Progress in Inflammation Research

Series Editor

Prof. Michael J. Parnham PhD
Director of Preclinical Discovery
Centre of Excellence in Macrolide Drug Discovery
GlaxoSmithKline Research Centre Zagreb Ltd.
Prilaz baruna Filipovića 29
HR-10000 Zagreb
Croatia

Advisory Board

G. Z. Feuerstein (Wyeth Research, Collegeville, PA, USA)
M. Pairet (Boehringer Ingelheim Pharma KG, Biberach a. d. Riss, Germany)
W. van Eden (Universiteit Utrecht, Utrecht, The Netherlands)

Forthcoming titles:

Angiogenesis in Inflammation: Mechanisms and Clinical Correlates, M.P. Seed, D.A. Walsh (Editors), 2008
New Therapeutic Targets in Rheumatoid Arthritis, P.-P. Tak (Editor), 2008
Inflammatory Cardiomyopathy (DCM) – Pathogenesis and Therapy, H.-P. Schultheiß, M. Noutsias (Editors), 2008
Matrix Metalloproteinases in Tissue Remodelling and Inflammation, V. Lagente, E. Boichot (Editors), 2008
Microarrays in Inflammation, A. Bosio, B. Gerstmayer (Editors), 2008
Occupational Asthma, T. Sigsgaard, D. Heederick (Editors), 2008
Nuclear Receptors and Inflammation, G.Z. Feuerstein, L.P. Freedman, C.K. Glass (Editors), 2009

(Already published titles see last page.)

Bone Morphogenetic Proteins:
From Local to Systemic Therapeutics

Slobodan Vukicevic
Kuber T. Sampath

Editors

Birkhäuser
Basel · Boston · Berlin

Editors

Slobodan Vukicevic
University of Zagreb
School of Medicine
Inst. Anatomy Drago Perovic
Salata 11
10000 Zagreb
Croatia

Kuber T. Sampath
Genzyme Corporation
One Mountain Road
Framingham, MA 01701-9322
USA

Library of Congress Control Number: 20089211606

Bibliographic information published by Die Deutsche Bibliothek
Die Deutsche Bibliothek lists this publication in the Deutsche Nationalbibliografie;
detailed bibliographic data is available in the internet at http://dnb.ddb.de

ISBN 978-3-7643-8551-4 Birkhäuser Verlag AG, Basel – Boston – Berlin

The publisher and editor can give no guarantee for the information on drug dosage and administration contained in this publication. The respective user must check its accuracy by consulting other sources of reference in each individual case. The use of registered names, trademarks etc. in this publication, even if not identified as such, does not imply that they are exempt from the relevant protective laws and regulations or free for general use.

This work is subject to copyright. All rights are reserved, whether the whole or part of the material is concerned, specifically the rights of translation, reprinting, re-use of illustrations, recitation, broadcasting, reproduction on microfilms or in other ways, and storage in data banks. For any kind of use, permission of the copyright owner must be obtained.

© 2008 Birkhäuser Verlag AG
Basel · Boston · Berlin
P.O. Box 133, CH-4010 Basel, Switzerland
Part of Springer Science+Business Media
Printed on acid-free paper produced from chlorine-free pulp. TCF ∞
Cover design: Markus Etterich, Basel
Cover illustration: Calcein green/ethidium bromide stained 100 μ thick vibratrome sections of fresh bovine cartilage explants showing the protective effect of OP-1 after mechanical injury. Top left: control cartilage; top right: normal cartilage exposed to OP-1; bottom left: cartilage that underwent 30 MPa impact injury; bottom right: cartilage that received an 30 MPa impact injury but was incubated with 100 ng/mL OP-1. Extensive superficial and middle zone chondrocyte death shown by orange ethidium bromide uptake is mitigated by the addition of OP-1 suggesting that this protein improves viability after sub-lethal mechanical injury. With friendly permission of Mark Hurtig.
Printed in Germany
ISBN 978-3-7643-8551-4   e-ISBN 978-3-7643-8552-1

9 8 7 6 5 4 3 2 1   www.birkhauser.ch

Contents

List of contributors vii

Preface xi

Slobodan Vukicevic and Kuber Sampath
Introduction 1

William F. McKay, Steven M. Peckham and Jeffrey M. Badura
Development of a novel compression-resistant carrier for
recombinant human bone morphogenetic protein-2 (rhBMP-2) and
preliminary clinical results 7

Michael Suk
Use of recombinant human BMP-2 in orthopedic trauma 25

Daniel B. Spagnoli
The application of recombinant human bone morphogenetic protein
on absorbable collagen sponge (rhBMP-2/ACS) to reconstruction of
maxillofacial bone defects 43

J. Kenneth Burkus
Clinical outcomes using rhBMP-2 in spinal fusion applications 71

Christina Sieber, Gerburg K. Schwaerzer and Petra Knaus
Bone morphogenetic protein signaling is fine-tuned on multiple levels 81

Daniel Graf and Aris N. Economides
Dissection of bone morphogenetic protein signaling using genome
engineering tools 115

Petra Seemann, Stefan Mundlos and Katarina Lehmann
Alterations of bone morphogenetic protein signaling pathway(s)
in skeletal diseases 141

Nandini Ghosh-Choudhury and Goutam Ghosh-Choudhury
Signaling crosstalk by bone morphogenetic proteins 161

*Stephen E. Harris, Wuchen Yang, Jelica Gluhak-Heinrich, Dayong Guo,
Xiao-Dong Chen, Marie A. Harris, Holger Kulessat, Brigid L.M. Hogan,
Alexander Lichtler, Barbara E. Kream, Jianhong Zhang, Jian Q. Feng,
Gregory R. Mundy, James Edwards and Yuji Mishina*
The role and mechanisms of bone morphogenetic protein 4 and 2
(BMP-4 and BMP-2) in postnatal skeletal development 179

*Jelica Gluhak-Heinrich, Dayong Guo, Wuchen Yang, Lilia E. Martinez,
Marie A. Harris, Holger Kulessat, Brigid L.M. Hogan, Alexander Lichtler,
Barbara Kream, Jianhong Zhang, Jian Q. Feng and Stephen E. Harris*
The role of bone morphogenetic protein 4 (BMP-4) in tooth development 199

Motoko Yanagita
Bone morphogenetic protein antagonists and kidney 213

Ugo Ripamonti, Jean-Claude Petit and June Teare
Induction of cementogenesis and periodontal ligament regeneration
by the bone morphogenetic proteins 233

*David J.J. de Gorter, Carola Krause, Clemens W.G.M. Löwik,
Rutger L. van Bezooijen and Peter ten Dijke*
Control of bone mass by sclerostin: inhibiting BMP- and
WNT-induced bone formation 257

Susan Chubinskaya, Mark Hurtig and David C. Rueger
Bone morphogenetic proteins in cartilage biology 277

*Slobodan Vukicevic, Petra Simic, Lovorka Grgurevic, Fran Borovecki
and Kuber Sampath*
Systemic administration of bone morphogenetic proteins 317

Index 339

List of contributors

Jeffrey M. Badura, Medtronic, Inc., 2600 Sofamor Danek Drive, Memphis, TN 38132, USA; e-mail: jeffrey.m.badura@medtronic.com

Rutger L. van Bezooijen, Department of Endocrinology and Metabolic Diseases, Leiden University Medical Centre, Albinusdreef 2, 2300 RC Leiden, The Netherlands; e-mail: r.l.van_bezooyen@lumc.nl

Fran Borovecki, Center for Functional Genomics, School of Medicine, University of Zagreb, 10000 Zagreb, Croatia

J. Kenneth Burkus, The Hughston Clinic, 6262 Veterans Parkway, Columbus, Georgia 31908-9517, USA; e-mail: jkb66@knology.net

Xiao-Dong Chen, University of Texas Health Science Center at San Antonio, 7703 Floyd Curl Dr., San Antonio, TX 78229, USA; e-mail: chenx4@uthscsa.edu

Susan Chubinskaya, Department of Biochemistry, Orthopedics and Section of Rheumatology (Department of Internal Medicine), Rush University Medical Center, Chicago, IL, 60612, USA; e-mail: schubins@rush.edu

Peter ten Dijke, Department of Molecular Cell Biology, Leiden University Medical Centre, Einthovenweg 20, 2300 RC Leiden, The Netherlands;
e-mail: p.ten_dijke@lumc.nl

Aris N. Economides, Genome Engineering Technologies, Regeneron Pharmaceuticals, Inc., 777 Old Saw River Road, Tarrytown, NY 10591, USA;
e-mail: aris.economides@regeneron.com

James Edwards, Vanderbilt University, Nashville, TN 37232, USA;
e-mail: james.edwards@vanderbilt.edu

Jian Q. Feng, Baylor College of Dentistry, Dallas, TX 75266, USA;
e-mail: jfeng@bcd.tamhsc.edu

List of contributors

Goutam Ghosh-Choudhury, Department of Medicine, University of Texas Health Science Center at San Antonio, 7703 Floyd Curl Drive, San Antonio, TX 78229, USA; e-mail: choudhuryg@uthscsa.edu

Nandini Ghosh-Choudhury, Department of Pathology, University of Texas Health Science Center at San Antonio, 7703 Floyd Curl Drive, San Antonio, TX 78229, USA; e-mail: choudhury@uthscsa.edu

Jelica Gluhak-Heinrich, University of Texas Health Science Center at San Antonio, 7703 Floyd Curl Dr., San Antonio, TX 78229, USA; e-mail: gluhak@uthscsa.edu

David J.J. de Gorter, Department of Molecular Cell Biology, Leiden University Medical Centre, Einthovenweg 20, 2300 RC Leiden, The Netherlands; e-mail: d.de_gorter@lumc.nl

Daniel Graf, Institute of Immunology, Biomedical Sciences Center 'Al. Fleming', 34 Al. Fleming Street, 166 72 Vari, Greece; e-mail: d.graf@fleming.gr

Lovorka Grgurevic, Laboratory for Mineralized Tissues, School of Medicine, University of Zagreb, 10000 Zagreb, Croatia

Dayong Guo, University of Texas Health Science Center at San Antonio, 7703 Floyd Curl Dr., San Antonio, TX 78229, and The University of Missouri at Kansas City, MO 64109, USA; e-mail: guod@umkc.edu

Marie A. Harris, University of Texas Health Science Center at San Antonio, 7703 Floyd Curl Dr., San Antonio, TX 78229, USA; e-mail: harrisma@uthscsa.edu

Stephen E. Harris, University of Texas Health Science Center at San Antonio, 7703 Floyd Curl Dr., San Antonio, TX 78229, USA: e-mail: harris@uthscsa.edu

Brigid L.M. Hogan, Duke University, Durham, N.C., USA; e-mail: b.hogan@cellbio.duke.edu

Mark Hurtig, Ontario Veterinary College, Guelph University, Guelph, Ontario, Canada N1G 2W1; e-mail: mhurtig@ovc.uoguelph.ca

Petra Knaus, Freie Universität Berlin, Institute for Chemistry/Biochemistry, Thielallee 63, 14195 Berlin, Germany; e-mail: knaus@chemie.fu-berlin.de

Barbara E. Kream, UCONN Health Center, Farmington, CT 06030, USA; e-mail: kream@nso1.uchc.edu

Carola Krause, Department of Molecular Cell Biology, Leiden University Medical Centre, Einthovenweg 20, 2300 RC Leiden, The Netherlands; e-mail: c.krause@lumc.nl

Holger Kulessa†

Katarina Lehmann, Institut für Medizinische Genetik, Universitätsmedizin Berlin, Charité, Augustenburger Platz 1, 13353 Berlin, Germany; e-mail: katarina.lehmann@charite.de

Alexander Lichtler, University of Connecticut Health Center, Farmington, CT 06032, USA; e-mail: lichtler@nso1.uchc.edu

Clemens W.G.M. Löwik, Department of Endocrinology and Metabolic Diseases, Leiden University Medical Centre, Albinusdreef 2, 2300 RC Leiden, The Netherlands; e-mail: c.w.g.m.lowik@lumc.nl

William F. McKay, Medtronic, Inc., 2600 Sofamor Danek Drive, Memphis, TN 38132, USA; e-mail: bill.mckay@medtronic.com

Lilia E. Martinez, The University of Texas Health Science Center at San Antonio, 7703 Floyd Curl Dr., San Antonio, TX 78229-3900, USA

Yuji Mishina, National Institute of Environmental Health Science, Research Triangle Park, NC 27709, USA; e-mail: mishina@niehs.nih.gov

Stefan Mundlos, Institut für Medizinische Genetik, Universitätsmedizin Berlin, Charité, Augustenburger Platz 1, 13353 Berlin, and Max-Planck-Institut für Molekulare Genetik, Ihnestr. 63–73, 14195 Berlin, Germany; e-mail: stefan.mundlos@charite.de

Gregory R. Mundy, Vanderbilt University, Nashville, TN 37232, USA; e-mail: gregory.r.mundy@vanderbilt.edu

Steven M. Peckham, Medtronic, Inc., 2600 Sofamor Danek Drive, Memphis, TN 38132, USA

Jean-Claude Petit, Bone Research Unit, MRC/University of the Witwatersrand, 7 York Road Medical School, 2193 Parktown Johannesburg, South Africa

Ugo Ripamonti, Bone Research Unit, MRC/University of the Witwatersrand, 7 York Road Medical School, 2193 Parktown Johannesburg, South Africa; e-mail: ugo.ripamonti@wits.ac.za

List of contributors

David C. Rueger, Stryker Biotech Division, Hopkinton, MA, 01748, USA; e-mail: davidrueger@msn.com

Kuber Sampath, Genzyme Corporation, Framingham, 01701-9322 Massachusetts, USA; e-mail: kuber.sampath@genzyme.com

Gerburg K. Schwaerzer, Freie Universität Berlin, Institute for Chemistry/Biochemistry, Thielallee 63, 14195 Berlin, Germany; e-mail: gerburg.schwaerzer@fu-berlin.de

Petra Seemann, Max-Planck-Institut für Molekulare Genetik, Research Group Development & Disease, Ihnestr. 63–73, 14195 Berlin, Germany; e-mail: seemann@molgen.mpg.de

Christina Sieber, Freie Universität Berlin, Institute for Chemistry/Biochemistry, Thielallee 63, 14195 Berlin, Germany; e-mail: tsieber@chemie.fu-berlin.de

Petra Simic, Laboratory for Mineralized Tissues, School of Medicine, University of Zagreb, 10000 Zagreb, Croatia

Daniel B. Spagnoli, University Oral & Maxillofacial Surgery, 7482 Waterside Crossing Blvd, Denver, NC 28037, USA; e-mail: dspagnoli@uomsnc.com

Michael Suk, University of Florida – Shands Jacksonville, 55 West 8th Street, ACC Building, 2nd Floor/Ortho, Jacksonville, FL 32209; e-mail: michael.suk@jax.ufl.edu

June Teare, Bone Research Unit, MRC/University of the Witwatersrand, 7 York Road Medical School, 2193 Parktown Johannesburg, South Africa

Slobodan Vukicevic, Laboratory for Mineralized Tissues, School of Medicine, University of Zagreb, 10000 Zagreb, Croatia; e-mail: vukicev@mef.hr

Motoko Yanagita, Kyoto University Graduate School of Medicine, Kyoto 606-8501, Japan; e-mail: motoy@kuhp.kyoto-u.ac.jp

Wuchen Yang, University of Texas Health Science Center at San Antonio, 7703 Floyd Curl Dr., San Antonio, TX 78229, USA; e-mail: yangw@uthscsa.edu

Jianhong Zhang, Vanderbilt University, Nashville, TN 37232, USA; e-mail: zhang01186@yahoo.com

Preface

We are happy to bring this 3rd volume covering the local and systemic use of bone morphogenetic proteins (BMPs) as a part of *Progress in Inflammation Research* edited by Michael J. Parnham, to our readers, including undergraduate students and basic and clinical scientists world-wide.

Since the original description of the remarkable potential of demineralized bone matrix to induce bone at an ectopic site it has taken more than three decades to bring BMPs to clinical use. In 2007, nearly one million patients were treated with BMP-2 or OP-1 (BMP-7) for spinal fusions, nonunions, acute fractions and maxillofacial reconstruction. This has led to the revolution of our understanding of the molecular processes responsible for bone regeneration, in particular when physiological mechanisms of fracture repair fail.

We appreciate very much the involvement of distinguished basic and clinical experts who contributed chapters and express our gratitude to them for their dedicated professionalism in finalizing this book. We acknowledge Mr. Branko Simat and Mrs. Morana Simat for their technical support in this project. We also appreciate the technical assistance from Ms. Ivancica Bastalic, Ms. Vera Kufner and Mr. Igor Erjavec. We thank Dr. Hans Detlef Klueber and Ms. Anke Brosius for their help in preparing and editing this book.

March 2008
Slobodan Vukicevic
Kuber T. Sampath

Introduction

Slobodan Vukicevic[1] and Kuber Sampath[2]

[1]Laboratory for Mineralized Tissues, School of Medicine, University of Zagreb, 10000 Zagreb, Croatia; [2]Genzyme Corporation, Framingham, 01701-9322 Massachusetts, USA

Bone has a remarkable potential to regenerate upon fracture. Marshall Urist discovered that non-living demineralized bone could become fully mineralized living functional bone when implanted at ectopic sites, and he coined this phenomenon 'bone by auto-induction' [1]. He hypothesized that the principle behind this regenerative potential resides in the extracellular matrix of bone and termed it 'bone morphogenetic proteins (BMPs)'. While the identity of BMP has been illusive for a long time, the demonstration that the proteins responsible for bone induction could be extracted from extracellular matrix of bone and assayed reproducibly for their activity by reconstituting them with an appropriate collagen scaffold and implanting at ectopic sites in rats permitted the molecular cloning and identification of '*bona fide*' BMPs [2–4]. This knowledge represents, for the first time, a prototype demonstration of tissue engineering *in vivo* and has been attributed to three biological components as prerequisites for bone tissue engineering: signaling molecules, responding cells and scaffold/microenvironment. The collagen carrier serves as substratum for migration and proliferation of mesenchymal stem cells (osteoprogenitors) and BMP signals their differentiation into fully vascularized functional bone [5, 6].

The BMPs represent a large family of proteins structurally related to transforming growth factor (TGF)-βs and activins, and are responsible for migration, proliferation and differentiation of several cell types [7]. Osteogenic devices containing BMP-7 (OP-1™) or BMP-2 (InFuse™) have now been developed commercially and approved by regulatory agencies as bone graft substitutes for the treatment of long bone nonunions, acute fractures and interbody fusion of vertebrae in humans [8, 9]. BMPs are the first therapeutic proteins approved for the use of tissue engineering in conjunction with a scaffold and biocompatible fixative devices in the arena of regenerative medicine. Here we provide extended clinical experience on the use of BMP-2-containing devices. Spagnoli summarizes the clinical achievements on the use of BMPs in maxillofacial bone defects; Burkus discusses the current clinical outcome in spinal fusion applications; and Suk reviews the clinical use of BMP-2 in orthopedic trauma – acute, delayed and nonunion fractures.

The biomaterial scaffold plays an important biological role as a component of BMP device to effect new bone formation [5]. The BMP devices used in clinical trials have all employed type I collagen, a natural component of bone and considered as the "gold standard" for comparison. Although BMP devices offer tremendous promise as bone graft substitutes, we are still faced with numerous challenges. There is a need for optimal delivery systems for BMPs with varying geometry and resorptive time, which depend on location and mechanical loading of the defects. The handling property of the device is also important, as the surgical approximation and retaining of the BMP device at the site of repair are required to affect a reproducible outcome. Furthermore, internal and/or external fixation has to be modernized according to the rate of osteogenesis induced by BMP devices and specific bone sites. McKay et al. describe the development of a novel compression-resistant carrier that contains both mineral and collagen as composites for BMP-2, and discuss its application in clinical setting.

BMPs elicit their biology by ligand-induced association of specific heterodimeric complexes of two related type I and type II serine/threonine kinase receptors at the cell surface [10]. Immediately upon forming the complexes, the constitutively active type II receptor kinase phosphorylates type I receptor and activates intracellular signaling by phosphorylating nuclear effector proteins known as Smads. There are three distinct subclasses of Smads and these can be divided into: signal transducing receptor-regulated Smads (R-Smads, i.e., Smad1, Smad5 and Smad8) and common mediator Smads (C-Smads, i.e., Smad4) and inhibitory Smads (I-Smads, i.e., Smad6 and Smad7). The phosphorylated BMPR-specific R-Smad1/5/8 forms heterodimeric complexes with C-Smad4 and then translocates to nucleus. Within the nucleus, R-Smad/C-Smad complexes act directly and/or in cooperation with other transcription factors, to regulate transcription of target genes. Inhibitory Smad6/7 specifically inhibits BMP signaling, thus inducing a negative regulation to keep the activation in balance. Ending of BMP signaling is achieved through ubiquitination and proteosome-mediated degradation of R-Smads. The identification, structure and mechanism of BMP receptor activation and Smad-mediated down signaling have been described in the previous editions. In this edition, we added chapters to bring out the advances made on the BMP signaling as it is fine-tuned at multiple levels (Knaus and colleagues), and its dissection using genomic approaches (Graf and Economides). Other chapters specifically discuss (1) the importance of mutation or alteration in the BMP receptor and downstream signaling pathways affecting skeletal disease (Seemann et al.), and (2) BMP signaling cross-talk between Smads/P13K/MAPKBMP mediated *via* Smad-binding elements (N. Ghosh-Choudhury and G. Ghosh-Choudhury). As BMP-2 and BMP-4 are very closely related BMPs, Harris et al. have contributed a chapter that describes the specific role of these two BMPs and molecular mechanism in postnatal skeletal development.

BMPs activity is tightly regulated by binding to the naturally occurring secreted soluble BMP antagonists at the extracellular space. BMP antagonists are secretory proteins that contain primary sequence allowing the formation of a knot structure representing a subfamily of cystine-knot family of proteins. Similar to the binding of TGF-β to α2-macroglobulin and small proteoglycan decorin, and activin to follistatin, BMPs bind to noggin, chordin, DAN/cereberus, gremlin, sclerostin and related USAG-1 [7]. The interplay between BMPs and their antagonists governs the BMP activity locally. The importance of two BMP antagonists, sclerostin and USAG-1 has been described in the context of osteoporosis (ten Dijke and colleagues) and kidney failure (Yanagita).

In addition to the bone-forming activity, several studies have demonstrated that BMPs, when applied alone or in combination with the appropriate scaffold onto chondral or osteochondral defects, are capable of inducing new articular cartilage formation *in vivo* [11]. Long-term preclinical studies, however, showed that the newly formed chondrocytes failed to maintain the cellular morphology and expression of articular cartilage phenotype over time. It is likely that providing BMPs intermittently or continuously for a sustainable time (instead of a one-time application at the beginning, as used to repair bone fractures) may help the regenerated cartilage to attain function under mechanical loading. Chubinskaya et al. describe advances made on the use of BMP in cartilage biology.

Another highly mineralized tissue in the body is the tooth. In several preclinical studies, BMP-2- and BMP-7-containing matrices have been shown to induce repair and regeneration of dental tissues including periodontium (periodontal ligament, cementum and alveolar bone) and dentin [12]. It is possible to regenerate the lost dentin and periodontal tissues in adults by applying a suitable scaffold and optimal concentration of specific BMPs at the repair site, provided that responding cells and vascular components are available. More details on the role of BMPs in maxillofacial and periodontal tissue repair and regeneration are presented by Ripamonti et al. Also added in this edition is the importance of BMP-4 in tooth development, offering specificity for potential application in dentin repair (Gluhak-Heinrich et al).

Given that BMPs induce new bone formation *in vivo*, promote the recruitment and growth of osteoblast progenitor, and maintain the expression of osteoblast phenotype in cultures, it is conceivable that providing an optimal dose of a BMP in the circulation may help trigger the osteogenic responses, as an endocrine signal, and restore the loss of bone mass and quality in postmenopausal osteoporosis and related metabolic bone diseases [13, 14]. Recent studies have shown that systemic administration of BMP-7 was able to restore the impaired remodeling associated with the aplastic bone disorder of renal osteodystrophy in a mouse model following 5/6 cortical ablation of the kidney [15]. This supports an endocrine role for BMP-7 since the highest level of its expression was found in the adult kidney. It is likely that BMPs can exert their tissue morphogenesis function both locally in an autocrine and paracrine manner during development, and systemically in an endocrine mode during growth and in the adult organism. In the final chapter, Vukicevic et al. describe the systemic use of BMP-6/7 for the prevention and restoration of bone mass in a model of osteoporosis, kidney regeneration in models of acute and chronic renal failure, liver regeneration, ischemic coronary infarction and stroke, and in a nude mouse model of different human cancers.

References

1. Urist MR (1965) Bone: Formation by autoinduction. *Science* 150: 893–899
2. Sampath TK, Reddi AH (1981) Dissociative extraction and reconstitution of extracellular matrix components involved in local bone differentiation. *Proc Natl Acad Sci USA* 78: 7599–7602
3. Wozney JM, Rosen V, Celeste AJ, Mitsock LM, Whitters MJ, Kriz RW, Hewick RM, Wang EA (1988) Novel regulators of bone formation: Molecular clones and activities. *Science* 242: 1528–1534
4. Ozkaynak E, Rueger DC, Drier EA, Corbett C, Ridge RJ, Sampath TK, Oppermann H (1990) OP-1 cDNA encodes an osteogenic protein in the TGF-beta family. *EMBO J* 9: 2085–2093
5. Reddi AH (1998) Role of morphogenetic proteins in skeletal *Tissue Eng*ineering and regeneration. *Nat Biotechnol* 16: 247–252
6. Reddi AH, Huggins C (1972) Biochemical sequences in the transformation of normal fibroblasts in adolescent rats. *Proc Natl Acad Sci USA* 69: 1601–1605
7. Martinovic S, Simic P, Borovecki F, Vukicevic S (2004) Biology of bone morphogenetic proteins. In: S Vukicevic, TK Sampath (eds): *Bone morphogenetic proteins: Regeneration of bone and beyond*. Birkhäuser, Basel, 45–73
8. Friedlaender GE, Perry CR, Cole JD, Cook SD, Cierny G, Muschler GF, Zych GA, Calhoun JH, LaForte AJ, Yin S (2001) Osteogenic protein-1 (bone morphogenetic protein-7) in the treatment of tibial nonunions. *J Bone Joint Surg* 83A: S151–158
9. Burkus JK, Gornet MF, Dickman CA, Zdeblick TA (2002) Anterior lumbar interbody fusion using rhBMP-2 with tapered interbody cages. *J Spinal Disord Tech* 15: 337–349
10. ten Dijke P, Miyazono K, Heldin CH (1996) Signaling *via* hetero-oligomeric complexes of type I and type II serine/threonine kinase receptors. *Curr Opin Cell Biol* 8: 139–145
11. Pecina M, Vukicevic S (2007) Biological aspects of bone, cartilage and tendon regeneration. *Int Orthop* 31: 719–720
12. Nakashima M (2005) Bone morphogenetic proteins in dentin regeneration for potential use in endodontic therapy. *Cytokine Growth Factor Rev* 16: 369–376
13. Vukicevic S, Basic V, Rogic D, Basic N, Shih MS, Shepard A, Jin D, Dattatreyamurty B, Jones W, Dorai H et al (1998) Osteogenic protein-1 (bone morphogenetic protein-7) reduces severity of injury after ischemic acute renal failure in rat. *J Clin Invest* 102: 202–214

14 Simic P, Buljan Culej J, Orlic I, Grgurevic L, Draca N, Spaventi R, Vukicevic S (2006) Systemically administered bone morphogenetic protein-6 restores bone in aged ovariectomized rats by increasing bone formation and suppressing bone resorption. *J Biol Chem* 281: 25509–25521

15 Lund RJ, Davies MR, Brown AJ, Hruska KA (2004) Successful treatment of an adynamic bone disorder with bone morphogenetic protein-7 in a renal ablation model. *J Am Soc Nephrol* 15: 359–369

Development of a novel compression-resistant carrier for recombinant human bone morphogenetic protein-2 (rhBMP-2) and preliminary clinical results

William F. McKay, Steven M. Peckham and Jeffrey M. Badura

Medtronic, Inc., 2600 Sofamor Danek Drive, Memphis, TN 38132, USA

Introduction

Recombinant human bone morphogenetic protein-2 (rhBMP-2) has been commercially available in the United States since July 2002 (INFUSE® Bone Graft/LT-CAGE® Lumbar Tapered Fusion Device, Medtronic, Inc., Memphis, TN). It was initially approved for use in interbody spinal fusions inside a threaded titanium interbody fusion device. Since then, it has been approved for two additional clinical indications: fresh tibial fractures and certain oral maxillofacial procedures (sinus elevation and extraction sockets). The approvals were based on Level I prospective, randomized clinical trials involving the absorbable collagen sponge (ACS) carrier and rhBMP-2 at a concentration of 1.5 mg/mL.

The interbody spinal fusion approval involved patients receiving single-level anterior interbody spinal fusion with dual threaded titanium lumbar tapered fusion cages (LT-CAGE® Device, Medtronic, Inc., Memphis, TN). The rhBMP-2/ACS was placed inside the cages where it was shielded from direct compressive forces. Equivalent fusion rates were achieved with rhBMP-2/ACS and iliac crest bone graft (94.5% *versus* 88.7%, respectively) [1]. The fresh tibial fracture trial involved internal fixation with an intramedullary nail (IM) without bone grafting as the current "standard of care" *versus* onlay grafting of rhBMP-2/ACS over the exposed fracture. The use of rhBMP-2/ACS resulted in a 41% reduction ($p < 0.001$) in the need for secondary surgical intervention to obtain fracture healing [2, 3]. The sinus elevation clinical trial demonstrated a similar level of new bone formation with rhBMP-2/ACS as compared to autograft, 10.2 mm *versus* 11.3 mm, respectively [4, 5]. Finally, rhBMP-2/ACS produced more bone in extraction sites than was formed in untreated standard of care control patients ($p \leq 0.05$) [6]. In both oral maxillofacial indications, the rhBMP-2/ACS was protected from soft tissue compressive forces.

The rhBMP-2 hydrated ACS sponge is soft and pliable and can be easily compressed during surgical implantation or from overlying soft tissue forces. However, it was effective in the FDA-approved clinical indications because it was used in relatively compression-free environments. Many other bone grafting indications require the bone graft to maintain space on its own, without the aid of a device or natural cavity, for an adequate volume of new bone to be formed. Examples include posterolateral spinal fusion procedures or treatment of large segmental bone defects in long bones or the mandible. The carrier is not required to carry weight-bearing loads, but it must resist compressive forces from overlying muscles or other soft tissues. The addition of a bulking agent, such as autograft, allograft, or calcium phosphate granules, to the ACS carrier results in a graft that is able to resist soft tissue compression in non-clinical and clinical applications [7–11]. The bulking agent helped maintain a space for the new bone formation to occur, but it was recognized that a carrier with inherent compression resistance would be more user friendly. Therefore, there was a need to develop a second-generation rhBMP-2 carrier that could be used in these more challenging clinical applications with better handling and clinical utility. This carrier was termed the compression-resistant matrix (CRM). The CRM carrier is a composite sponge consisting of cross-linked Type I bovine collagen impregnated with biphasic calcium phosphate ceramic granules [15% hydroxyapatite (HA) and 85% β-tricalcium phosphate (β-TCP)] (Fig. 1). This chapter focuses on the identification, characterization, and preclinical testing of the rhBMP-2/CRM bone graft substitute.

Identifying the need for a compression-resistant rhBMP-2 carrier

One of the first preclinical spinal fusion studies examining the use of rhBMP-2/ACS was conducted in a rabbit posterolateral fusion model. In a posterolateral fusion technique, muscle and soft tissue is removed from the spinous processes and lamina to prepare a bleeding bone bed onto which the bone graft is placed. Upon surgical closure, the overlying soft tissues subject the bone graft to some compressive forces. In the current study, the bone formation and fusion success of rhBMP-2/ACS alone was compared to autograft bone using radiographic, biomechanical, and histological analyses [12]. Furthermore, a low dose and high dose of rhBMP-2 were examined and compared, 0.7 mg and 2.7 mg, respectively. Results demonstrated that all animals (100%) treated with rhBMP-2/ACS achieved solid fusion *via* manual palpation and radiography, while only 42% of the autograft group achieved solid fusion. Histology demonstrated normal mature bone growth in the rhBMP-2-treated animals. In addition, the high-dose rhBMP-2 group demonstrated greater trabecular bone growth than the low-dose group.

This study led to further investigation of rhBMP-2/ACS in other spinal fusion models, including that of a non-human primate anterior lumbar interbody fusion (ALIF). This study examined and compared a buffer/ACS control to two doses of rhBMP-2 on the ACS carrier [13]. At the time of surgery, the ACS was soaked with a buffer solution, 0.75 mg/mL or 1.5 mg/mL of rhBMP-2 solution, and placed within a threaded titanium cylindrical cage prior to cage insertion. CT scans at 12

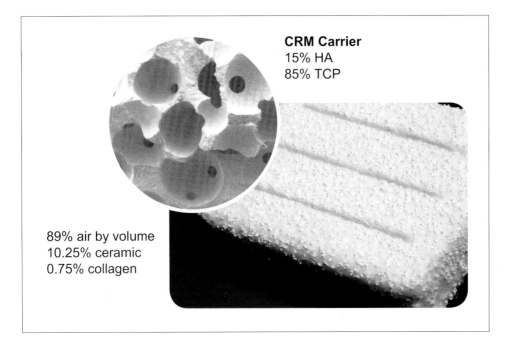

Figure 1
Composition of the compression-resistant matrix (CRM) carrier.

and 24 weeks and histology at sacrifice were used to evaluate the fusion efficacy. All animals treated with rhBMP-2/ACS achieved solid fusion through the cage. The buffer/ACS group demonstrated some bone growth into the outer edges of the cage but ultimately developed pseudoarthrosis. Histology demonstrated that the rhBMP-2/ACS treatment induced normal, mature trabecular bone.

The successful fusion results in the rabbit posterolateral model and non-human primate interbody fusion model led to the investigation of rhBMP-2/ACS in a non-human primate posterolateral fusion model. In this study, posterolateral fusion was performed at L4–L5 using different rhBMP-2 carrier formulations and rhBMP-2 doses [14]. This study demonstrated that in six animals receiving rhBMP-2/ACS alone, only one treated with the high dose (8 mg rhBMP-2 at 1.7 mg/mL) achieved fusion. It was noted that this fusion mass was more robust at the transverse processes and narrower at the center of the intertransverse space. In the other animals treated with rhBMP-2/ACS alone, including the 2 mg and 4 mg doses, some bone formation was observed between the transverse processes. However, this bone growth was not continuous and led to pseudoarthrosis, as assessed by manual palpation. The minimal fusion mass volume and non-contiguous bone formation between adjacent transverse processes led the investigators to believe that soft tissue forces from the surrounding muscles compressed the sponge, resulting in limited and inconsistent bone formation. To test this theory, another group was examined in which 9 mg rhBMP-2 on ACS was covered by a protective polyethylene mesh shield. Results from this group confirmed that the shield protected the malleable rhBMP-2/ACS implant from compressive forces, allowing for more robust fusion masses to form.

Results from this early preclinical efficacy work demonstrated that the non-human primate was the most challenging animal model for inducing reproducible spinal fusion. In particular, the non-human primate posterolateral fusion environment not only required higher doses of rhBMP-2 than the rabbit model, but, to be successful, required protection from compressive soft tissue forces as well.

This early work also demonstrated two primary reasons why the non-human primate most accurately models human fusion potential. First, the timing required for bone formation and fusion in primates is most similar to humans. Thus, the appropriate rhBMP-2 doses and carrier resorption for inducing reproducible *de novo* bone formation should be examined in a non-human primate model before deciding on a clinical formulation. Second, as reproduced in the non-human primate intertransverse posterolateral fusion environment, compressive forces caused by the surrounding soft tissue can affect overall fusion. Thus, the physical characteristics (i.e., compression resistance) of the rhBMP-2 carrier could ultimately influence fusion success. These important preclinical investigations were the catalyst for designing a second-generation rhBMP-2 carrier, specifically for the intertransverse posterolateral fusion environment.

In vitro and *in vivo* characterization testing of the CRM carrier

Based on previous preclinical characterization work with the first-generation ACS carrier, certain design requirements were established for the CRM carrier. First, the CRM carrier had to resist compressive forces created by surrounding soft tissue in the posterolateral fusion environment. Second, the CRM carrier had to hydrate and retain, or incorporate, the applied rhBMP-2 upon soaking and compression. Finally, the CRM carrier had to release rhBMP-2 at the site of implantation, allowing *de novo* bone formation to occur.

One study was conducted to compare the compression resistance of the first-generation ACS carrier to the novel CRM carrier [15]. The ACS and CRM were hydrated with water, as instructed, prior to testing, mimicking an rhBMP-2-soaked pre-implantation state. Load was applied with an MTS 810 Materials Test System at a constant rate of 0.1 mm/s for a total displacement of 5 mm (Fig. 2). Results demonstrated that the ACS began to resist compression at 3.5 mm of displacement and the CRM carrier resisted compression immediately upon loading. The average compressive loads at 5 mm of displacement were 48.4 N and 489.2 N for the ACS and

Figure 2
Images from the compression testing of the absorbable collagen sponge (ACS) (A, B) and CRM (C, D) carriers.

CRM, respectively (Fig. 3). This test clearly demonstrated that the CRM resisted greater compressive forces than the ACS at a constant displacement rate, suggesting that the CRM would be capable of resisting soft tissue compressive forces observed in the posterolateral fusion environment.

Another important *in vitro* study examined the incorporation of the rhBMP-2 into the CRM carrier where incorporation is defined as the amount of rhBMP-2 remaining in the carrier after expressing as much fluid as possible from the carrier [16]. The testing material was prepared by applying 2.5 mL 4.0 mg/mL rhBMP-2 solution to a 5.0-mL dry piece of CRM, resulting in a bulk rhBMP-2 concentration of 2.0 mg/mL, or 2.0 mg rhBMP-2 per volume of CRM carrier. The rhBMP-2 solution was then allowed to soak on the CRM carrier for 5 or 60 min. Following the proper soak time, manual compression or centrifugation was performed to express the maximal amount of fluid from the carrier. Manual compression involved insertion of the soaked carrier into a syringe and compressing to half its original length. Centrifugation was performed at 1500 rpm for 5 min. The rhBMP-2 concentration

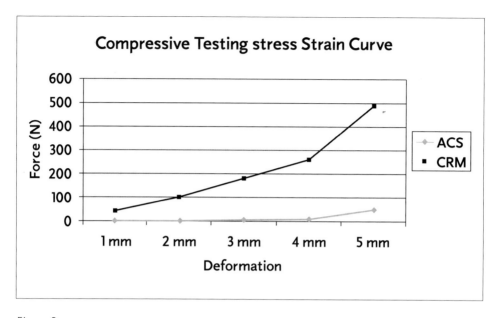

Figure 3
Stress-strain curve from the compression testing of the ACS and CRM carriers, with CRM demonstrating a 100 times greater max force at 5 mm of displacement compared to ACS.

of expressate was then measured using RP-HPLC analysis. Results demonstrated that following manual compression, 82.19% and 89.96% ($p=0.0605$) of rhBMP-2 was retained within the carrier for the 5- and 60-min soak times, respectively. However, following centrifugation, 36.09% and 56.95% ($p=0.0001$) of rhBMP-2 was retained within the carrier for the 5- and 60-min soak times, respectively. It is important to note that both the manual compression and centrifugation expression techniques represented worst-case scenarios for compression of the rhBMP-2/CRM that should not be recreated clinically. The results from this *in vitro* incorporation study demonstrated retention of the rhBMP-2 by the CRM even at the short soak times and high compressive loads.

The next step in the preclinical development of rhBMP-2/CRM involved a study investigating the localization of rhBMP-2 at the site of implantation in a rabbit posterolateral fusion model [17]. ^{125}I-labeled rhBMP-2 was applied to biphasic calcium phosphate (BCP) ceramic granules (60% HA and 40% β-TCP) and a BCP (5% HA and 95% β-TCP) ceramic-collagen composite sponge carrier to examine area under the curve (AUC), mean residence time (MRT), initial retention/incorporation, and terminal retention half-life of the rhBMP-2 for 36 days postoperatively. Results demonstrated that the AUC was 988%×day for BCP and 1070%×day for the composite sponge and the MRT was 10.2 days for BCP and

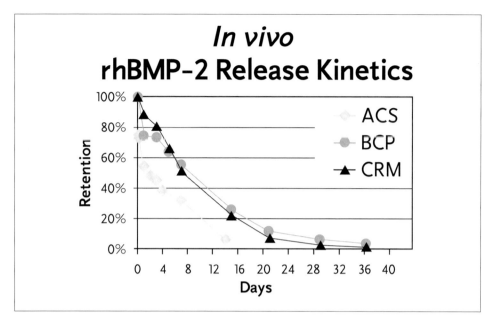

Figure 4
In vivo *rhBMP-2 release kinetics for the ACS, CRM, and BCP carriers. From [17].*

7.6 days for the composite sponge. Neither of these differences was statistically significant. Initial retention/incorporation was statistically higher for the composite sponge compared to BCP (96.8% *versus* 86.0%). On the other hand, animals receiving rhBMP-2/BCP showed a longer terminal retention half-life compared to the rhBMP-2/composite sponge, 7.5 *versus* 4.5 days, respectively. Finally, systemic catabolism and elimination of the rhBMP-2 was extensive and the systemic detection of rhBMP-2 was negligible. These data demonstrated that in a common rabbit posterolateral fusion model, both the BCP composite sponge and BCP granule carriers were successful at retaining rhBMP-2 at the site of implantation. Furthermore, the BCP composite sponge retained the rhBMP-2 at the implantation site for up to 5 weeks, approximately 3 weeks longer than the first-generation ACS carrier (Fig. 4).

Efficacy testing of CRM and optimization of rhBMP-2/CRM

After determining that a compression-resistant rhBMP-2 carrier was ideal and necessary for fusion success in the posterolateral environment, a series of preclinical efficacy studies were performed to optimize fusion results with the carrier. More

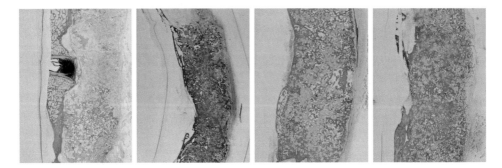

Figure 5
Sagittal histology images for the biphasic calcium phosphate (BCP) carrier loaded with 0, 6, 9, or 12 mg (A–D) of rhBMP-2.

specifically, the ceramic composition (i.e., percentage HA and percentage β-TCP) of the CRM and the appropriate rhBMP-2 bulk concentration for reproducible fusion were investigated.

The initial work was performed by soaking blocks of BCP (60% HA and 40% β-TCP) ceramic with rhBMP-2 solution, resulting in bulk rhBMP-2 concentrations of 0, 1.35, 2.0, and 2.7 mg/mL, or 0, 6, 9, or 12 mg rhBMP-2, respectively [18]. These groups were compared to primates treated with iliac crest bone graft (ICBG). The animals were killed at 24 weeks and fusion was assessed using CT scans, manual palpation, and histology. Manual palpation indicated that all animals receiving ICBG did not achieve fusion. On the other hand, all animals receiving the ceramic blocks, with or without rhBMP-2, obtained fusion as determined *via* manual palpation. In addition, CT scans indicated the blocks loaded with rhBMP-2 resulted in complete graft incorporation. Histology demonstrated that animals receiving 0 mg rhBMP-2 on the composite sponge contained fibrous tissue, predominantly, surrounding residual ceramic and very little bone formation, which extended from the decorticated transverse processes (Fig. 5). The 1.35 mg/mL rhBMP-2 group contained more variable amounts of bone formation compared to the 2.0 mg/mL and 2.7 mg/mL treated animals. In addition, both the 2.0 mg/mL and 2.7 mg/mL groups demonstrated comparable bone formation, indicating the possibility of an efficacious dose threshold level for rhBMP-2 on the ceramic blocks.

One limitation of the BCP block carrier was that a significant amount of residual ceramic was left unremodeled at the 24-week sacrifice time [18]. Ceramic is radio-opaque, which makes it difficult to distinguish between new bone growth and residual ceramic, especially on plain film radiographs. Ultimately, the objective was to have a ceramic carrier capable of maintaining space for an adequate fusion mass to form with remodeling taking place over time and eventually being replaced by new bone formation. To accomplish this, a series of rabbit and non-human primate studies were conducted to investigate the overall impact of decreasing the amount of slow-resorbing HA and increasing the amount of faster-resorbing β-TCP.

In one study, Suh and colleagues [19] examined two different ceramic-based carriers in a validated rabbit posterolateral fusion model. This study evaluated rhBMP-2 soaked onto a ceramic-collagen composite sponge containing a 5% HA and 95% β-TCP ceramic granules imbedded in collagen and loose BCP ceramic granules consisting of 60% HA and 40% β-TCP. Twenty-eight skeletally mature New Zealand white rabbits, $n = 14$/group, underwent L5–L6 posterolateral spinal fusions and were killed at 5 weeks. A bulk rhBMP-2 concentration of 0.29 mg/mL was used for both carriers. All animals showed a 100% radiographic fusion rate. Furthermore, the 5:95 HA:TCP composite sponge group was less radio-opaque compared to the 60:40 HA:TCP granule carrier on plain film radiographs. Histological analysis demonstrated normal new bone formation in both groups. Finally, it was noted that the subjective handling properties of the ceramic-collagen composite sponge were much improved compared to those of the loose granules of the BCP granule carrier.

Suh and colleagues [19] performed another study that examined two different ceramic-collagen composite sponges in a non-human primate posterolateral fusion model. Six skeletally mature rhesus monkeys underwent L4–L5 posterolateral spinal fusion. The animals were divided into three investigational groups, which examined: (1) a BCP ceramic (15% HA and 85% β-TCP)/collagen composite sponge with 2.1 mg/mL rhBMP-2 ($n=2$); (2) a BCP ceramic (5% HA and 95% β-TCP)/collagen composite sponge with 2.1 mg/mL rhBMP-2 ($n=2$); and (3) a BCP ceramic (5% HA and 95% β-TCP)/collagen composite sponge with 1.0 mg/mL rhBMP-2 ($n=2$). It should be noted that the 1.0 mg/mL group had the same overall rhBMP-2 dose as the 2.1 mg/mL groups, but delivered on twice the volume of carrier. The animals were killed at 6 months. All animals treated with 2.1 mg/mL rhBMP-2, independent of ceramic composition, resulted in 100% fusion. The group treated with only 1.0 mg/mL rhBMP-2 formed some new bone, but did not achieve solid fusion. The fusion masses in the animals treated with the 15:85 HA:TCP composite sponge were larger than those treated with the relatively faster-degrading 5:95 HA:TCP composite sponge. Histological analysis revealed the fusion masses contained normal bone formation and minimal residual ceramic.

These studies performed by Suh and colleagues highlighted a number of key findings, including the preferred subjective handling of a composite sponge compared to loose granules, a 15:85 HA:β-TCP BCP composition capable of supporting new bone formation while resorbing over time, and determination that the bulk concentration of rhBMP-2 (i.e., milligrams of rhBMP-2 to volume of carrier) when placed on a composite sponge carrier are more important to fusion success than the overall dose of rhBMP-2 supplied at the implantation site.

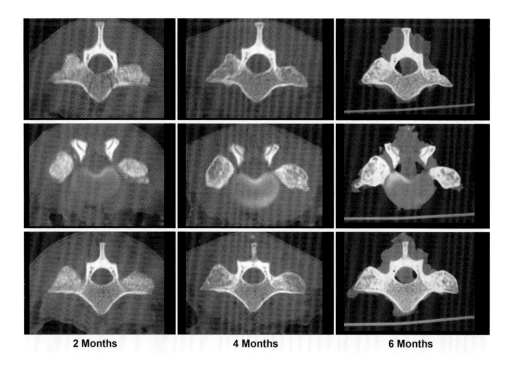

Figure 6
CT scans at 2, 4, and 6 months from a non-human primate treated with rhBMP-2/CRM at the 2.0 mg/mL bulk concentration [20].

The final preclinical study was performed to verify the efficacy of the 2.0 mg/mL rhBMP-2 bulk concentration on the CRM carrier prior to initiating the pivotal investigational device exemption (IDE) trial [20]. In this study, nine rhesus monkeys were treated with either rhBMP-2/CRM at 2.0 mg/mL bulk concentration ($n=3$), rhBMP-2/CRM at 0.6 mg/mL bulk concentration ($n=3$), or rhBMP-2/ACS at 1.5 mg/mL wrapped around dry CRM ($n=3$). CT scans at 2, 4, and 6 months postoperatively and histology results at sacrifice indicated that rhBMP-2/CRM at 2.0 mg/mL bulk concentration and 1.5 mg/mL rhBMP-2/ACS wrapped around dry CRM were successful at inducing 100% fusion (Figs 6 and 7). The fusion masses were also comparable in size and shape to the original implanted graft. None of the animals receiving 0.6 mg/mL rhBMP-2/CRM demonstrated fusion at any time, which indicated that the bulk concentration was too low to induce fusion. The most important aspect of this study was that it verified that rhBMP-2 on the CRM carrier at 2.0 mg/mL bulk concentration was effective at achieving reproducible fusion in the challenging primate intertransverse posterolateral environment.

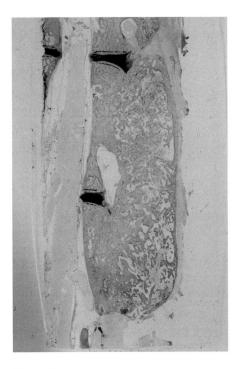

Figure 7
Sagittal histology image for rhBMP-2/CRM. Dense trabecular bone and minimal residual ceramic are present in this fusion mass [20].

Prospective randomized clinical trial

Today, a significant portion of spinal fusion procedures are performed using a posterolateral fusion technique, and the majority of these procedures are done using internal fixation with rods and pedicle screws. rhBMP-2/CRM was investigated in an IDE multicenter [21] prospective randomized instrumented (CD HORIZON® rod and screw fixation system) single-level posterolateral fusion clinical trial comparing rhBMP-2/CRM to iliac crest autograft (Fig. 8). Any local bone obtained from the procedure was discarded in both groups.

A total of 463 patients were enrolled into the study with 224 patients randomized to the autograft group and 239 to the rhBMP-2/CRM group [21, 22]. No difference in patient demographics (age, weight, gender, workers compensation, litigation, tobacco use, alcohol use, and prior back surgery) was found except for a statistically higher incidence of litigation in the control group (2.5% vs. 6.7%, $p = 0.042$). Both the operative time (2.5 vs. 2.9 days, respectively) and blood loss

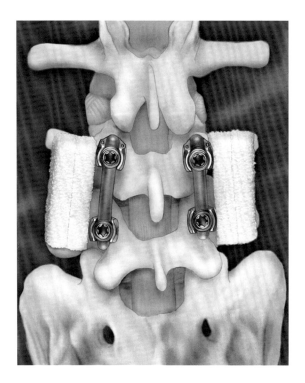

Figure 8
Schematic of rhBMP-2/CRM implanted in the posterolateral environment with supplemental posterior fixation

(343 vs. 448 mL, respectively) were statistically less in the rhBMP-2/CRM group ($p<0.001$). Hospital stay duration was the same for both groups.

Clinical outcome was recorded preoperatively and at 1.5, 3, 6, 12, 24 months postoperatively. Independent radiographic assessment was done at 6, 12, and 24 months postoperatively. At the time of the drafting of this chapter, all patients had reached the 12-month follow-up time point and 282 (61%) of the patients had reached 24 months.

Clinical outcome was assessed using the Oswestry pain questionnaire. Significant improvement in Oswestry scores was found in both groups postoperatively ($p<0.001$), but there was no difference between the two study groups (Fig. 9). Similarly, no difference was found between the two groups for SF-36 physical component or reduction in low back and leg pain scores. Iliac crest donor site pain was present, even up to 24 months, in 54% of the autogenous control patients, with an average pain score of 5.2 based on a 20-point pain scale.

Fusion assessment was conducted by two independent blinded radiologists using both the plain film radiographs and thin slice CT scans. A third independent,

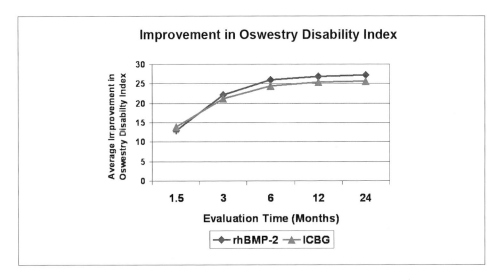

Figure 9
Improvement in Oswestry over the 24-month study.

blinded radiologist was utilized as an adjudicator if the first two radiologists did not agree. The protocol definition criteria for fusion were that there had to be bilateral bridging trabecular bone, and less that 3-mm translation and less than 5 degrees of angulation on the lateral bending films. In addition, under the FDA protocol, if the patient was revised for any reason, even if the radiologist had determined them to be fused, the patient was to be classified as a failed fusion or pseudoarthrosis.

At all time points (6, 12, and 24 months), the rhBMP-2/CRM had higher fusion rates than the autograft control group, but the difference was not statistically significant. Preliminarily, at 24 months, the fusions rates were 86.8% and 94.9% for the autograft and rhBMP-2/CRM groups, respectively. When assessing for the presence of bridging bone using only the thin slice CTs, the rhBMP-2/CRM group was found to have statistically higher rates at all time points, 56.4% vs. 74.4% at 6 months ($p<0.001$), 71.3% vs. 86.9% at 12 months ($p<0.001$), and 82.6 vs. 93.6% at 24 months ($p<0.024$) for autograft and rhBMP-2/CRM, respectively (Fig. 10). Figure 11 shows sample radiographs taken at 12 months postoperative of a patient who received rhBMP-2/CRM.

Discussion and conclusions

In the INFUSE® Bone Graft kit, the ACS is the FDA-approved carrier for delivery of the rhBMP-2 solution. The ACS carrier is hydrated with a fixed volume of rhBMP-2

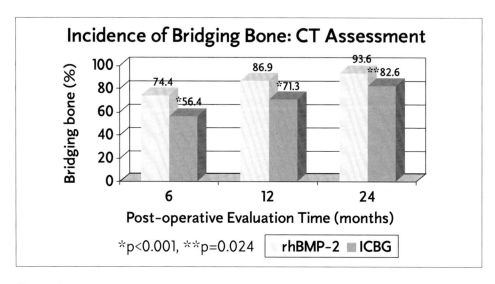

Figure 10
Presence of bridging bone on thin slice CT scans.

solution at the time of surgery and it absorbs the volume of the rhBMP-2 solution placed onto it. The hydrated ACS sponge is soft and pliable and can be easily compressed during surgical implantation or from overlying soft tissue forces. rhBMP-2 at a 1.5 mg/mL concentration delivered on the ACS sponge carrier was FDA approved in four surgical indications for which it was clinically studied (anterior lumbar interbody spinal fusion, treatment of fresh tibial fractures, sinus elevation procedures and the filling of extraction sockets). These surgical applications involve placement of bone graft material in relatively compression-free environments, such as inside a metal fusion cage or a sinus cavity, where pressure on the ACS carrier is minimal. The rhBMP-2/ACS was not placed in a compression-free environment in the tibial fracture study; it was only used as an onlay graft to deliver rhBMP-2 to the fracture site. Its effectiveness was not dependent on forming new bone throughout the ACS carrier, but within the fracture.

Although the ACS carrier was shown to be an effective rhBMP-2 carrier for these indications, the addition of bulking agents to the ACS sponge was found to be required in more mechanically challenging environments where the ACS is subjected to forces from surrounding soft tissues [11, 14]. Therefore, a second-generation carrier, CRM, was developed for use in these implantation environments. Compression resistance was achieved by the addition of resorbable calcium phosphate granules to the collagen sponge in a quantity sufficient to provide a certain degree of compression resistance yet still maintain some flexibility after hydration with rhBMP-2 solution. The CRM carrier demonstrates rapid binding of the rhBMP-2

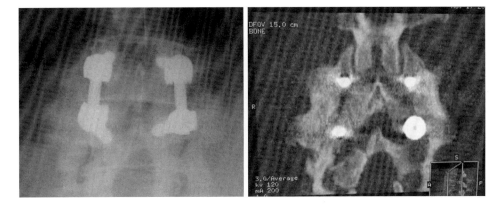

Figure 11
Sample AP radiograph (A) and coronal CT scan (B) of an rhBMP-2/CRM patient 12 months postoperatively.

solution intraoperatively (at a minimum of 5 min) and sustains rhBMP-2 delivery at the surgical site for up to 5 weeks postoperatively [16, 17]. The composition of the calcium phosphate was also carefully chosen to match its resorption profile to the rate of new bone formation [20].

Preclinical testing in non-human primates demonstrated the CRM carrier to be more effective at inducing consistent bone formation and fusion compared to autograft and the ACS carrier alone in mechanically challenging environments, such as a posterolateral fusion model. The high rate of fusion in an instrumented posterolateral environment with rhBMP-2/CRM was subsequently confirmed in a multicenter, prospective, randomized Level I clinical trial [21]. The culmination of preclinical and clinical data suggests that the CRM carrier may successfully expand the use of rhBMP-2 into more challenging bone grafting applications, such as intertransverse posterolateral fusion.

References

1 Burkus JK, Gornet MF, Dickman, Zdeblick TA (2002) Anterior interbody fusion using rhBMP-2 with tapered interbody cages. *J Spin Disord Tech* 15: 337–349
2 BESTT Study Group, Govender S, Csimma C, Genant HK, Valentin-Opran A (2002) Recombinant human bone morphogenetic protein-2 for treatment of open tibial fractures: A prospective, controlled, randomized study of four hundred and fifty patients. *J Bone Joint Surg* 84–A: 2123–2134
3 Swiontkowski MF, Aro HT, Donell S, Esterhai JL, Goulet J, Jones A, Kregor PJ, Nordsletten L, Paiement G, Patel A (2006) Recombinant human bone morphogenetic protein-

2 in open tibial fractures: A subgroup analysis of data combined from two prospective randomized studies. *J Bone Joint Surg Am* 88: 1258–1265

4   Boyne PJ, Lilly LC, Marx RE, Moy PK, Nevins M, Spagnoli DB, Triplett RG (2005) *De novo* bone induction by recombinant human bone morphogenetic protein-2 (rhBMP-2) in maxillary sinus floor augmentation. *J Oral Maxillofac Surg* 63: 1693–1707

5   INFUSE® Bone Graft for Certain Oral Maxillofacial and Dental Regenerative Uses [package insert] (2007) Memphis, TN: Medtronic

6   Fiorellini JP, Howell TH, Cochran D, Malmquist J, Lilly LC, Spagnoli D, Toljanic J, Jones A, Nevins M (2005) Randomized study evaluating recombinant human bone morphogenetic protein-2 for extraction socket augmentation. *J Periodontol* 76: 605–613

7   Glassman SD, Carreon LY, Djurasovic M, Campbell MJ, Puno RM, Johnson JR, Dimar JR (2007) Posterolateral lumbar spine fusion with INFUSE® Bone Graft. *Spine J* 7: 44–49

8   Singh K, Smucker JD, Boden SD (2006) Use of recombinant human bone morphogenetic protein-2 as an adjunct in posterolateral lumbar spine fusion: A prospective CT-scan analysis at one and two years. *J Spinal Disord Tech* 19: 416–423

9   Jones AL, Bucholz RW, Bosse MJ, Mirza SK, Lyon TR, Webb LX, Pollak AN, Golden JD, Valentin-Opran A; BMP-2 Evaluation in Surgery for Tibial Trauma-Allograft (BESTT-ALL) Study Group (2006) Recombinant human BMP-2 and allograft compared with autogenous bone graft for reconstruction of diaphyseal tibial fractures with cortical defects: A randomized controlled trial. *J Bone Joint Surg* 88-A: 1431–1441

10 Schwartz ND, Hicks BM (2006) Segmental bone defects treated using recombinant human bone morphogenetic protein. *J Orthop* 3: e2

11 Akamaru T, Suh D, Boden S, Kim HS, Minamide A, Louis-Ugbo J (2003) Simple carrier matrix modifications can enhance delivery of recombinant human bone morphogenetic protein-2 for posterolateral spinal fusion. *Spine* 28: 429–434

12 Schimandle JH, Boden, SD, Hutton, WC (1995) Experimental spinal fusion with recombinant human bone morphogenetic protein-2. *Spine* 20: 1326–1337

13 Boden SD, Martin GJ, Horton WC, Truss TL, Sandhu HS (1998) Laparoscopic anterior spinal arthrodesis with rhbMP-2 in a titanium interbody threaded cage. *J Spin Disord* 11: 95–101

14 Martin GJ, Boden SD, Morone MA, Moskovitz PA (1999) Posterolateral intertransverse process spinal arthrodesis with rhBMP-2 in a nonhuman primate: Important lessons learned regarding dose, carrier, and safety. *J Spin Disord* 12: 179–186

15 Medtronic, Inc (2000) *Compare compressive resistance of absorbable collagen sponge and the calcium phosphate/collagen composite.* Memphis, TN: Medtronic; Internal Technical Report – TS00-076 (PC-0013)

16 Medtronic, Inc (2007) *An examination of the rhBMP-2 incorporation properties using compression resistant matrix: An* in vitro *study comparing commercial and clinical rhBMP-2/buffer formulations.* Memphis, TN: Medtronic; Internal Technical Report – PC-0666.

17 Louis-Ugbo J, Kim HS, Boden SD, Mayr MT, Li RC, Seeherman H, D'Augusta D, Blake

C, Jiao A, Peckham S (2002) Retention of ^{125}I-labeled recombinant human bone morphogenetic protein-2 by biphasic calcium phosphate or a composite sponge in a rabbit posterolateral spine arthrodesis model. *J Orthop Res* 20: 1050–1059
18 Boden SD, Martin GJ, Morone MA, Louis-Ugbo J, Moskovitz PA (1999) Posterolateral lumbar intertransverse process spine arthrodesis with recombinant human bone morphogenetic protein-2/hydroxyapatite-tricalcium phosphate after laminectomy in the nonhuman primate. *Spine* 24: 1179–1185
19 Suh DY, Boden SD, Louı-Ugbo J, Mayr M, Murakami H, Kim HS, Minamide A, Hutton WC (2002) Delivery of recombinant human bone morphogenetic protein-2 using a compression-resistant matrix in posterolateral spine fusion in the rabbit and in the non-human primate. *Spine* 27: 353–360
20 Barnes B, Boden SD, Louis-Ugbo J, Tomak PR, Park JS, Park MS, Minamide A (2005) Lower Dose of rhBMP-2 achieves spine fusion when combined with an osteoconductive bulking agent in non-human primates. *Spine* 30: 1127–1133
21 Dimar JR, Glassman SD, Burkus JK, Pryor P, Hardacker J, Boden SD (2006) *Evaluation of rhBMP-2/ceramic matrix as an ICBG replacement in posterolateral fusions: A multicenter Level I clinical study.* International Society for the Study of the Lumbar Spine. Bergen, 58
22 Dimar JR, Glassman SD, Burkus JK, Carreon LY (2006) Clinical outcomes and fusion success at 2 years of single-level instrumented posterolateral fusions with recombinant human bone morphogenetic protein-2/compression resistant matrix *versus* iliac crest bone graft. *Spine* 31: 2534–2539

Use of recombinant human BMP-2 in orthopedic trauma

Michael Suk

University of Florida – Shands Jacksonville, Jacksonville, FL 32209, USA

Introduction

It has been more than 40 years since Dr. Marshall Urist's initial discovery of the osteoinductive activity of bone morphogenetic protein (BMP) [1, 2]. Today, as a direct result of that discovery, orthopedic surgeons have the ability to control osteogenesis through the local application of BMP in their patients. Osteoinduction is defined as the ability of a protein to mediate and induce bone formation in an extraossesous site. BMPs are the family of osteoinductive proteins that can be extracted from bone matrix and are responsible for embryonic skeletal formation, skeletal regeneration and bone healing. The first publications on the clinical use of extracts of human BMP from allograft bone matrix began in the late 1980s. This was followed in 1988 by the isolation of the DNA for individual BMP molecules from a purified extract and the subsequent recombinant production of different BMPs [3]. This recombinant production process is similar to the method used to manufacture human insulin and other protein-based drugs (such as erythropoietin and interferon); it enables the reproducible production of a single BMP, at a known concentration and at very high purity. The availability of recombinant human bone morphogenetic protein-2 (rhBMP-2, also known as dibotermin alfa) permitted the large-scale testing of different carriers and concentrations in preclinical animal models, prior to starting clinical studies and gaining regulatory approval of a synthetic rhBMP-2 product.

The osteoinductive potential of rhBMP-2 has been widely studied in different bone healing environments. Preclinical and clinical research has demonstrated that rhBMP-2, carried on an absorbable collagen sponge (ACS), can induce new bone formation when implanted surgically. For example, rhBMP-2/ACS has been shown in animal models to heal critical-size bone defects and accelerate fracture healing. Clinical studies have also demonstrated the efficacy of rhBMP-2/ACS to improve fracture healing in open tibia fractures. These data resulted in EMEA (European Agency for the Evaluation of Medicinal Products) and FDA (Food and Drug Administration) approval of the rhBMP-2/ACS implant for use in treating open tibia fractures (rhBMP-2/ACS is sold by Medtronic under the trade name InductOs in Europe and INFUSE® Bone Graft in the United States). Each kit contains sterile, freeze-dried rhBMP-2 powder, which is reconstituted with sterile water (to a concentration of 1.50 mg/mL) and then that solution is soaked onto the ACS for 15 min prior to implantation (Fig. 1). This same product has also gained approval in both spine fusion and oral/maxillofacial indications (which will not be discussed in this review). rhBMP-2 is the most researched and published bone graft material and is arguably one of the most significant advances in orthopedics. What follows is a summary of the data published or presented by surgeons on their usage of BMP to treat orthopedic trauma patients.

Clinical use of BMP extracts

As Director of the UCLA Bone Research Laboratory, Dr. Urist was able to secure sponsored research grants to investigate the physiology of bone formation and its relationship to BMP (spanning from 1970 to 1994) [4]. Throughout the 1970s, pre-clinical research in Dr. Urist's Lab demonstrated the involvement of BMP in the bone formation cascade of chemotaxis, cell-specific proliferation, stem cell differentiation, callus formation and ultimately bone formation (see Table 1). BMP was shown to induce both endochondral (through a cartilage intermediate) and direct (intramembranous) bone formation. In both cases, woven bone forms and then remodels and becomes populated with bone and bone marrow. Proof that BMP could be extracted from demineralized bone matrix and still retain its ability to form bone came in 1977 [5]. Preclinical experiments on the bone induction potential of extracts of human BMP (hBMP) from bone matrix followed. These extracts of hBMP could then be soaked onto various carriers and used to treat patients.

Surgeons from UCLA have published several case reports on the use of extracts of hBMP, starting in 1988 with a report on 12 nonunion patients [6]. Union was obtained in 11 out of the 12 nonunions, at an average time to union of 4.7 months. This was followed by a report of extracts of hBMP implanted along with autograft in 6 patients with traumatic 3–17-cm tibial defects [7]. The tibia were stabilized with external fixation and function was restored in all 6 patients. A third publication reported on 4 patients with severely deformed nonunions of the distal metaphyseal tibia who were successfully treated by a combination of internal fixation and implantation of absorbable gelatin capsules filled with extracts of hBMP [8]. Before the hBMP implantation, patients had an average of 5.8 surgical procedures. The group next reported on the successful union of 24 out of 25 difficult, previously unresponsive femoral and tibial nonunions after treatment involving allograft bone soaked in extracts of hBMP [9]. This same combination of allograft bone soaked in extracts of hBMP was reported to successfully treat 14 out of 15

Table 1 - BMP and the bone formation cascade.

Chemotaxis	BMP stimulates the migration of mesenchymal stem cells into the area of implantation
Angiogenesis	BMP stimulates osteoblasts to release growth factors (VEGF), which cause the development of new blood vessels
Proliferation	BMP causes a mesenchymal stem cell to divide and proliferate into two daughter cells of the original cell
Differentiation of chondrocytes	Under avascular conditions, low doses of BMP induce mesenchymal stem cells to differentiate into chondrocytes
Differentiation of osteoblasts	Higher doses of BMP induce mesenchymal stem cells to differentiate into osteoblasts
Endochondral bone formation	Over a range of concentrations, BMP initiates a series of events which result in cartilage formation, removal of the cartilage and replacement by bone formation
Direct bone formation	Higher concentrations of BMP cause bone to form at the same time as cartilage formation, suggesting direct (intramembranous) ossification

atrophic femoral nonunions [10] and then 24 out of 30 femoral nonunions [11]. All together, this group reported on the treatment of 90 patients with extracts of hBMP and a high success rate of healing. These treatments were only limited by the limited supply of cadaver bone from which hBMP is extracted, clearly establishing the need for a way to synthetically produce an unlimited supply of BMP for the treatment of patients.

Preclinical studies with rhBMP-2

Preclinical animal studies have demonstrated that rhBMP-2/ACS can induce bone and repair large, segmental critical-sized defects in rat femora, rabbit radii and ulnae, dog radii, and nonhuman primate radii [12–17]. The segmental critical-sized defect is an appropriate animal model to simulate the nonunion healing environment. The induced bone biologically and structurally integrates with the pre-existing bone, and remodels physiologically, i.e., consistent with the biomechanical forces placed on it. In addition, the rhBMP-2/ACS-induced bone can repair itself following fracture, in a manner indistinguishable from native bone. Separate studies demonstrated that rhBMP-2/ACS can accelerate healing in rabbit and goat long bone fracture models [18, 19]. Radiographic, biomechanical, and histological evaluation of the induced bone indicates that it is appropriate for the anatomic site where it forms, and functions biologically and biomechanically as native bone. Application of rhBMP-2/ACS results in the induction of normal bone locally at the site of implantation. The bone induced by rhBMP-2/ACS remodels and assumes the structure appropriate to its location and function, as would be expected from host bone.

Clinical studies with rhBMP-2/ACS

Clinical studies on rhBMP-2/ACS have involved three different treatments regimes: (1) the acute treatment of open fractures, (2) staged bone grafting of fracture with cortical bone loss and (3) bone grafting of a nonunion. Each of these regimes represents a different stage in the attempt to heal a fracture. Acute treatment occurs within the first few weeks following the initial fracture (at definitive wound closure) and is intended to aid the healing response in fractures with small amounts of bone loss. If there is significant cortical bone loss, then the typical treatment involves the initial reduction of the fracture and a second staged bone grafting at 3–12 weeks after the initial fracture. This staged approach, first proposed by Gustilo in the mid-1980s [20, 21], maximizes the potential for a successful outcome by allowing the soft tissue swelling and inflammation to subside prior to adding the bone graft material. Without this staged bone grafting procedure, the cortical bone loss will not heal and the defect will proceed on to a nonunion situation. If the fracture site shows no visibly progressive signs of healing and/or if the internal fixation hardware becomes loose or breaks, then the patient has a nonunion. Surgical treatment of this nonunion may involve the need for a bone graft during that procedure. Throughout the history of orthopedics, surgeons have tried to use a variety of natural and synthetic materials to repair nonunions. Only recently have they been successful at treating these challenging patients with rhBMP-2/ACS.

Acute treatment of open fractures

An open tibia fracture can be a life altering and devastating event, requiring a multi-year treatment and rehabilitation regime. Usually the result of high-energy trauma, open tibia fractures continue to be associated with high rates of nonunion. Many patients require multiple additional surgical treatments (i.e., secondary interventions) to finally achieve healing. A large observational study (Lower Extremity Assessment Project; LEAP), conducted in the United States and involving 545 patients with severe leg injuries, found that the rate of nonunion among grade IIIB fractures to be as high as 28.9% [22]. The LEAP study also found that 19.1% of these patients required at least one secondary intervention during their treatment and that 10.9% of patients still had unhealed fractures after 24 months of treatment. This and other

Figure 1
A 30-year-old male motorcycle accident patient with a grade IIIB open tibia fracture.

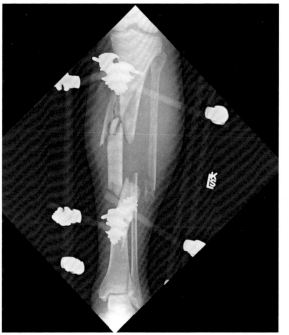 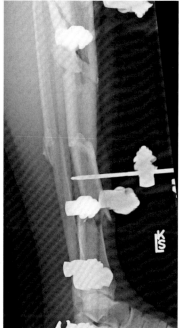

Figures 2 and 3
Antero-posterior (AP) and lateral radiographs.

studies establish the clear need for a biological adjunct to improve healing in open tibia fractures. Figures 1–3 present Case 1, a 30-year-old male motorcycle accident patient with a grade IIIB open tibia fracture.

A prospective safety study of tibial fractures treated with rhBMP-2/ACS (0.43 mg/mL) has been completed [23]. That preliminary safety study involved 12 patients treated with either an intramedullary nail or external fixation. The median time between injury and rhBMP-2/ACS implantation was 4 days (range, 2–6 days). Upon completion of the 9-month follow-up, 9 of the 12 patients had healed without further intervention. The conclusion of that study was that rhBMP-2/ACS was safe and well tolerated as an adjunct to treatment of a tibial fracture. This study set the design criteria for a larger, multicenter clinical trial.

A prospective, randomized clinical study of 450 patients with an open tibial fracture was carried out at 49 centers to evaluate the safety and efficacy of rhBMP-2/ACS [24]. Patients were randomized to one of three groups: (1) standard treatment only, which included intramedullary (IM) nail fixation and routine soft-tissue management (the control group), (2) standard treatment and an implant containing 0.75 mg/mL rhBMP-2, or (3) standard treatment and an implant containing 1.50 mg/mL rhBMP-2. This was designed to be a dosing study to select the proper concentration of rhBMP-2/ACS. Gustilo-Anderson types I–IIIB fractures were studied. Patients were followed for 12 months, with assessments at 6, 10, 14, 20, 26, 39, and 50 weeks after treatment. The rhBMP-2/ACS was placed as an onlay over the fracture at the time of definitive wound closure. The primary endpoint measurement was the proportion of patients who required a secondary intervention within 12 months of wound closure. The 1.50 mg/mL improved overall clinical success by 28% over the control (65% success vs 47%), and reduced the risk of secondary interventions by 44% (26% vs 46%). Adding 1.50 mg/mL to the fracture site accelerated the rate of clinical healing, with 50% of patients being healed 39 days faster than the IM nail alone. Significantly more patients were clinically healed at 14–52 weeks follow-up when 1.50 mg/mL rhBMP-2/ACS was added ($p<0.01$). Surprisingly, the incidence of infection was 45% lower (24% vs 44%, $p=0.047$) in patients with severe open fractures (grade IIIA and IIIB) when 1.50 mg/mL rhBMP-2/ACS was added to the fracture site. This study helped establish the approved concentration of 1.50 mg/mL for the use of rhBMP-2/ACS in acute open tibia fractures. Figures 4–7 present Case 2, with grade IIIA open tibia.

Concurrent with the study above, a 60-patient prospective randomized study involving the same protocol and open tibia fractures was also completed in the United States. A subgroup analysis of data combined from these two studies has been published by Swiontkowski et al. [25]. From the combined data set, 131 patients with Gustilo-Andersen type IIIA or IIIB fractures were analyzed. The use of 1.50 mg/mL rhBMP-2 resulted in a 90% risk reduction in the need for autograft procedure to treat delayed union of these severe fractures. It also significantly decreased the number of invasive secondary interventions that were performed to promote fracture healing. The rate of infection was also significantly lower in patients treated with rhBMP-2 than it was in the control patients, representing a 48% risk reduction in this complication. This analysis further establishes the clinical efficacy of rhBMP-2/ACS for the acute treatment of severe open tibia fractures.

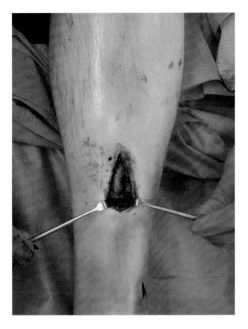
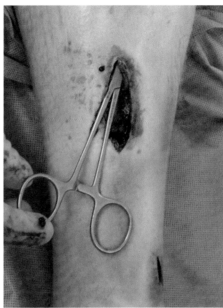

Figures 4 and 5
Grade IIIA open tibia.

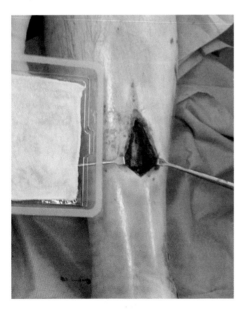
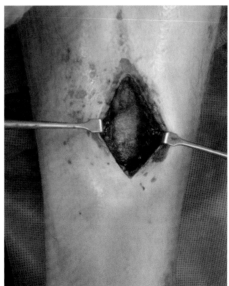

Figures 6 and 7
Placement of rhBMP-2 collagen sponge.

Staged treatment of fractures

The treatment of open tibia fractures associated with substantial cortical bone loss often involves the use of a staged reconstruction. First, the fracture is reduced, a thorough irrigation and debridement is performed, and the soft-tissue injury is treated (Fig. 8). No bone graft is applied during this acute phase of treatment, with bone grafting at a later date once the inflammation is calmed down and adequate/healthy soft-tissue coverage is established (Figs 9, 10). The use of autologous bone graft harvested from the iliac crest is associated with high healing rates, but can result in donor-site morbidity and pain. In addition, autograft bone may be unfeasible or unavailable in certain patients. The need for an alternative to autograft harvest and transplantation has lead to clinical studies involving the use of rhBMP-2/ACS in the staged reconstruction of open tibia fractures with cortical bone loss.

A prospective, randomized study has been conducted using rhBMP-2/ACS combined with allograft bone to treat tibia fractures with traumatic cortical bone loss [26]. Allograft bone was added to supplement the volume that was missing from the cortical bone loss. Adult patients undergoing staged bone grafting between 6 and 12 weeks after injury for diaphyseal tibia fractures (either open or closed injuries) with cortical bone defects between 1 and 5 cm in length and 50% to 100% circumferential bone loss were included. Thirty patients were randomly assigned to two groups: the control group received autogenous bone graft harvested from the iliac crest ($n=15$, autograft group). The treatment group received allograft (in the form of cancellous bone chips) implanted within the bone defect and rhBMP-2/ACS (1.50 mg/mL) applied as an onlay circumferentially around/across the defect site ($n=15$; rhBMP-2/allograft group). All patients were followed for 12 months or, at a minimum, through completion of healing, as assessed by the clinical investigators. Ten of the 15 patients in the autograft group and 13 of the 15 patients in the rhBMP-2/allograft group were deemed to be healed. A statistically significant decrease in estimated blood loss was observed in the rhBMP-2/allograft group compared with the autograft group ($p=0.0073$). Mean length of surgery was also lower in the rhBMP-2/allograft group, but this difference was not statistically significant ($p=0.4309$). On the basis of the study results, rhBMP-2/allograft may be a reasonable alternative to autogenous bone grafting in cases of tibial fractures with traumatic diaphyseal bone loss. The staged bone grafting of a segmental defect is in many ways very similar to the bone grafting of a nonunion site, in that a segmental defect will not heal on its own without the bone graft procedure.

Surgeons at Walter Reed Army Hospital conducted a retrospective review of 59 consecutive patients with type III open tibia fractures and segmental cortical bone loss (defect lengths ranged from 2 to 10 cm) sustained during Operation Iraqi Freedom. These data were presented at the Orthopedic Trauma Association's (OTA) annual meeting in Ottawa, Canada in October 2005 [27]. Patients treated with

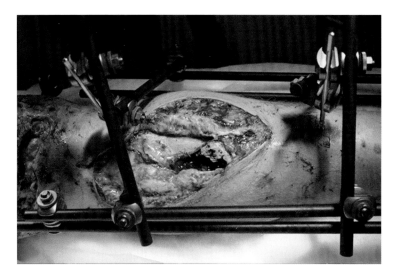

Figure 8
Staged treatment of grade IIIB open tibia fracture.

primary fixation, INFUSE® Bone Graft and supplemental allograft (placed from 6 to 29 days post injury) were reviewed. Patients were treated with either external fixation (ringed or spanning) or internal fixation. Definitive union was observed in 54 of 59 fractures (92%). The rhBMP-2/ACS implant in combination with allograft bone proved to be effective in the supplemental treatment of severe open lower extremity fractures with severe bone loss.

Surgeons in Fort Wayne, Indiana performed a retrospective analysis of patients treated with INFUSE® Bone Graft in combination with two different bone graft substitutes (calcium sulfate or calcium phosphate) to repair 19 segmental cortical bone defects in 18 patients [28]. There were 11 males and 7 females. There were 9 femur fractures, 6 tibial fractures, 2 clavicle fractures, 1 humerus fracture, and 1 ulna fracture. Ten defects were 100% circumferential, while 9 were partial defects. Defect length averaged 4.75 cm, ranging from 1.5 to 8.0 cm. Open fractures occurred in 14 patients. Bony union occurred in 16 of 19 bone defects, with a union rate of 84%. Average time to union was 8.4 months (range 3.5–13.5 months). Failure was noted in 3 patients. Two of these patients were treated early on in the study with a calcium sulfate product in association with the rhBMP-2/ACS implant, and these patients had premature resorption of the ceramic graft material. The third failed patient had fixation failure at 6 weeks due to non-compliance. No infections were reported. No clinical reactions from the rhBMP-2/ACS implant were noted. The rhBMP-2/ACS implant has the capability to heal critically sized bone defects in a variety of patients, with a success rate of 84% in this study. This study also suggests

Table 2 - Combined union success for staged bone grafting studies.

	Success of iliac crest autograft bone	Success of rhBMP-2/ACS + osteoconductive filler
Jones et al. [26]	10/15 (66.7%)	13/15 (86.7%)
Kuklo et al. [27]	–	54/59 (92%)
Schwartz et al. [28]	–	16/19 (84%)
Total	10/15 (66.7%)	83/93 (89%)

that rhBMP-2/ACS should not be used in combination with fast resorbing bone void filler materials, such as calcium sulfate-based ceramics, as they may resorb too quickly. Table 2 summarizes these studies on staged bone grafting studies. Figures 9–14 show Case 1.

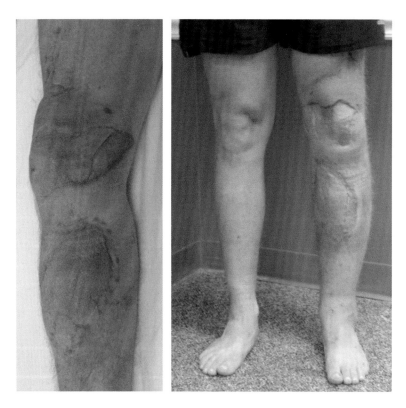

Figures 9 and 10
Soft tissue coverage at 10 weeks.

 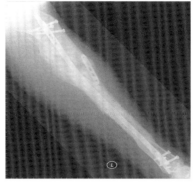

Figures 11 and 12
AP and lateral radiographs at 10 weeks post op.

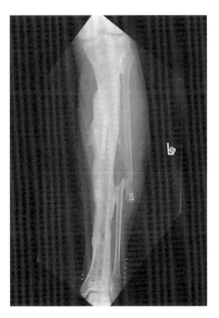 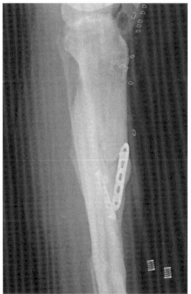

Figures 13 and 14
AP and lateral radiograph 1 year post op.

Treatment of nonunions

The successful surgical treatment of fracture nonunions involves three crucial steps. First, proper reduction and fixation of the fractured bone provides the mechanical stability required for callus formation and remodeling. This step may involve the surgical removal of broken/failed fixation hardware and implantation of new hardware. Second, debridement clears the site of fibrous scar tissue that may impede bone growth, and it exposes fresh bony surfaces, resulting in a local environment suitable for bone formation. Third, bone grafting enhances the osteogenic potential of the site, supplying the materials necessary to induce bone formation. The current gold standard for the treatment of nonunions is the transplantation of autologous bone harvested from the iliac crest, proximal tibia or distal femur. Despite the reportedly high success rate for autograft, patients often need to be treated two or three times before achieving a successful union (Fig. 4).

In the past few years, there have been a number of retrospective reviews by surgeons on their use of INFUSE® Bone Graft to treat patients with fracture nonunions. After obtaining IRB ethics approval, Hicks et al. [29] conducted a retrospective review of 54 nonunions treated with rhBMP-2/ACS either alone or in combination with a calcium phosphate bone void filler material. The nonunions occurred in 14 femurs, 17 tibias, 9 humeri, 8 clavicles, 2 ankles and 4 in the forearm. The average follow up was 13.9 months. Of the 54 patients, 44 (81%) achieved radiographic union and there were no complications associated with the use of the rhBMP-2 observed. Surgeons from Orlando have presented a retrospective review of 98 patients who received INFUSE® Bone Graft, and compared the healing rates of their younger patients to their Medicare aged (<60 years old) patients [30]. That analysis included both primary fractures and nonunions treated with rhBMP-2/ACS. The majority of the treatments occurred in the femur (23), tibia (22) and ankle (19). The authors concluded that both groups healed at a similar rate, and, overall, 94 patients were considered a clinical success. Ringler et al. [31] conducted a review of 20 nonunion patients treated with INFUSE® Bone Graft combined with either calcium phosphate, allograft or autograft bone. Non-unions occurred in the femur, tibia and humerus. At 12 months, 100% of the rhBMP-2 treated patients were considered healed by radiographic bridging bone and clinical success. Surgeons from the Vanderbilt University reported on 101 nonunion patients treated during a 5-year period at their institution [32]. The bone grafting material was decided by the treating physician; 83 patients were treated with iliac crest autograft and 22 patients were treated with INFUSE® Bone Graft combined with allograft chips. Success was determined by radiographic review by an independent orthopedic surgeon. At 6 months, 71/83 (86%) of the autograft and 17/22 (77%) of the rhBMP-2-treated patients were considered to have successful healing. The authors found no difference in the odds of a complication occurring between the two groups. Combined, these four reviews suggest that rhBMP-2/ACS might be a reasonable alternative to autograft bone in the treatment of a nonunion (Tab. 3).

Figures 15–26 present Case 3, a 26-year-old male pedestrian who sustained a Grade IIIA open tibia fracture.

As part of a 'compassionate use' case series (prior to regulatory approval), Richards et al. [33] reported on the use of rhBMP-2/ACS in 7 patients with congenital

Table 3 - Combined clinical success of nonunion treatment

	Success of rhBMP-2/ACS
Hicks et al. [29]	44/54 (81%)
Cole et al. [30]	94/98 (96%)
Ringler et al. [31]	20/20 (100%)
Obremskey et al. [32]	17/22 (77%)
Total	175/194 (90%)

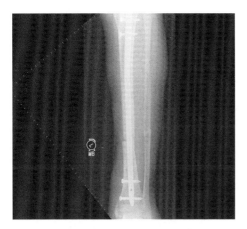 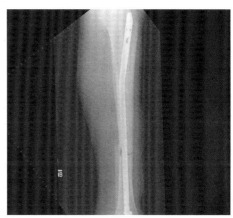

Figures 15 and 16
A 26-year-old male with a grade IIIA open tibia fracture.

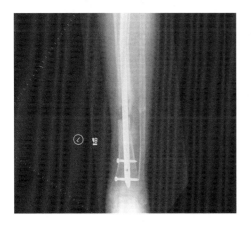 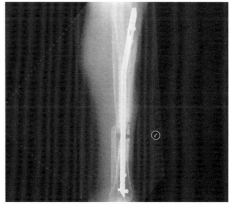

Figures 17 and 18
At 3 month post op.

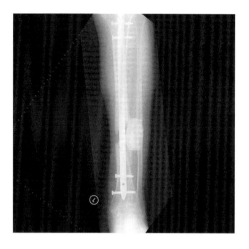
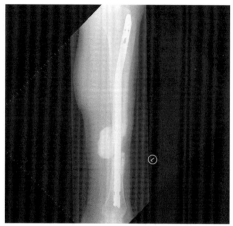

Figures 19 and 20
At 1 year post index; 3 months post exchange nailing.

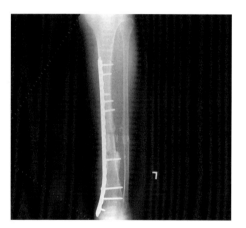
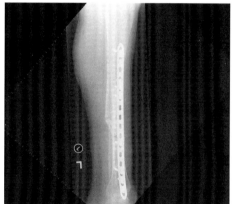

Figures 21 and 22
Non-union repair with allograft and rhBMP-2.

pseudarthrosis of the tibia. The mean age was 7.5 years (range 2–14 years). The rhBMP-2/ACS was implanted in combination with autograft bone. Follow-up averaged 3.6 years and there were no adverse effects from the rhBMP-2 noted. Radiographic and clinical union was achieved in 5 of 7 patients (71%), at a mean of 29 weeks post operation. This study demonstrated that rhBMP-2 was well tolerated in this patient population and that it may be beneficial in increasing union rates in patients with congenital pseudarthrosis.

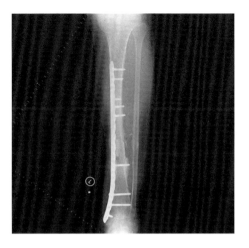
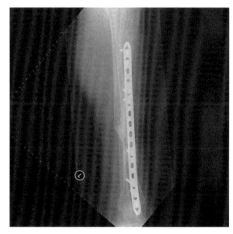

Figures 23 and 24
At 6 months post op.

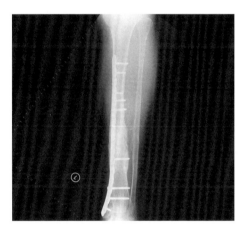
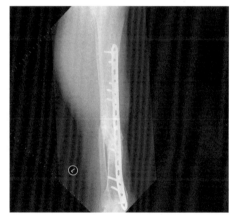

Figures 25 and 26
At 1 year post op.

References

1 Urist MR (1965) Bone: Formation by autoinduction. *Science* 150: 839–839
2 Urist MR, Strates BS (1097) Bone morphogenetic protein. *J Dent Res* 50: 1392–1406
3 Wozney JM, Rosen V, Celeste AJ, Mitsock LM, Whitters MJ, Kriz RW, Hewick RM, Wang EA (1988) Novel regulators of bone formation: molecular clones and activities. *Science* 242: 1528–1534

4   Reddi AH, Marshall R (2003) Urist: A renaissance scientist and orthopaedic surgeon. *J Bone Joint Surg* 85-A: 3–7
5   Urist MR, Granstein R, Nogami H, Svenson L, Murphy R (1977) Transmembrane bone morphogenesis across multiple-walled diffusion chambers. *Arch Surg* 122: 612–619
6   Johnson EE, Urist MR, Finerman GAM (1988) Bone morphogenetic protein augmentation grafting of resistant femoral nonunions: A preliminary report. *Clin Orthop Relat Res* 230: 257–265
7   Johnson EE, Urist MR, Finerman GA (1988) Repair of segmental defects of the tibia with cancellous bone grafts augmented with human bone morphogenetic protein. A preliminary report. *Clin Orthop Relat Res* 236: 249–257
8   Johnson EE, Urist MR, Finerman GAM (1990) Distal metaphyseal tibial nonunion: Deformity and bone loss treated by open reduction, internal fixation, and human bone morphogenetic protein (hBMP). *Clin Orthop Relat Res* 250: 234–240
9   Johnson EE, Urist MR, Finerman GAM (1992) Resistant nonunions and partial or complete segmental defects of long bones: Treatment with implants of a composite of human bone morphogenetic protein (BMP) and autolyzed, antigen-extracted, allogeneic (AAA) bone. *Clin Orthop Relat Res* 277: 229–237
10 Johnson EE, Urist MR (1998) One-stage lengthening of femoral nonunion augmented with human bone morphogenetic protein. *Clin Orthop Relat Res* 347: 105–116
11 Johnson EE, Urist MR (2000) Human bone morphogenetic protein allografting for reconstruction of femoral nonunion. *Clin Orthop Relat Res* 371: 61–74
12 Zabka AG, Pluhar GE, Edwards RB 3rd, Manley PA, Hayashi K, Heiner JP, Kalscheur VL, Seeherman HJ, Markel (2001) Histomorphometric description of allograft bone remodeling and union in a canine segmental femoral defect model: a comparison of rhBMP-2, cancellous bone graft, and absorbable collagen sponge. *J Orthop Res* 19: 318–327
13 Sciadini MF, Johnson KD (2000) Evaluation of recombinant human bone morphogenetic protein-2 as a bone-graft substitute in a canine segmental defect model. *J Orthop Res* 18: 289–302
14 Hollinger JO, Schmitt JM, Buck DC, Shannon R, Joh SP, Zegzula HD, Wozney J (1998) Recombinant human bone morphogenetic protein-2 and collagen for bone regeneration. *J Biomed Mater Res* 43: 356–364
15 Li RH, Wozney JM (2001) Delivering on the promise of bone morphogenetic proteins. *Trends Biotechnol* 19: 255–265
16 Cook SD, Patron LP, Brown S (2005) Evaluation of INFUSE bone graft in a canine critical size defect: effect of sponge placement on healing. *Transactions of the Orthopaedic Trauma Association Annual Meeting*, October 20–22, 2005, Ottawa, Ontario, Canada, 18: Paper no. 36
17 Jones C, Badura J, Marotta J (2007) Examination of extending rhBMP-2/absorbable collagen sponge (ACS) with allograft and ceramic in a canine critical-sized segmental defect. *Transactions of the Orthopaedic Research Society Annual Meeting*, February 11–14, 2007, San Diego, California, USA, 53: Poster no. 354

18 Bouxsein ML, Turek TJ, Blake CA, D'Augusta D, Li X, Stevens M, Seeherman HJ, Wozney JM (2001) Recombinant human bone morphogenetic protein-2 accelerates healing in a rabbit ulnar osteotomy model. *J Bone Joint Surg Am* 83-A: 1219–1230
19 Welch RD, Jones AL, Bucholz RW, Reinert CM, Tjia JS, Pierce WA, Wozney JM, Li XJ (1998) Effect of recombinant human bone morphogenetic protein-2 on fracture healing in a goat tibial fracture model. *J Bone Miner Res* 13: 1483–1490
20 Gustilo RB, Mendoza RM, Williams DN (1984) Problems in the management of type III (severe) open fractures: a new classification of type III open fractures. *J Trauma* 24: 742–746
21 Gustilo RB, Anderson JT (1976) Prevention of infection in the treatment of one thousand and twenty-five open fractures of long bones: retrospective and prospective analyses. *J Bone Joint Surg Am* 58: 453–458
22 Bosse MJ, MacKenzie EJ, Kellam JF, Burgess AR, Webb LX, Swiontkowski MF, Sanders RW, Jones AL, McAndrew MP, Patterson BM, McCarthy ML, Travison TG, Castillo RC (2002) An analysis of outcomes of reconstruction or amputation of leg-threatening injuries. *N Engl J Med* 347: 1924–1931
23 Riedel GE, Valentin-Opran A (1999) Clinical evaluation of rhBMP-2/ACS in orthopedic trauma: a progress report. *Orthopedics* 22: 663–665
24 Govender S, Csimma C, Genant HK, Valentin-Opran A, Amit Y, Arbel R, Aro H, Atar D, Bishay M, Börner MG et al (2002) Recombinant human bone morphogenetic protein-2 for treatment of open tibial fractures: a prospective, controlled, randomized study of four hundred and fifty patients. *J Bone Joint Surg Am* 84-A: 2123–2134
25 Swiontkowski MF, Aro HT, Donell S, Esterhai JL, Goulet J, Jones A, Kregor PJ, Nordsletten L, Paiement G, Patel A (2006) Recombinant human bone morphogenetic protein-2 in open tibial fractures. A subgroup analysis of data combined from two prospective randomized studies. *J Bone Joint Surg Am* 88: 1258–1265
26 Jones AL, Bucholz RW, Bosse MJ, Mirza SK, Lyon TR, Webb LX, Pollak AN, Golden JD, Valentin-Opran A (2006) Recombinant human BMP-2 and allograft compared with autogenous bone graft for reconstruction of diaphyseal tibial fractures with cortical defects. A randomized, controlled trial. *J Bone Joint Surg Am* 88: 1431–1441
27 Kuklo TR, Groth AT, Andersen RC, Islinger RB (2005) Use of rhBMP-2 for open segmental tibia fractures in Iraq. *Transactions from the Orthopaedic Trauma Association Annual Meeting*, Ottawa, Ontario, 20–22 October 2005, Paper no: 49
28 Schwartz ND, Hicks BM (2006) Segmental bone defects treated using recombinant human bone morphogenetic protein. *J Orthopaedics* 3(2) e2, accessed on May 15, 2007 at http://www.jortho.org/2006/3/2/e2/index.htm
29 Hicks MB (2006) BMP-2 and its use in nonunion and malunions *Transactions from the Orthopaedic Trauma Association annual meeting*, Phoenix, AZ, 5–7 October 2006, Poster no: 27
30 Cole JD, Nguyen S, Lamb DL (2006) Review of healing with rhBMP-2/ACS in the medicare-aged population. *Transactions from the Orthopaedic Trauma Association annual meeting*, Phoenix, AZ, 5–7 October 2006, Poster no: 102

31. Ringler JR, Endres TJ, Jones CB (2007) Clinical outcomes for long-bone nonunions implanted with bone morphogenetic protein. *Transactions from the American Academy of Orthopaedic Surgeons annual meeting*, San Diego, CA, 14–18 February 2007, Poster no: 508
32. Obremskey Wt, Kregor PJ, Shuler FD, Tressler MA (2006) Bone morphogenic proteins compared to iliac crest bone graft in long bone nonunions. *Transactions from the Orthopaedic Trauma Association annual meeting*, Phoenix, AZ, 5–7 October 2006, Paper no: 15
33. Richards BS, Johnston CE, Welch RD, Shrader MW (2006) Use of rhBMP-2 in congenital pseudarthorsis of the tibia. *Transactions from the American Academy of Orthopaedic Surgeons annual meeting*, Chicago, IL, 22–26 March 2006, Poster no: P252

The application of recombinant human bone morphogenetic protein on absorbable collagen sponge (rhBMP-2/ACS) to reconstruction of maxillofacial bone defects

Daniel B. Spagnoli

Uniersity Oral and Maxillofacial Surgery, Charlotte, NC, USA

Introduction

The facial skeleton supports a diverse variety of functions, and thus its morphology is equally diverse. In addition, the facial bones are altered from normal by a wide variety of processes including trauma, infection, cysts, neoplasms, congenital defects, developmental deformities, periodontal disease, tooth extraction, atrophy, and edentulous bone loss. The potential for successful reconstruction of these defects is based upon appropriate host graft interactions. Host characteristics that may influence success include the patient's metabolic status, age, tissue vascularity, availability of stem cells, and a host site uncompromised by infection or excessive scar. Graft characteristics should include a lack of immune response, osteogenic potential, appropriate geometry to support vascular and cellular migration, osteoconduction, and a chemotactic and osteoinductive influence on host stem cells, and it must mature to sustain physiological forces.

The traditional approach to facial bone reconstruction has relied on autogenous grafts alone or in combination with various alloplastic or allograft matrices, with or without supporting bone plates, meshes, membranes or osteotomies. When the appropriate stability, host, and graft characteristics are present one can achieve a reconstruction that restores osseous structural morphology and regenerates vascular cellular bone that is capable of supporting function and osseointegration. With the realization that dental implants not only support a functional restoration of the dentition, but they also preserve bone or prevent edentulous bone loss, a greater number of dentists are recommending implants for the restoration of missing teeth. This, together with trends toward prosthetic-driven implant placement, has led to an increase in the number of bone grafts of the facial skeleton beyond the traditional reconstruction of traumatic and pathological defects. The increased number of grafts, together with the desire to decrease morbidity for all grafts and improve patient acceptance of bone grafts, has lead to the pursuit of alternative grafts that can provide the characteristics of morphologically correct structurally sound bone, with viable trabeculae, a vascular marrow space, and a cellular stroma [1].

Since the discovery of the osteoinductive properties of demineralized bone matrix (DBM) by Marshall Urist in 1965, there has been extensive research regarding the role of bone morphogenetic proteins (BMPs) in the regeneration of skeletal defects. Further research led to the isolation of individual BMPs that could stimulate mesenchymal stem cells to form bone, and recombinant technologies are now used to clone and manufacture large quantities of specific BMPs [4–8]. Of the available BMPs, Cheng et al. [9] have shown that BMP-2, 6, and 9 have the most osteogenic activity. However, only rhBMP-2 has been used in the extensive pre-clinical and clinical research (discussed in this chapter) required to obtain FDA approval for grafting of the facial skeleton. In March 2007, Infuse Bone Graft, a combination of recombinant human (rh) BMP-2 and an absorbable collagen sponge (ACS) carrier capable of delivering an rhBMP-2 concentration of 1.5 mg/ml, was approved by the FDA for grafting specific facial defects. The approval was based upon research that defined safety and efficacy of Infuse and states that INFUSE® Bone Graft is indicated as an alternative to autogenous bone graft for sinus augmentations, and for localized alveolar ridge augmentations for defects associated with extraction sockets.

The INFUSE® Bone Graft consists of the two components, rhBMP-2 placed on an ACS. These components must be used as a system for the prescribed indication. The BMP solution component must not be used without the carrier/scaffold component or with a carrier/scaffold component different from the one described in the package insert [10].

Preclinical research

Preclinical research established that rhBMP-2 is a safe and efficacious form of treatment for a variety of simulated critical size defects in craniofacial bones. A study published in 1991 by Toriumi et al. [11] established that 3-cm critical-sized continuity defects in dog mandibles stabilized by bone plates were reconstructed reliably by rhBMP-2 in an inactive dog bone matrix carrier. The rhBMP-2 group included 12 dogs all of which formed bone of sufficient volume density and strength to permit removal of the bone plates by 10 weeks post reconstruction. Histology showed minimal evidence of residual carrier, and normal osteoblast-rich completely incorporated bone. The only deficiency revealed by the study was less than normal bone width, a problem the authors assigned to a lack of space maintenance by the carrier. These animals did not have any infections, fistulas, dehiscence, heterotopic bone, or antibodies to rhBMP-2. In contrast, no treatment and carrier-only controls showed minimal bone formation, they did not achieve a stable union, and a number of control animals developed infections and fistulas.

Subsequent studies showed that critical-sized mandibular continuity defects in porcine and subhuman primate models (*Macaca fascicularis*) could also be reconstructed with rhBMP-2. In the porcine model, the defects were stabilized with reconstruction plates, and, in the *Macaca fascicularis*, the defects were stabilized by an orthopedic mesh [12–15]. In these studies, the rhBMP-2 was delivered on an ACS carrier. In both of these models, the defects were completely reconstructed *via de novo* bone formation similar to intramembranous ossification. The animals did not have ectopic bone formation or other morbidity. In addition, when older *Macaca fascicularis* monkeys were studied (equivalent to 75–80-year-old human men) with the same dose of rhBMP-2 used in middle aged animals, there was no discernable difference in the amount or quality of bone formation [16]. A clinical case report in the veterinary literature has shown successful reconstruction of a golden retriever mandible after hemimandibulectomy resection of an atypical squamous cell carcinoma. The mandible was reconstructed using a reconstruction plate together with a monocortical rib tray and rhBMP-2 in a collagen-tricalcium phosphate sponge (TCP). The dog had an excellent outcome without significant complications. The cosmetic form as well as the functional ability to eat drink, and grasp objects were restored. A bone biopsy at 1 year revealed complete bone formation, with resorption of the TCP, and ongoing remodeling [17].

Boyne et al. [18] studied simulated bilateral alveolar clefts in *Macaca mulatta* animals of 1.5 years of age (similar to 5-year-old humans). In this study, he grafted the cleft on one side with rhBMP-2 in an ACS carrier, and contralateral clefts were grafted with autogenous bone. Both study groups formed bone of similar thickness and there were no histomorphometric differences; however, the rhBMP-2 group formed a normal cortical surface by 3 months and the autogenous group did not. Other areas of clinical interest studied by preclinical animal studies include successful grafting of porcine critical-sized cranial defects [19], and a variety of studies of experimental defects of the alveolar ridge (for a comprehensive review of this subject see Wikesjo et al. [20]). In a study of large saddle defects of hound dog alveolar ridges, rhBMP-2/ACS was found to be equivalent to guided bone regeneration [21], and a number of studies have identified the benefits of adding space maintenance or compression resistant substrates to the rhBMP-2 graft to achieve maintenance of graft volume when tight tissue pressure is present [22–24].

Implant placement in the posterior maxillary alveolar ridge following sinus lift bone grafts has become a mainstay oral rehabilitation procedure. Preclinical maxillary sinus lift studies with rhBMP-2/ACS grafts have been completed in rabbit goat and subhuman primate models. In the rabbit, Wada et al. [25] compared autogenous iliac crest grafts to rhBMP-2/ACS with findings of no histometric or histological differences in the grafts at 8 weeks post-op. Kirker-Head et al. [26] developed a goat model for sinus lift evaluation and studied rhBMP-2/ACS grafts *versus* buffer/ACS controls. They showed that the rhBMP-2/ACS grafts maintained their volume and developed mineralized bone. In contrast, the control grafts were resorbed within 4 weeks. Nevins et al. [27] used the same model to complete an efficacy, safety, and technical feasibility study. Their results showed that rhBMP-

2/ACS grafts produced substantive new bone formation without adverse sequelae. Hanisch et al. [28] extended sinus lift research to subhuman primates (cynomolgus monkeys) by comparing rhBMP-2/ACS sinus lift grafts to ACS plus vehicle controls. Their study showed that significantly greater vertical bone height was achieved by the BMP cohort, but bone density and quality were similar.

The immediate placement of dental implants following tooth extraction is an area of dentistry with significant interest, because it potentially decreases treatment time and provides immediate stabilization of the alveolar bone [29]. In most cases it is considered a contraindication to place immediate implants in extraction sites with buccal wall defects or horizontal gaps between the implants and alveolar wall of ≥3 mm [30]. Thus, it would be beneficial if modifications to immediate implant placement protocols were available to address these deficiencies. Hall et al. [31] have shown that it is possible to adsorb rhBMP-2 onto titanium porous oxide implant surfaces and grow bone attached to the implants in a rat ectopic (subcutaneous ventral thoracic pouch) model. Sumner et al. [32] tested hydroxyapatite/TCP-coated implants treated with rhBMP-2 implanted into canine humeri with 3-mm defects between the implants and bone osteotomy surface. They tested three doses of rhBMP2 applied to the implants, 100, 400, and 800 µg, and compared these to non-BMP-coated controls. They showed that rhBMP2 treatment of the implants led to gap healing, and a 3.5-fold increase in bone ingrowth compared to controls. They also showed that bone ingrowth was inversely related to dose, such that the 100 mg cohort grew more bone than the 800 µg cohort. An alternative approach to treating gaps or defects at immediate implant sites would be to grout the space or defect with rhBMP-2 on a carrier. Wikesjo et al. [33] treated 5-mm peri-implant alveolar ridge gaps with rhBMP-2 in a calcium phosphate cement and compared these to controls that only received the cement. They showed significant bone fill, vertical height gain, and bone implant interface in the experimental group and limited bone growth in the control group. They concluded that the rhBMP-2/calcium phosphate cement graft in conjunction with immediate dental implant placement would be an effective alveolar socket protocol.

Clinical research

Safety and efficacy information from preclinical research played an essential role in designing human clinical trials to assess the feasibility, safety, and efficacy of maxillofacial applications of rhBMP-2. It was determined that initial studies would be performed in alveolar ridge extraction sockets, and deficient maxillas requiring sinus lift bone grafts. These sites were chosen for a number of reasons, including a wealth of preclinical information, ease of measurement, and the need to return to these sites to place dental implants. The placement of dental implants in rhBMP-2 grafts and control sites provides core biopsies of bone histology, bone growth data *via* tetracycline florescence microscopy, radiographic bone loading response data, and implant integration and functional loading data. To obtain the necessary information a series of studies were performed. Initial alveolar ridge and sinus lift studies were designed to assess safety and feasibility. The Phase II studies continued the safety assessment while studying dose response and controls. The Phase III pivotal study was the largest study and was developed to verify safety and efficacy in a large diverse adult population using the sinus lift model.

The Phase I alveolar ridge extraction socket preservation, and ridge augmentation study were reported in two phases by two separate reports. The initial paper by Howell et al. [34] reported on a two-center 12-patient study that enrolled three socket preservation and three socket augmentation patients for each site. The surgical treatment included extraction of teeth where required and preparation of socket or alveolar defect sites by debridement and multiple perforations of the bone with a number two round burr to expose the marrow space. The graft was rhBMP-2/ACS at a concentration of 0.43 mg/ml. Cut portions of the BMP-loaded ACS were placed into the defect to fill the space without voids and then a final layer of the rhBMP-2/ACS was placed over the defect. All study sites were closed with suture, but some did not achieve primary closure. The amount of study material placed in each defect was recorded so that any variation in bone growth response could be assessed. The total dose of rhBMP-2 per patient varied from 0.12 to 0.44 mg/ml. There were no safety issues identified or differences between extraction socket and augmentation patients. A few patients exhibited transient erythema at graft sites, but no patients had mucosal dehiscence or infection. As noted above, primary wound closure over the graft was not achieved at every site, but all of the sites healed without loss of the device or infection. Vital signs, hematology, serum chemistry, and urinalysis were normal for all patients. No patients developed antibodies to rhBMP-2, bovine collagen type 1 or human collagen type 1. It was also determined that it was feasible to use the graft because it was easy to prepare, it conformed to the defect, and it was cohesive and easy to handle. Methods of graft site assessment were evaluated during this study and it was determined that CT scan evaluations of width, height and bone density were an effective assessment tool. The evaluation of rhBMP-2 biological activity is limited in this study by the lack of a control. However, in the extraction socket cohort, five of six patients had good maintenance of socket height. CT scan analysis showed loss of the lamina dura in extraction sockets, and the cortical plate at augmentation sites, suggesting incorporation of the grafts into adjacent bone. CT scans also showed a linear dose-response relationship between the amount of rhBMP-2 graft and the bone height in the socket preservation patients. CT analysis of the augmentation cohort failed to show any significant bone regeneration. The authors felt that the lack of bone growth at augmentation sites was possibly related to an insufficient dose of rhBMP-2, or lack of space maintenance at the graft site. This report showed that it was feasible to graft with rhBMP-2 and that with this concentration short-term safety was excellent. It also suggests that clinical efficacy may require higher doses and space maintenance at graft sites. A second phase of the above study led to the procurement of bone biopsies and placement of dental implants in available sites. The second phase reported by Cochran et al. [35] followed all 12 patients for 3 years; 6 extraction socket patients and 4 augmentation patients received implants at study sites, and all implants integrated and remained functionally loaded. Histological assessment of bone core biopsies from implant sites revealed bone with cellular marrow spaces and trabeculae formed of woven and lamellar bone. They did not identify any long-term safety issues with these patients, and concluded that rhBMP-2/ACS at 4.3 mg/ml is a safe graft for human alveolar ridge extraction socket or augmentation sites.

An additional multicenter open label safety and feasibility study was completed by Boyne et al. [36] with 12 patients who required sinus lift augmentation of the posterior maxilla for dental implants. These patients were grafted with 0.43 mg/ml rhBMP-2/ACS, the same concentration used in the socket preservation and alveolar ridge augmentation study noted above. The total dose of rhBMP-2 used to graft the sinuses varied from 1.77 to 3.40 mg, with the higher dose representing almost an eightfold increase above the highest dose delivered in the socket study. Safety assessment revealed that the most typical adverse events were facial edema, oral erythema, pain, and rhinitis, all within what patients normally experience with this surgery. The patients did not have clinically significant alterations in complete blood counts, blood chemistries, or urinalysis. None of the patients developed antibodies to rhBMP-2 or human type 1 collagen; however, one patient developed antibodies to bovine type 1 collagen with no associated clinical symptoms. The surgeons participating in the study evaluated the rhBMP-2/ACS graft material for cohesiveness, form, handling, placement, volume, ease of placement, and time of preparation to ascertain if it is feasible to implant this graft in sinus lift applications. The scores by the surgeons showed that the graft was easy to use. This study also evaluated the use of periapical films *versus* CT scans for the evaluation of bone grafts in the maxillary sinus and determined that CT scans were the best method of evaluating these grafts. One patient in the study was not evaluable due to a sinus mucus retention cyst, but the remaining 11 patients evaluated by CT scans all grew bone. The mean bone height response above baseline was 8.51 mm with a range of 2.28 mm to 15.73 mm as shown in Table 1 below.

Of the 12 patients in this study, 11 received implants, and 8 of the 11 patients developed adequate bone for placement of implants of ideal size. The three patients who did not receive ideal implants were treated using the sinus lift in fracture method, suggesting that either exposing the rhBMP-2 to the sinus membrane is important or possibly the weight of the infractured bone compromised the space maintenance potential of the ACS. Implants were placed at either 19 weeks or between 24 and 27 weeks, and trephine core biopsies were obtained. All specimens revealed trabecular bone and no residual collagen. As expected, the bone from the

Table 1 - Sinus lift grafts on 11 patients with 0.43 mg/ml rhBMP-2/ACS.

Patient no.	Mean height			Height response		
	Baseline	Week 16	Mean	Rater 1	Rater 2	Rater 3
1	9.40	20.53	11.13	10.50	12.30	10.60
2	5.35	13.70	8.35	8.95	10.05	6.05
3	2.35	18.08	15.73	15.20	16.35	15.65
4	2.42	6.15	3.73	3.80	3.10	4.30
5	3.83	10.38	6.55	5.65	6.45	7.55
6	1.63	9.63	8.00	7.75	7.65	8.60
7	9.37	11.65	2.28	3.45	1.10	2.30
8	3.57	9.70	6.14	6.15	6.20	6.05
9	7.98	16.17	8.18	7.70	9.05	7.80
10	1.62	10.28	8.67	8.45	8.50	9.05
11	2.03	16.87	14.83	14.20	15.30	15.00

later biopsies was more mature. The authors concluded that safety and feasibility criteria were met, and that future studies to evaluate higher concentrations of rhBMP-2/ACS were needed to determine if greater amounts of bone formation were possible. The safety and efficacy of higher doses of rhBMP-2 were evaluated in Phase II clinical trials. Fiorellini et al. [37] reported on a multicenter (eight-site) randomized double-masked, controlled clinical trial evaluating four treatments of extraction sockets with at least 50% buccal wall defects. Two sequential cohorts of 40 patients were randomized in a double-masked manner to receive 0.75 mg/ml or 1.5 mg/ml rhBMP-2/ACS, placebo (ACS) alone, or no treatment to the socket in a 2:1:1 ratio. The mean total dose of rhBMP-2 implanted per socket was 0.9 mg in the 0.75 mg/ml group and 1.9 mg in the 1.5 mg/ml group.

Safety assessment revealed that oral edema and erythema were more common in the rhBMP-2 groups; however, clinical safety profiles were not significantly different between the treatment groups. Serum chemistry and hematology evaluations of the patients did not reveal any clinically significant trends. No patients formed antibodies to rhBMP-2 or human type 1 collagen. Three patients had antibodies to bovine type 1 collagen at baseline. Two of these patients were in the no treatment control group and thus were not exposed to collagen in this study. Eight patients developed antibodies to bovine type 1 collagen, 4 of whom normalized within 4 months and none of them had clinical adverse events that differed significantly from controls. Bone augmentation was significantly greater in sockets treated with 1.5 mg/ml rhBMP-2/ACS than sockets treated with 0.75 mg/ml rhBMP-2/ACS or

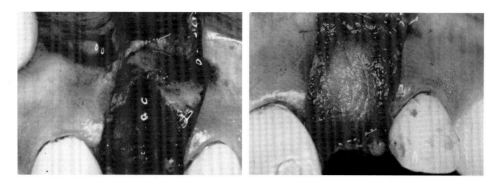

Figure 1
Buccal wall defect exposed and after graft with rhBMP-2/ACS.

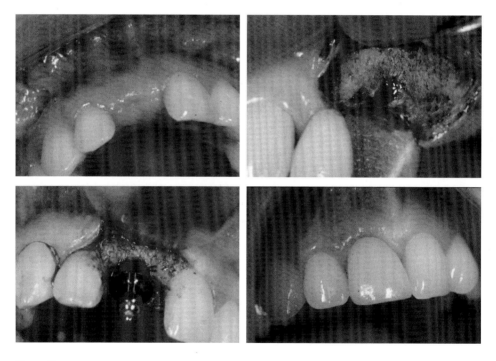

Figure 2
Buccal wall defect regenerated with new bone above and implant placed and restored below.

controls. The adequacy of bone present for implant placement was approximately three times greater in the 1.5 mg/ml rhBMP-2/ACS group than no treatment or placebo controls.

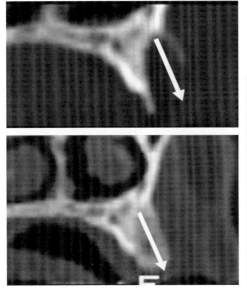
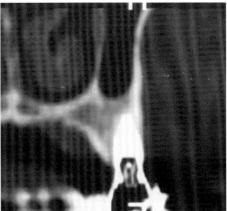

Figure 3
Buccal wall defect after extraction and rhBMP-2/ACS placement above, and after graft maturation and implant placement below.

Histology samples showed mostly trabecular bone with evidence for remodeling from woven to lamellar bone and active osteoblasts as common features.

The authors of this study concluded that 1.5 mg/ml rhBMP-2/ACS was safe at the doses provided to these patients, and had a significant effect on bone regeneration in extraction sockets with buccal wall defects.

Further evaluation of the safety and efficacy of escalating doses of rhBMP-2/ACS was obtained from another Phase II study designed to evaluate bone regeneration in human posterior maxillary sites that required a sinus lift bone graft and dental implants. This study reported by Boyne et al. [38] in 2005 was a controlled six-investigative site study involving patients with ≤6 mm (confirmed by CT scan) of the posterior maxillary bone at proposed implant sites. This 48-patient study was divided into two cohorts of 24 patients each and randomized in a 2:1 ratio to receive either rhBMP-2/ACS (16 patients) or bone graft (8 patients). The rhBMP-2/ACS dose for the first cohort was 0.75 mg/ml and for the second cohort was 1.5 mg/ml. Bone graft for both cohorts was defined as the standard of care bone graft method for the individual investigator's site, but had to consist of autogenous bone alone or mixed with allogeneic bone. The volume of study treatment grafted per sinus was mostly determined by the size of the antral void created by the surgeon at each sinus site. The mean total graft volume per sinus was 6.9 ml for the bone graft group, 11.9 ml

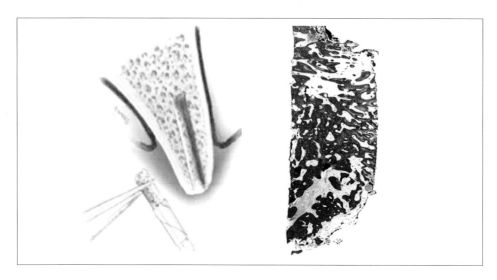

Figure 4
Trephine biopsy technique on the left and trephine histology H&F 2× on the right.

for the 0.75 mg/ml group, and 13.8 ml for the 1.5 mg/ml group. The per patient dose of rhBMP-2, which reflects bilateral sinus lifts in some patients, ranged from a low of 5.2 mg to a high of 48 mg with a mean total dose per patient for the 0.75 mg/ml cohort of 8.9 mg and for the 1.5 mg/ml cohort of 20.8 mg. The total dose of rhBMP-2 received by many of these patients was significantly greater than the doses received in previous studies. The patients did not experience any significant variation in hematology or blood chemistry, and the clinical adverse event experience of the patients was very comparable except for a few exceptions. Edema, rash, and pain were more common in the autogenous bone graft group with the symptoms found in association with the donor site. Less frequent but observed harvest site findings included sensory loss and gait disturbance. Patients that received the 1.5 mg/ml rhBMP-2/ACS graft had a significantly greater amount of facial edema, but no other clinical problems associated with the edema. No patients had antibodies to rhBMP-2 prior to the study and no bone graft or 0.75 mg/ml patients developed them during the study. Two patients (12%) in the 1.5 mg/ml group developed anti-rhBMP-2 antibodies, but the responses were transient with no clinical manifestations of an immune response or neutralizing effects on the activity of rhBMP-2. Bone height gain above baseline was not significantly different for the three study groups. The mean height gain was 9.47 mm for the 0.75 mg/ml group, 10.16 mm for the 1.5 mg/ml group and 11.29 mm for the bone graft group. The gain in bone width was significantly greater for the bone graft group at the ridge crest, but not at the midpoint or apex region of the grafts. One of the most interesting findings in this study was related to changes in bone density when the groups were compared

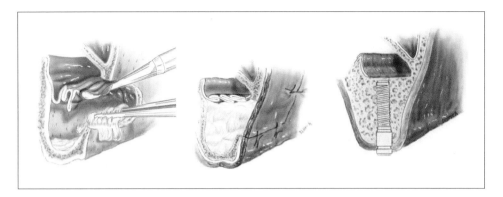

Figure 5
Sinus lift graft treatment sequence.

by CT scan over time. At 4 months, the graft density for the bone graft group was significantly denser than either rhBMP-2/ACS group, and the 1.5 mg/ml group was significantly denser than the 0.75 mg/ml group. By 6 months post grafting, both rhBMP-2 groups had densities greater than the bone graft group, suggesting substantial maturation of the rhBMP-2 grafts between 4 and 6 months. The density change information is also relevant to understanding the different mechanisms of healing that occur between bone grafts and rhBMP-2. In bone grafts, one transfers a tissue that is already mineralized and the graft heals by progressive resorption to expose native BMPs and other cytokines followed by cell differentiation bone deposition, remodeling and maturation. In contrast, rhBMP-2 leads to *de novo* bone formation following a chemotaxis cell differentiation pathway with formation of new bone trabeculae followed by remodeling and maturation. Trephine burr core graft biopsies from implant placement sites showed no significant differences between the three groups in the histological parameters evaluated. However, all the trabeculae of the rhBMP-2 grafts were formed by viable bone, and some areas in the bone graft group had residual non-viable autogenous or allogeneic graft remnants. The proportion of patients who received dental implants that were functionally loaded and remained so for 36 months was not significantly different for the three groups and was highest in the 1.5 mg/ml group at 76%, lowest in the bone graft group at 62%, and in the middle in the 0.75 mg/ml group at 67%.

This study showed the safety of rhBMP-2/ACS at high dose, and unequivocal evidence of substantial *de novo* bone regeneration with dimensions and density similar to the standard of care bone grafts. In addition, the rhBMP-2-derived bone demonstrated normal healing and functional properties by successfully supporting osseointegration and functional loading of dental implants.

Evaluation of the above results by the investigators led to the conclusion that the 1.5 mg/ml rhBMP-2/ACS grafts had safety similar to bone grafts, and eliminated

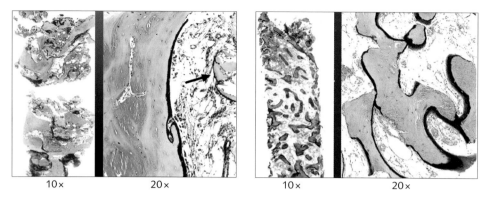

Figure 6
Left core biopsy of autogenous tibial plateau/DBM graft from sinus lift obtained 6 months post grafting. Note viable bone trabeculae and non-viable residual graft (arrow). Right core biopsy of rhBMP-2/ACS. Note all the bone is viable.

donor site morbidity. They also determined that this concentration was the most effective and appropriate concentration of rhBMP-2 for use in a Phase III pivotal study.

The pivotal study (G. Triplett, in preperation) was a multicenter (21-site), open-label, randomized (1:1), and parallel evaluation of sinus lift sites with ≤6 mm of native bone height. The study population included 160 total subjects randomized to receive 1.5 mg/ml rhBMP-2/ACS ($n=82$) or bone graft ($n=78$). Bone graft was defined as autogenous bone alone or mixed with allogeneic bone.

The sinus lift sites were prepared by a sinus window technique designed to preserve and expose the vascularity of the sinus membrane to the graft as shown in Figure 5. The mean volume of study treatment implanted per sinus was 8.3 ml in the bone graft group, and 12.9 ml in the rhBMP-2 group. The mean dose of rhBMP-2 implanted per sinus was 19.4 mg with a standard deviation of 5.9 mg. Both study groups developed a significant amount of new bone above base line. The mean increase in bone height in the rhBMP-2/ACS group was 7.83 mm, and in the bone graft group it was 9.46 mm.

The safety data collected in this study parallels that reported in the Phase II sinus lift clinical trial. Adverse events were generally consistent with the effects of the surgical treatment. Facial edema was more prominent in the rhBMP-2/ACS group, but it does not cause other clinical morbidity or affect outcomes. The bone graft patients had significantly more occurrences of pain, edema, erythema, arthralgia and gait disturbance. The evaluation of immune response also paralleled the results of the Phase II study, with rhBMP-2 antibodies in none of the bone graft patients and 2% of the rhBMP-2 patients. None of the patients with antibodies had a clinical response or neutralizing effect of the biological activity of rhBMP-2. Only 29% of the patients

Figure 7
Sinus lift window, top left. Note vascularity of sinus membrane and maxillary bone. Top right, AhBMP-2 is applied to the ACS. Below, the rhBMP-2/ACS is prepared on the left and placed in the sinus on the right.

in the rhBMP-2/ACS group developed treatment emergent antibodies to bovine type I collagen, and 32% of the bone graft patients developed these antibodies.

The intent-to-treat analysis of the study population was designed to study subject success based upon functional loading of the intended implants placed into grafted study sites with maintenance of functional loading for 6 months. The subject success rate was significantly higher for the bone graft group (91%) than the rhBMP-2/ACS group (79%). Interestingly, both groups exceeded the target success rate of 73% that was derived from a literature review of published success rates for dental implants placed into autogenous bone sinus lift graft sites. The difference in success was partially affected by a greater number of patients who withdrew from the study or were lost to follow up in the rhBMP-2 group. Seven subjects in the rhBMP-2/ACS group withdrew at various functional loading time points *versus* two that withdrew from the bone graft group, all prior to the 6-month evaluation point. A number of the rhBMP-2 patients who grew significant bone height were categorized as failures because in some cases investigators had to perform a secondary bone graft augmentation to obtain extra width at the ridge crest prior to implant placement. The placement of rhBMP-2/ACS in this study was limited to placement within the sinus lift space, and ridge width augmentation was not part of the protocol. Thus, even though a patient grew adequate bone within the grafted sinus for implant placement, the need for any secondary graft, even if it was only one of their treatment planned sites, was outside of the protocol and the patient was deemed a failure. When functionally loaded dental implants were used as a unit of analysis there was no significant difference in implant success rates between bone graft and rhBMP-2/ACS. Dental implant success rates were 83% for both treatment groups after 6 months of functional loading ($p = 1.00$). The success rates decreased to 82% and 79% for the bone graft and rhBMP-2/ACS groups by 12 months of functional loading (p = 0.491), and then did not change significantly to the final 24-month evaluation. This evaluation thus showed no significant difference between implant survivals in newly induced bone in the two treatment groups.

Maxillary sinus lift pivotal study case

Comparison of bone graft and rhBMP-2/ACS group core bone biopsy specimens obtained at the time of implant placement (approximately 8.6 months post graft) by qualitative histological assessment showed no inflammatory cell infiltration or residual collagen matrix. In addition there were no significant differences in trabecular bone assessment, presence of active osteoblasts and osteoclasts, or bone marrow components. Li et al. [39] in a further assessment of the biopsy specimens from this study have shown that the osteogenic pathway of bone regeneration is different when bone graft specimens are compared to rhBMP-2/ACS specimens. This histological study showed that bone graft induces woven bone formation on its surface, forming a woven bone graft complex. This complex undergoes osteoconductive creeping substitution by osteoclastic resorption followed by osteoblastic formation of new lamellar bone. This process results in the formation of viable trabeculae in some areas, but in some areas the bone graft becomes isolated or encapsulated resulting in residual non-viable bone. In contrast, rhBMP-2/ACS promotes an osteoinductive process characterized by chemotactic mediated migration proliferation and differentiation of mesenchymal stem cells into the osteoblast lineage. The histology shows that preosteoblasts differentiate to form osteoblast clusters that produce osteoid matrix. The preosteoblasts and osteoblast clusters are located in close association with vascular spaces and a connective tissue matrix. Foci of woven bone trabeculae form and connect to form a bone trabecular network. The woven bone trabecular network is remodeled by osteoclastic resorption of woven bone and

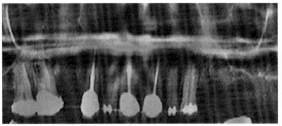

Edentulous left posterior maxilla   Stent to identify study sites

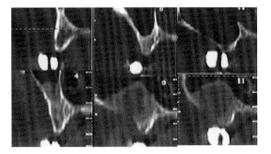

Figure 8
CT scan above, identification of site 8 and 11 with ≤6 mm of basal bone. Below, 6 months after graft with substantial bone regeneration.

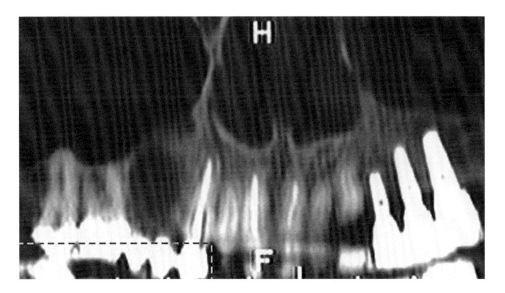

Figure 9
Restored implants. Note significant bone density and intimate implant bone interface.

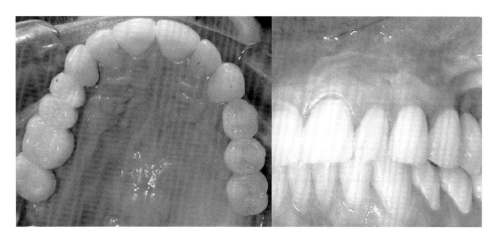

Figure 10
Restored implants with appropriate occlusion and emergence profile.

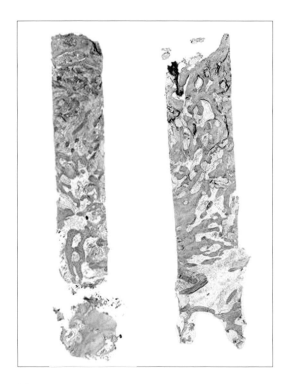

Figure 11
Left, autogenous bone and right, rhBMP-2/ACS trephine core biopsy specimens, 10×. Note equivalent bone trabecular pattern and features.

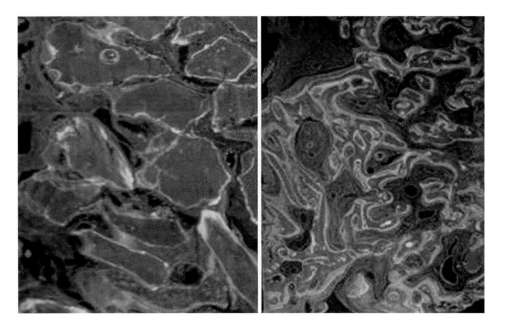

Figure 12
Tetracycline label of autogenous bone on the left and rhBMP-2/ACS on the right, 10×. New bone develops on the surface of autogenous grafts. New bone develops de novo from the core to the surface of the trabeculae with rhBMP-2/ACS.

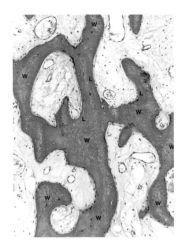

Figure 13
Characteristics of rhBMP-2/ACS derived bone, 20×. All trabeculae are viable and interconnected. The marrow space is very vascular and densely populated with stellate mesenchymal cells.

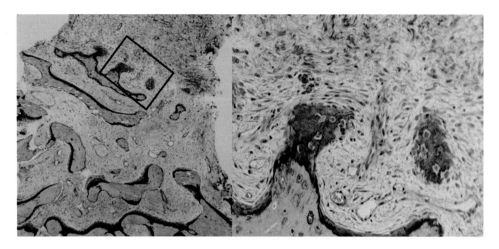

Figure 14
Right: De novo *bone formation in an area near the surface of a graft, 10×.* Left: *The outlined area is shown, 20×. Note preosteoblasts and the condensation of osteoblasts forming new osteoid in areas coupled to blood vessels.*

osteoblastic formation of mature connected lamellar bone trabeculae. In contrast to the bone graft specimens, all of the resultant bone is viable. This histological study of maxillary sinus grafts has thus shown that rhBMP-2/ACS leads to *de novo* bone formation *via* an intramembranous pathway.

Clinical applications

The preclinical studies and clinical research discussed above have shown that rhBMP-2/ACS can safely regenerate vascular viable trabecular bone in the maxillofacial skeleton, and the bone generated by this graft responds favorably to functional loads such as those generated by implant borne prosthesis. It is important to note that the characteristics of the ACS, such as its benign tissue response, ability to concentrate the rhBMP-2 in the grafted location for an appropriate period of time, and its complete replacement by host tissue are important components of this graft, and that rhBMP-2 should not be used clinically without the ACS. Although the ACS has many beneficial properties, it is not designed to substantially maintain space at graft sites. Thus, surgeons planning grafts that use rhBMP-2/ACS as the chemotactic/inductive components of their graft should apply it together with other surgical techniques similar to those already used in conjunction with cortical cancellous particulate grafts. A few of many possible techniques are described in the clinical application section below.

A clinical report by Chin et al. [40] has shown that 1.5 mg/ml rhBMP-2/ACS (infuse) alone or in combination with distraction osteogenesis can successfully repair alveolar and facial clefts in children without the need for autogenous bone procurement. The authors report bone consolidation across the cleft in 49 of 50 clefts treated without any systemic or graft site adverse events. A clinical paper by Herford et al. [41] has also documented successful treatment of alveolar clefts in children, but with a reduced dose of rhBMP-2/ACS of 1.0 mg/ml. This report also provides evidence of clinical success when large maxillary or mandibular atrophic, tumor, or trauma defects are treated with rhBMP-2/ACS in combination with bone plates or meshes, and in some defects in combination with autogenous bone. Eight patients with alveolar clefts at our site have been treated with 1.5 mg/ml rhBMP-2/ACS (infuse). The first of these patients was an adult with an untreated alveolar cleft and the six were children followed by our cleft palate team. One child was transferred to our team after an unsuccessful attempt to graft her cleft with autogenous bone from the iliac crest. None of the patients had significant swelling or other adverse events. All of the children recovered well and were discharged to home on the day of surgery, except one who was kept overnight due to simultaneous treatment of multiple mandibular cysts. None of the grafted sites developed a dehiscence or infection, and soft tissue healing appeared mature within 1 week. All of the patients achieved complete bone union at the cleft site, without signs of adjacent bone or tooth resorption. Canine teeth have shown natural root development and migration through the bone. The parents of the children and the adult patient all reported that discomfort associated with the cleft graft was minimal and narcotic analgesics were not required. The parents of the child previously treated with an autogenous iliac crest graft stated that the elimination of the pain associated with the iliac crest graft completely changed their child's experience and helped alleviate the fear of surgery she had developed from the former treatment.

Large maxillary or mandibular defects treated at our site with rhBMP-2/ACS (infuse) have been treated using infuse alone or in combination with allogeneic bone or autogenous bone marrow. Large mandibular reconstruction sites have been stabilized with bone plates or in some defects that cross the mandibular midline; we have used perforated cadaveric mandibles as a tray to contain the graft and have achieved a functional and esthetic facial form. Large maxillary or mandibular defects that are the result of benign tumors or trauma are similarly stabilized with bone plates or titanium mesh, and treated with infuse combined with freeze-dried demineralized or mineralized particulate bone, or particulate ceramic. Large defects that have been irradiated are treated with hyperbaric oxygen (HBO) of approximately 40 pre-op treatments and then a combination of infuse and autogenous marrow followed by 10–20 post-op HBO treatments [42]. The marrow provides a source of osteoblasts that may contribute to osteogenesis, as well as mesenchymal stem cells and bone marrow capillary pericytes, both of which are capable of maturing to form osteoblasts under the influence of chemotactic and inductive signals of BMP. In addition,

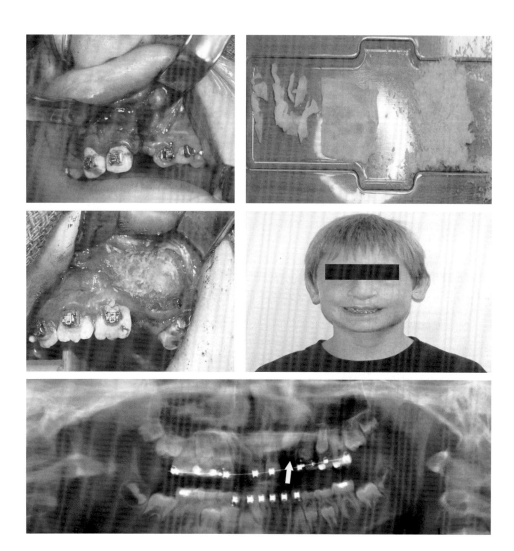

Figure 15
Top left: Alveolar cleft with nasal floor and palate repaired. Top right: 1.5 mg/ml rhBMP-2/ACS with DBM. Middle left: Graft placed. Middle right: One week post-op with no swelling. Bottom: Alveolar cleft reconstructed (arrow) at 4 months with bone continuity and normal canine tooth migration. Note the normal roots on adjacent teeth.

pericytes are tolerant of a hypoxic environment and may survive transplantation. The use of infuse together with autogenous marrow or other alloplastic/allogeneic matrices is technique sensitive, and one must respect the role of rhBMP-2 chemotaxis, and induction as well as the principle of an appropriate host environment with mesenchymal stem cells and blood vessels available to facilitate bone regeneration.

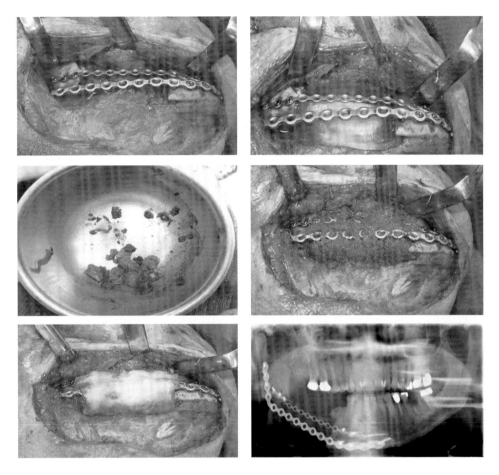

Figure 16
Top left: Mandible resected due to osteoradionecrosis and stabilized with reconstruction plate and superior tension band/soft tissue tenting plate. Top right: 1.5 mg/ml rhBMP-2/ACS placed at medial soft tissue interface. Middle left: Autogenous bone marrow procured by iliac crest trephine technique. Middle right: Core of the defect filled with a 1:1 mixture of marrow and rhBMP-2/ACS. Bottom left: rhBMP-2/ACS covers the graft and provides interface to superficial tissue. Bottom right: 2 months post-op panorex. Note early bone formation and fracture of tension band plate.

We find it important to line the periphery of graft site defects with the 1.5 mg/ml rhBMP-2/ACS infuse providing the research defined concentration at the interface between the defect and surrounding tissues where it can have the most significant chemotactic and vascular field effect. We then particulate additional infuse sponge (approximately 3–4-mm cubes) and blend in an approximate 1:1 ratio with the

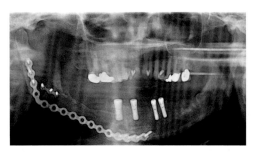
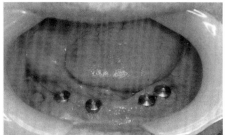

Figure 17
Bottom left: 6 months post-op panorex, the bone graft is mature, tension band plate partially removed, and dental implants placed. Bottom right: Alveolar ridge is restored and dental implants are placed.

chosen matrix or graft extender to be placed in the core of the defect. This approach assures a maximal rhBMP-2 chemotactic effect from the surface of the defect, and improves graft geometry/bulk together with the inductive properties of rhBMP-2 in the core of the defect. Using this approach, the rhBMP-2 concentration is maintained within the therapeutic range defined in the maxillary sinus lift dose response study of 0.75–1.5 mg/ml rhBMP-2/ACS. This concept places the higher 1.5 mg/ml rhBMP-2/ACS concentration at the periphery where signal diffusion and chemotaxis can affect a larger tissue and vascular field, thus achieving rapid angiogenesis and substantial population of the graft with inducible cells. The core of the graft maintains an approximate rhBMP-2 concentration of 0.75 mg/ml together with the benefits of improved osteoconduction due to the graft geometry and architecture provided by the matrix.

The approach of using a composite graft as discussed above can also be used to extend the infuse graft when treating maxillary sinus lift sites. Using this approach has maintained the efficacy of rhBMP-2/ACS while by reducing total dose, and thus decreasing side effects such as swelling. As noted earlier, autogenous bone grafts placed in maxillary sinus lift sites together with allogeneic particulate bone or alloplastic materials tend to outperform autogenous grafts alone because the additional matrix is more resistant to resorption. We have evaluated Infuse rhBMP-2 together with freeze-dried particulate demineralized or mineralized bone allograft, as well as alloplastic ceramics including Cgraft® and Mastergraft Matrix® at sinus lift sites. We have completed a preliminary assessment of 20 consecutive patients with 30 maxillary sinus lifts and 90 implants placed into sites that had ≤6 mm of native bone prior to grafting (C. Chandler, D. Misiek, X.J. Li, D. Cook, and D. Spagnoli, in preparation). All of the sinus lifts were prepared by the sinus window technique used in the dose response and pivotal studies using the graft placement techniques discussed above. None of the patients experienced infection, wound dehiscence or

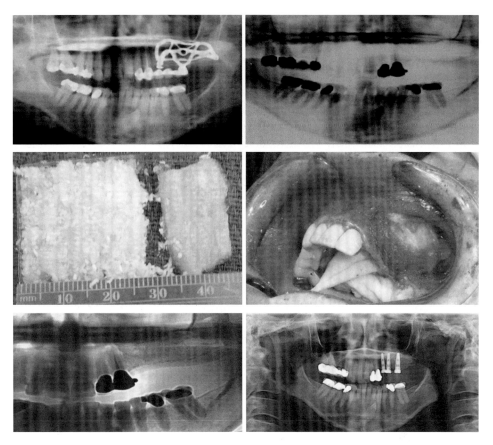

Figure 18
Top left: Patient with failing maxillary subperiosteal implant. Top right: Panorex after removal of the subperiosteal implant. Note less than 1 mm of crestal bone. Middle left: rhBMP-2/ACS (Infuse) and 1:1 mixture of infuse with DBM. Middle right: Graft placed in sinus lift site. Bottom left: Panorex of graft site 4 months post-op. Bottom right: Panorex of graft site 6 months post-op with implants.

sinus side effects. All of the sites grew sufficient bone for implant placement and implants were placed on average at 6 months after the graft. All of the implants achieved initial integration, but one implant in a patient that had six implants failed during the temporization stage. This patient had infuse/particulate DBM grafts to both sinuses, and the histology showed excellent formation of viable bone trabeculae. The reason for implant failure was due to early off axis loads from a partial denture on implants placed for fixed restorations. We have obtained representative core trephine samples from implant sites of each of the composite grafts noted above. Histological assessment of these grafts shows that they all grow significant

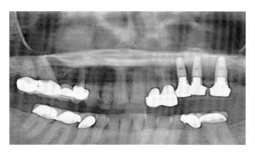
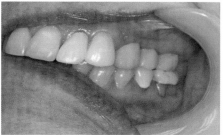
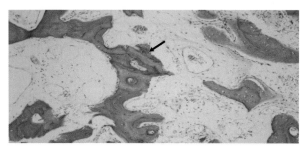
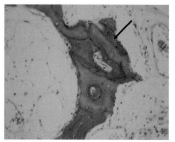

Figure 19
Top left: Panorex of graft site 9 months post-op with implants restored. Note significant increase in bone density. Top right: Functional restoration of implants with proper emergence profile and occlusion. Bottom left: 20× representative histological specimen from trephine biopsy of implant site. Note viable bone trabeculae and vascular marrow space with minimal residual allograft. Bottom right: 40× view of area in parenthesis on the left. Note normal viable trabecular bone with a small area of residual non-viable graft (arrow) embedded in the specimen.

viable trabecular bone similar to rhBMP-2/ACS alone, with the difference being that there is more residual matrix in the ceramic specimens than the allogeneic bone specimens.

Conclusion

The information discussed in this review covers much, but not all, of the important research that has led to the US FDA approval of rhBMP-2/ACS Infuse for oral, maxillofacial, and dental bone reconstruction. This research has provided compelling evidence that rhBMP-2/ACS is a safe and effective graft substance when used to treat maxillary alveolar ridge buccal wall extraction site defects, and when used as a graft for maxillary sinus lift augmentations. Additional clinical experience in treating congenital cleft defects of the maxillary alveolus, and various other tumors or trauma-related defects are provided as examples of the potential extended application of rhBMP-2/ACS. The graft placement techniques described above, although useful now, will ultimately be replaced as improved or enhanced when matrices become available for combination with rhBMP-2ACS [43, 44]. Protein-signaled bone grafting with rhBMP-2 has the potential to significantly enhance patient care due to its efficacy and propensity to accelerate wound healing, and by the reduction of donor site morbidity and invasiveness. We have already witnessed significant psychological benefits for patients that fear bone graft procurement.

References

1 Spagnoli DB, Gollehon SG, Misiek DJ (2004) Preprosthetic and reconstructive surgery. In: M Miloro (ed): *Petersons principles of oral and maxillfacial surgery*. BC Decker, Hamilton, 157–187
2 Urist MR (1965) Bone: Formation by autoinduction. *Science* 150: 893
3 Urist MR, Mikulski A, Lietze A (1979) Solubilized and insolubilized bone morphogenetic protein. *Proc Natl Acad Sci USA* 76: 1828
4 Wozney JM, Rosen V, Celeste AJ, Mitsock LM, Whitters MJ, Kriz RW, Hewick RM, Wang EA (1988) Novel regulators of bone formation: Molecular clones and activities. *Science* 242: 1528–1534
5 Wang EA, Rosen V, Cordes P, Hewick RM, Kriz MJ, Luxenberg DP, Sibley BS, Wozney JM (1988) Purification and characterization of distinct bone-inducing factors. *Proc Natl Acad Sci USA* 85: 9484–9488
6 Wang EA, Rosen V, D'Alessandro JS, Bauduy M, Cordes P, Harada T, Israel DI, Hewick RM, Kerns KM, LaPan P et al (1990) Recombinant human bone morphogenetic protein induces bone formation. *Proc Natl Acad Sci USA* 87: 2220–2224
7 Wozney JM (1993) Bone morphogenetic proteins and their gene expression. In: M Noda (ed): *Cellular and molecular biology of bone*. Academic Press, San Diego, 131–167
8 Wozney JM (1995) BMPs: Roles in bone development and repair. In: *Proceedings of the Portland Bone Symposium*, 2–5 August; Portland, OR
9 Cheng H, Jiang W, Phillips FM, Haydon RC, Peng Y, Zhou L, Luu HH, An N, Breyer B, Vanichakarn P et al (2003) Osteogenic activity of the fourteen types of human bone morphogenetic proteins (BMPs). *J Bone Joint Surg* AM 85-A: 1544–1552
10 Medtronic package insert no M704819B001 (2007) INFUSE® Bone graft for certain oral maxillofacial and dental regenerative uses.
11 Toriumi DM, Kolter HS, Luxenberg DP, Holtrop ME, Wang EA (1991) Mandibular reconstruction with a recombinant bone-inducing factor. Functional, histologic, and biomechanical evaluation. *Arch Otolarynogol Head Neck Surg* 117: 1110–1112
12 Boyne PJ (1996) Animal studies of application of rhBMP-2 in maxillofacial reconstruction. *Bone* 19 (1 Suppl): 83S-92S
13 Boyne PJ, Nakamura A, Shabahang S (1999) Evaluation of the long-term effect of func-

tion on rhBMP-2 regenerated hemimandibulectomy defects. *Br J Oral Maxillofac Surg* 37: 344–352

14 Boyne PJ (2001) Application of bone morphogenetic proteins in the treatment of clinical oral and maxillofacial osseous defects. *J Bone Joint Surg Am* 83-A (Suppl 1): S146–150

15 Carstens MH, Chin M (2005) *In situ* osteogenesis: Regeneration of 10-cm mandibular defect in porcine model using recombinant human bone morphogenetic protein-2 (rhBMP-2) and helistat absorbable collagen sponge. *J Craniofac Surg* 16: 1033–1042

16 Boyne PJ, Salina S, Nakamura A, Audia F, Shabahang S (2006) Bone regeneration using rhBMP-2 induction in hemimandibulectomy type defects of elderly sub-human primates. *Cell Tissue Bank* 7: 1–10

17 Boudrieau RJ, Mitchell SL, Seeherman H (2004) Mandibular reconstruction of a partial hemimandibulectomy in a dog with severe malocclusion. *Vet Surg* 33: 119–130

18 Boyne PJ, Nath R, Nakamura A (1988) Human recombinant BMP-2 in osseous reconstruction of simulated cleft palate defects. *Br J Oral Maxillofac Surg* 36: 84–90

19 Chang SC, Wei FC, Chuang H (2003) *Ex vivo* gene therapy in autologous critical-sized craniofacial bone regeneration. *Plast Reconstr Surg* 112: 141–150

20 Wikesjo UM, Sorensen RG, Wozney JM (2001) Augmentation of alveolar bone and dental implant osseointegration: Clinical implications of studies with rhBMP-2. A comprehensive review. *J Bone Joint Surg* AM 83: S136–S145

21 Jovanovic SA, Hunt DR, Bernard GW, Spiekermann H, Nishimura R, Wozney JM, Wikesjö UME (2003) Long-term functional loading of dental implants in rhBMP-2 induced bone. A histologic study in the canine ridge augmentation model. *Clin Oral Implants Res* 14: 793–803

22 Barboza EP, Duarte MEL, Geolás L, Sorensen RG, Riedel GE, Wikesjö UME (2000) Ridge augmentation following implantation of recombinant human bone morphogenetic protein-2 in the dog, *J Periodontol* 71: 488–496

23 Hanisch O, Sorensen RG, Kinoshita A, Spiekermann H, Wozney JM, Wikesjö UME (2003) Effect of recombinant human bone morphogenetic protein-2 in dehiscence defects with non-submerged immediate implants: An experimental study in cynomolgus monkeys. *J Periodontol* 74: 648–657

24 Barboza EP, Caúla AL, Caúla FO, de Souza RO, Neto LG, Sorensen RG, Li XJ, Wikesjö UME (2004) Effect of recombinant human bone morphogenetic protein-2 in an absorbable collagen sponge with space-providing biomaterials on the augmentation of chronic alveolar ridge defects. *J Periodontol* 75: 702–708

25 Wada K, Niimi A, Watanabe K, Sawai T, Ueda M (2001) Maxillary sinus floor augmentation in rabbits: A comparative histologic-histomorphometric study between rhBMP-2 and autogenous bone. *Int J Periodontics Restorative Dent* 21: 253–263

26 Kirker-Head CA, Nevins M, Palmer R, Nevins ML, Schelling SH (1997) A new animal model for maxillary sinus floor augmentation: Evaluation parameters. *Int J Oral Maxillofac Implants* 12: 403–411

27 Nevins M, Kirker-Head C, Nevins M, Wozney JA, Palmer R, Graham D (1996) Bone

formation in the goat maxillary sinus induced by absorbable collagen sponge implants impregnated with recombinant human bone morphogenetic protein-2. *Int J Periodontics Restorative Dent* 16: 9

28 Hanisch O, Tatskis DN, Rohrer MD, Wohrle PS, Wozney JM, Wikesjo UM (1997) Bone formation and osseointegration simulated by rhBMP-2 following subantral augmentation procedures in nonhuman primates. *Int J Oral Maxillofac Implants* 12: 785–792

29 Bogaerde LV, Rangert B, Wendelhag I (2005) Immediate/early function of Branemark system® TiUnite™ implants in fresh extraction sockets in maxillae and posterior mandibles: An 18–month prospective clinical study. *Clin Implant Dent Relat Res* 7: S121–S130

30 Hammerle CH, Chen ST, Wilson TG (2004) Consensus statements and recommended clinical procedures regarding the placement of implants in extraction sockets. *Int J Oral Maxillofac Implants* 19 (Suppl): 26–28

31 Hall J, Sorensen RG, Wozney JM Wikesjo UM (2007) Bone formation at rhBMP-2 coated titanium implants in the rat ectopic model. *J Clin Periodontol* 34: 444–451

32 Sumner DR, Turner TM, Urban RM, Turek T, Seeherman H, Wozney JM (2004) Locally delivered rhBMP-2 enhances bone ingrowth and gap healing in a canine model. *J Orthop Res* 22: 58–65

33 Wikesjo UM, Sorensen RG, Kinoshita A (2002) rhBMP-2/aBSM induces significant vertical alveolar ridge augmentation and dental implant osseointegration. *Clin Implant Dent Relat Res* 4: 174–182

34 Howell TH, Fiorellini J, Jones A, Alder M, Nummikoski P, Lazaro M, Lilly L, Cochran D (1997) A feasibility study evaluating rhBMP-2/absorbable collagen sponge device for local alveolar ridge preservation or augmentation. *Int J Periodontics Restorative Dent* 17: 125–139

35 Cochran DL, Jones AA, Lilly LC, Fiorellini JP, Howell H (2000) Evaluation of recombinant human bone morphogenetic protein-2 in oral applications including the use of endosseous implants: 3-year results of a pilot study in humans. *J Periodontol* 71: 1241–1257

36 Boyne PJ, Marx RE, Nevins M, Triplett G, Lazaro E, Lilly LC, Alder M, Nummikoski P (1997) A feasibility study evaluating rhBMP-2/absorbable collagen sponge for maxillary sinus floor augmentation. *Int J Periodontics Restorative Dent* 17: 11–25

37 Fiorellini JP, Howell TH, Cochran D, Malmquist J, Lilly LC, Spagnoli D, Toljanic J, Jones A, Nevins M (2005) Randomized study evaluating recombinant human bone morphogenetic protein-2 for extraction socket augmentation. *J Periodontol* 76: 605–613

38 Boyne PJ, Lilly LC, Marx RE, Moy PK, Nevins M, Spagnoli DB, Triplett RG (2005) *De novo* bone induction by recombinant human bone morphogenetic protein-2 (rhBMP-2) in maxillary sinus floor augmentation. *J Oral Maxillofac Surg* 63: 1693–1707

39 Li XJ, Boyne P, Lilly L, Spagnoli DB (2007) Different osteogenic pathways between rhBMP-2/ACS and autogenous bone graft in 190 maxillary sinus floor augmentation surgeries. *J Oral Maxillofac Surg* 65 (9 Suppl 2): 36

40 Chin M, Ng T, Tom WK, Carstens M (2005) Repair of alveolar clefts with recombinant

human bone morphogenetic protein (rhBMP-2) in patients with clefts. *J Craniofac Surg* 16: 778–789

41 Hereford AS, Boyne PJ, Williams RP (2007) Clinical applications of rhBMP-2 in maxillofacial surgery. *California Dental Assoc J* 35: 335–341

42 Marx RE, Arnes JR (1982) The use of hyperbaric oxygen as an adjunct in the treatment of osteoradionecrosis of the mandible. *J Oral Maxillofac Surg* 40: 412–418

43 Seeherman H, Wozney JM (2005) Delivery of bone morphogenetic proteins for orthopedic tissue regeneration. *Cytokine Growth Factor Rev* 16: 329–345

44 Wikesjo UM, Polimeni G, Qahash M (2005) Tissue engineering with recombinant human bone morphogenetic protein-2 for alveolar augmentation and oral implant osseointegration: experimental observations and clinical perspectives. *Clin Implant Dent Relat Res* 7: 112–119

Clinical outcomes using rhBMP-2 in spinal fusion applications

J. Kenneth Burkus

Staff Physician, Spine Service, The Hughston Clinic, P.C., 6262 Veterans Parkway, Columbus, Georgia 31908-9517, USA

Introduction

Bone morphogenetic proteins (BMPs) are an integral part of new bone formation and are capable of inducing the entire bone formation cascade [1, 2]. It is this unique property that allows these proteins, when they are combined with a suitable carrier, to be used as a bone graft replacement. The new bone formation occurs in four distinct phases: recruitment and proliferation, differentiation, calcification, and maturation [3]. During the recruitment and proliferation phase, undifferentiated mesenchymal cells are attracted to the site by chemotaxis. These stem cells divide and increase in number. During the differentiation phase, the mesenchymal stem cells are transformed into osteoblasts. During the calcification phase, the osteoblasts produce matrix, generate callous and form new bone. During the maturation phase, the newly formed bone remodels into trabecular bone and increases in vascularity.

Spine fusion, creating new bone formation across a spinal motion segment, occurs in a challenging healing environment. During the spinal fusion procedure, bioactive materials are placed in varying anatomic positions spanning an intervertebral motion segment [4]. A successful fusion induces new bone formation that bridges an anatomic region of the spine that normally does not support viable bone. It is a complex process that does not always heal successfully. This process is, in part, a race among resorption of the graft material, cellular apoptosis, and the formation of new bone growing through the graft connecting the two adjacent, mobile vertebral bodies. BMPs function as a differentiation factor and act on mesenchymal stem cells to form bone [2, 5]. When used at an optimized concentration and with an appropriate carrier, BMPs can be a successful bone graft replacement [6–8].

Administration of rhBMP-2

The rhBMP-2 acts as a local agent, not as a systemic agent. The process of osteoinduction using rhBMP-2 requires surgical implantation. To apply rhBMP-2 locally, the protein is applied to a resorbable biological matrix material. The matrix serves several functions: (1) it helps to deliver the rhBMP-2 to the surgical site by improving the handling characteristics for easy placement; (2) it helps to augment retention of the BMP at the site; and (3) it provides a compatible environment for bone induction [9]. The concentration and dose used in human clinical trials have been formulated to the specific site of implantation and to the specific carriers.

Preclinical testing

Rigorous preclinical testing was undertaken to establish the appropriate dose, concentration, and carrier of each spinal application. The osteoinductive capability of rhBMP-2 was first tested in rats [2]. RhBMP-2 was able to cause ectopic bone to form successfully in a soft tissue pouch. After the rodent model, spinal fusions were conducted in rabbits, dogs, sheep, and finally in non-human primates to test the ability of rhBMP-2 to consistently achieve a solid fusion [6, 10, 11]. The performance of the carrier was also tested for its ability to restrict bone formation to the surgical implantation site. From these studies, it was found that healing is species-specific. The concentration of rhBMP-2 required to induce bone formation varied from species to species [8, 10]. Higher level animal species, such as the rhesus monkey, required a considerably higher concentration and dose of rhBMP-2 than lower species to successfully form bone. The time required to achieve a fusion was also longer in higher species.

Clinical application of rhBMP-2 in human trials

The largest number of clinical BMP trials on the human spine has involved rhBMP-2 [12–21]. For lumbar and cervical anterior interbody fusions, the concentration of rhBMP-2 was always constant at 1.5 mg rhBMP-2/mL. An absorbable collagen sponge was used as the carrier in all interbody studies. The sponge was never used as a stand-alone implant because it could not resist the physiological loads applied to the disc space. The sponge was always used in conjunction with an interbody spacer that bears the compressive loads across the vertebral interspace [22].

Posterior and posterolateral spinal fusion represent a more difficult healing environment when compared with the interbody space. To overcome the inherent challenges of posterolateral spinal applications, two distinct concentrations of rhBMP-2 (1.5 mg/mL and 2.0 mg/mL) and three different carriers were studied [13, 21, 23–25].

Anterior lumbar interbody spine fusion

Several large, prospective, multicenter, human clinical studies of patients undergoing anterior lumbar interbody fusion (ALIF) using rhBMP-2 have been completed

[18, 26]. In each of these studies, the interbody fusion was performed using either threaded, cylindrical allograft bone dowels [16, 19] or threaded titanium cages [15, 17] as the intradiscal implant. These interbody fusion cages were not supplemented with any other form of spinal fixation or instrumentation [22]. In all of these studies, the same concentration of rhBMP-2 and absorbable collagen carrier was used. When combined with rhBMP-2 on a collagen carrier (rhBMP-/ACS), these two distinct interbody fusion implants have been shown to decrease pain, improve clinical functional outcomes, and improve rates of fusion in patients after ALIF surgery; rhBMP-2 was used at a concentration of 1.5 mg/mL in all of these studies. Importantly, the total dose of rhBMP-2 that the individual patient received varied from 4.2 to 12 mg depending on the size of the implants used. In most cases, rhBMP-/ACS was placed only within the interbody construct at a volume consistent with its internal volume.

In a study of 131 patients who were treated with ALIF using cortical allograft implants, the patients treated with rhBMP-2 had statistically superior outcomes ($p<0.05$) with regard to length of surgery, blood loss, and hospital stay to those in patients treated with autograft [19]. Fusion rates were statistically superior in the rhBMP-2-treated group to those in the autograft group at all time points studied ($p<0.05$). Enhanced incorporation of the allograft bone dowel with the host's bone could be seen in patients treated with rhBMP-2 [27]. Importantly, clinical outcomes were improved in the rhBMP-2-treated group when compared with the autograft group; average Oswestry Disability Index scores, SF-36 Health Survey scores, low-back and leg-pain scores at 6, 12, and 24 months were statistically superior in the rhBMP-2-treated group. The need for supplemental surgery over time was also less in these patients.

An analysis of the outcomes in 277 patients who underwent stand-alone ALIF with the tapered titanium fusion cages used with rhBMP-2 on an absorbable collagen sponge was reported in 2003 [17]. As in other reports, the patients treated with rhBMP-2 had statistically superior outcomes to those in patients treated with autograft with regard to length of surgery, blood loss, hospital stay, reoperation rate, and median time to return to work. Oswestry Disability Index scores and the Physical Component Summary scores and Pain Index of the SF-36 at 3, 6, 12, and 24 months showed statistically superior outcomes in the rhBMP-2 group ($p \leq 0.0053$). Similarly, fusion rates at 24 months were statistically superior in the rhBMP-2 group ($p = 0.022$).

Additionally, a long-term follow-up study of patients treated with stand-alone cages has been completed to ascertain if the use of rhBMP-2 maintains anterior intervertebral spinal fusion and sustains improvements in clinical outcomes and reduction of pain. In this unpublished study presented at the Scoliosis Research Society in 2007, 134 patients with a minimum follow-up of 4 years showed high rates of fusion at 48 months (99%), low rates of additional surgery, and sustained improvements in clinical outcomes as measured by average Oswestry Disability

Index scores, SF-36 Physical Component Summary scores, low-back and leg-pain scores and work status.

Anterior cervical interbody spine fusion

In a prospective clinical trial, 33 patients with one- and two-level degenerative cervical disc disease were treated with anterior cervical interbody fusion using cortical allografts and anterior plate fixation [12]. Patients were randomly divided into two groups. The control group received iliac crest bone graft, and the investigational group received rhBMP-2 on a collagen-soaked sponge. The concentration of rhBMP-2 (1.5 mg/mL) and carrier were the same as that used in the ALIF studies. However, because of significantly smaller interbody construct, the graft volume was markedly lower than the interbody studies at 0.4 cc containing 0.6 mg rhBMP-2.

Postoperatively, both groups showed significant improvement in functional clinical outcomes and in neck and arm pain. At 24 months, the rhBMP-2-treated group had mean improvement scores superior to that of the control group in neck disability (52.7 *versus* 36.9; $p<0.03$) and in arm pain (14 *versus* 8.5 on a 20-point scale; $p<0.03$). Both groups had fusion rates of 100%. No patient had extension of new bone formation into the spinal canal.

Posterolateral lumbar spine fusion

Because preclinical non-human primate studies demonstrated that a bulking agent needed to be added to the absorbable collagen sponge carrier, carrier and growth factor concentration were optimized to consistently induce bone formation in the posterolateral human spine [6]. In the posterior spine area, soft tissue compression and rapid resorption of the collagen scaffold had resulted in thin fusion masses. These issues required the development of compression-resistant carriers with longer resorption times. An appropriate bulking agent was sought to enhance the volume of the posterolateral fusion mass when the collagen sponge was used to deliver and retain rhBMP-2 at the fusion site [23, 28]. The addition of a compression-resistant osteoconductive material to the rhBMP-2 carrier maintains a scaffold for fusion mass development and prevents thinning of the fusion mass by the forces exerted by the surrounding posterior spinal muscles. The bulking agent must be able to be remodeled and replaced by host bone; its resorption profile should match the rate of new bone formation. Autogenous grafts have also been used as bulking agents for the rhBMP-2-soaked (1.5 mg/mL) sponge in one- and two-level lumbar fusions [25, 29]. Preclinical studies also suggested that a higher concentration of rhBMP-2 (2 mg per unit of carrier volume) was necessary to achieve consistent fusion outcomes when an alternative carrier was used. Four prospective, randomized clinical studies have been completed to assess the carrier, concentration, and dose of rhBMP-2 in posterolateral fusion applications ([13, 20], and unpublished data presented at the Annual Meeting of the Orthopaedic Research Society in 2007).

Ceramic granules wrapped in an rhBMP-2-soaked collagen sponge

Preclinical animal studies show that the addition of ceramic granules to the collagen sponge offers enough soft tissue compression resistance to produce a fusion mass similar in size to the original matrix at the time of implantation [23]. This surgical application involves preparing the absorbable collagen sponge with the 1.5 mg/mL concentration of rhBMP-2 and involves no autogenous bone grafts. The sponge is wrapped around 5.0 cc of ceramic granules and is implanted posteriorly in the lumbar spine. Biphasic ceramics have been widely used as bone graft expanders [30]. Various ratios of hydroxyapatite and tricalcium phosphate (HA:TCP) were tested before selecting the 15:85 composition found in MASTERGRAFT® Granules (Medtronic Sofamor Danek, Memphis, TN). When used with rhBMP-2, this ratio appeared to have the best resorption profile relative to new bone formation. Graft options with faster resorption profiles such as with a pure beta-TCP or calcium sulfate may provide different results.

A prospective, randomized, multicenter clinical pilot study of 46 patients has been completed using this technique, and their results were reported at the 2007 Annual Meeting of the Orthopaedic Research Society. Outcomes for all patients were significantly improved by 6 weeks from preoperative values ($p = 0.091$). However, there was a trend towards greater improvement in functional outcomes and back and leg pain scores in the rhBMP-2 group than in the autograft group. By 24 months, fusion was evident in 94.7% of the investigational patients compared with 70.0% of the autograft-treated patients ($p = 0.091$) (Fig. 1). At 24 months, clinical and radiographic success was higher in the rhBMP-2-treated patients and the morbidity associated with iliac crest bone graft harvesting experienced by more than 70% of the autograft-treated patients was avoided. RhBMP-2 on a collagen sponge combined with ceramic granules was an effective autograft replacement for one-level instrumented posterolateral fusion in the lumbar spine.

Ceramic carrier with high dose rhBMP-2

In a pilot study, 25 patients were entered into a multicenter prospective study [13]. A higher dose of rhBMP-2 was used in these study patients than in the interbody trials. In this posterolateral spinal application, a solution containing 40 mg rhBMP-2 was applied to ceramic granules and placed in the posterolateral aspect of the spine. The composition of the ceramic granules was 60% hydroxyapatite/40% β-

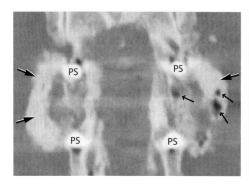 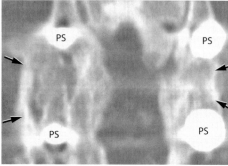

Figure 1
(A) Frontal plane reconstructed thin-slice 1-mm computed tomography (CT) scan obtained within 24 h of surgery. Minute residual air pockets appear as dark shadows overlying the graft (small arrows). The MASTERGRAFT ceramic granules (large arrows) are seen overlying the transverse processes in the posterolateral area of the spine between the fourth and fifth lumbar vertebra. (B) At 24 months after surgery, frontal plane reconstructed CT scan shows that the ceramic granules have been replaced by mature, trabeculated bone (arrows) spanning the transverse processes in the posterolateral region of the spine. PS: pedicle screw.

tricalcium phosphate. All patients (18/18) who received rhBMP-2 achieved a solid spine fusion rigorously determined by plain radiographs and thin-cut computed tomography (CT) scans. The control autograft-treated group had a fusion success rate of only 40% (2/5). Patients treated with rhBMP-2 also demonstrated faster and greater improvement in clinical outcomes when compared with those in the control group treated with autograft.

An expanded prospective, randomized, multicenter study using this ceramic carrier and the higher concentration of rhBMP-2 has been conducted outside of the United States (unpublished observations). In the first study, which was reported by Alexander at the 2005 Annual Meeting of the American Academy of Orthopaedic Surgeons, 97 patients underwent one- or two-level lumbar spinal fusions (48 investigational and 49 control patients). Of these patients, 29 underwent two-level-only posterolateral fusion surgery (12 investigational and 17 control patients). Again, the fusion data, reported by Alexander at this same meeting, showed that the rates of new bone formation approached 100%.

Compression resistant matrix high-dose BMP

A third alternative carrier was formulated to maintain the graft volume and resist the *in vivo* compressive muscle forces. A compression-resistant carrier (CRM) was developed, consisting of a biphasic ceramic (15% hydroxyapatite and 85% trical-

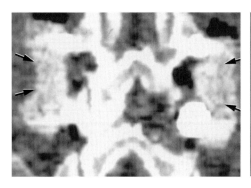 

Figure 2
(A) Frontal plane reconstructed thin-slice 1-mm CT scan obtained within 24 h of surgery. The compression resistant matrix (CRM) block of ceramic and collagen is seen spanning the transverse processes between L4 and L5. The ceramic granules (arrows) are seen distinctly. (B) At 24 months after surgery, the CRM block has been replaced by a robust fusion mass composed of mature trabeculated bone (arrows).

cium phosphate) suspended in a collagen matrix. The use of this carrier with the higher dose rhBMP-2 (2.0 mg/mL) has been evaluated in a prospective, randomized, multicenter study with an enrollment of 463 patients [20]. Each investigational patient received a 20-cc matrix carrier combined with 40 mg rhBMP-2. Radiographic and clinical outcomes show that the fusion rates in the rhBMP-2-treated group are similar to those seen in the anterior interbody studies [15, 17]. At 24 months, the rhBMP-2/matrix group of patients showed similar clinical outcomes but higher rates of fusion when compared with the autograft group for a single-level instrumented posterolateral fusion (Fig. 2).

With the high rate of successful fusion found in one-level posterolateral lumbar applications, a second prospective study (unpublished observations) was undertaken to assess the ability of this same concentration, dose, and carrier to induce new bone formation and fusion over two levels. Adding local autogenous bone graft to the posterolateral fusion site increased the graft volume. Thirty patients underwent an instrumented posterolateral lumbar spinal fusion at two adjacent levels using local bone graft as a graft volume expander in addition to the rhBMP-2 (2.0 mg/mL) on the matrix carrier. Follow-up is ongoing, but early results are promising (Fig. 3).

Conclusions

These clinical and radiographic studies confirm the results found in previous animal studies that rhBMP-2 is a safe and effective combination for patients who are undergoing cervical and lumbar fusion surgery. The results show that an increased

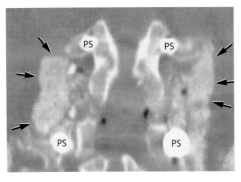 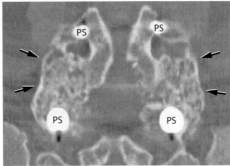

Figure 3
(A) Frontal plane reconstructed thin-slice 1-mm CT scan taken after surgery shows the CRM block (arrows) extending over two spinal levels between L4 and the sacrum. The ceramic is distinct from the surrounding soft tissues. (B) At 12 months after surgery, a cortical rim of bone is seen surrounding the graft (arrows). Most of the ceramic matrix has been resorbed and replaced by trabecular bone. There is solid fusion over two levels.

fusion rate was associated with improved functional outcomes and a decrease in pain scores at statistically significant levels. Long-term follow-up studies show that these improved outcomes are maintained over time.

References

1 Urist MR (1965) Bone formation by autoinduction. *Science* 150: 893–899
2 Wozney JM (2002) Overview of bone morphogenetic proteins. *Spine* 27 (16 Suppl 1): S2–8
3 Boden SD (2002) Overview of the biology of lumbar spine fusion and principles for selecting a bone graft substitute. *Spine* 27 (Suppl 1): S26–31
4 Vaccaro AR, Chiba K, Heller JG, Patel TC, Thalgott JS, Truumees E, Fischgrund JS, Craig MR, Berta SC, Wang JC, North American Spine Society for Contemporary Concepts in Spine Care (2002) Bone grafting alternatives in spinal surgery. *Spine J* 2: 206–215
5 Cheng H, Jiang W, Phillips FM, Haydon RC, Peng Y, Zhou L, Luu HH, An N, Breyer B, Vanichakarn P et al (2003) Osteogenic activity of the fourteen types of human bone morphogenetic proteins (BMPs). *J Bone Joint Surg Am* 85: 1544–1552
6 Boden SD, Schimandle JH, Hutton WC (1995) The 1995 Volvo award in basic sciences. The use of an osteoconductive growth factor for lumbar spinal fusion. Part II: Study of dose, carrier, and species. *Spine* 20: 2633–2644
7 McKay B, Sandhu HS (2002) Use of recombinant human bone morphogenetic protein-2 in spinal fusion applications. *Spine* 27 (16 Suppl 1): S66–S85

8 Sandhu HS, Kanim LE, Kabo JM, Toth JM, Zeegen EN, Liu D, Delamarter RB, Dawson EG (1996) Effective doses of recombinant human bone morphogenetic protein-2 in experimental spinal fusion. *Spine* 21: 2115–2122

9 Winn SR, Uludag H, Hollinger JO (1999) Carrier systems for bone morphogenetic proteins. *Clin Orthop* 46: 193–202

10 Sandhu HS, Toth JM, Diwan AD, Seim HB, Kanim LE, Kabo JM, Turner AS (2002) Histologic evaluation of the efficacy of rhBMP-2 compared with autograft bone in sheep spinal anterior interbody fusion. *Spine* 27: 567–575

11 Schimandle JH, Boden SD, Hutton WC (1995) Experimental spine fusion with recombinant human bone morphogenetic protein-2. *Spine* 20: 1326–1337

12 Baskin DS, Ryan P, Sonntag V, Westmark R, Widmayer MA (2003) A prospective, randomized, controlled cervical fusion study using recombinant human bone morphogenetic protein-2 with the CORNERSTONE-SR Allograft Ring and the ATLANTIS anterior cervical plate. *Spine* 28: 1219–1224

13 Boden SD, Kang J, Sandhu H, Heller JG (2002) Use of recombinant human bone morphogenetic protein-2 to achieve posterolateral lumbar spine fusion in humans: A prospective, randomized clinical pilot trial. 2002 Volvo Award in Clinical Studies. *Spine* 27: 2662–2673

14 Boden SD, Zdeblick TA Sandhu HS, Heim SE (2000) The use of rhBMP-2 in interbody fusion cages. *Spine* 25: 376–381

15 Burkus JK, Gornet MF, Dickman C, Zdeblick TA (2002) Anterior lumbar interbody fusion using rhBMP-2 with tapered interbody cages. *J Spinal Disord Tech* 15: 337–349

16 Burkus JK, Transfeldt EE, Kitchel SH, Watkins RG, Balderston RA (2002) Clinical and radiographic outcomes of anterior lumbar interbody fusion using recombinant human bone morphogenetic protein-2. *Spine* 27: 2396–2408

17 Burkus JK, Heim SE, Gornet MF, Zdeblick TA (2003) Is INFUSE bone graft superior to autograft bone? An integrated analysis of clinical trials using the LT-CAGE lumbar tapered fusion device. *J Spinal Disord Tech* 16: 113–122

18 Burkus JK, Gornet MF, Schuler TC (1999) An analysis of clinical trials using rhBMP-2 as a bone graft replacement in stand-alone lumbar interbody fusions. *Orthopedics* 22: 669–671

19 Burkus JK, Sandhu HS, Gornet MF, Longley MC (2005) Use of rhBMP-2 in combination with structural cortical allografts: Clinical and radiographic outcomes in anterior lumbar spinal surgery *J Bone Joint Surg Am* 87: 1205–1212

20 Dimar JR, Glassman SD, Burkus JK, Carreon LY (2006) Clinical outcomes and fusion success at 2 years of single-level instrumented posterolateral fusions with recombinant human bone morphogenetic protein-2/compression resistant matrix *versus* iliac crest bone graft. *Spine* 31: 2534–2539

21 Kleeman TJ, Ahn UM, Talbot-Kleeman A (2001) Laparoscopic anterior lumbar interbody fusion with rhBMP-2. A prospective study of clinical and radiographic outcomes. *Spine* 26: 2751–2756

22 Burkus JK (2003) Stand-alone anterior lumbar interbody fusion constructs: Effect of interbody design, bone graft and bone morphogenetic protein on clinical and radiographic outcomes. In: K Lewandrowski, MJ Yaszemski, AA White, DJ Trantolo, DL Wise (eds): *Advances in spinal fusion: Clinical applications of basic science, molecular biology, biomechanics, and engineering.* Marcel Dekker, New York, 69–84
23 Akamaru T, Suh D, Boden S, Kim HS, Minamide A, Louis-Ugbo J (2003) Simple carrier matrix modifications can enhance delivery of recombinant human bone morphogenetic protein-2 for posterolateral spine fusion. *Spine* 28: 429–434
24 Martin GJ Jr, Boden SD, Titus L, Scarborough NL (1999) New formulations of demineralized bone matrix as a more effective graft alternative in experimental posterolateral lumbar spine arthrodesis. *Spine* 24: 637–645
25 Singh K, Smucker JD, Boden SD (2006) Use of recombinant human bone morphogenetic protein-2 as an adjunct in posterolateral lumbar spine fusion. A prospective CT-scan analysis at one and two years. *J Spinal Disord Tech* 19: 416–423
26 Burkus JK (2004) Bone morphogenetic proteins in anterior lumbar interbody fusion: Old techniques – New technologies. *J Neurosurg (Spine 1)* 3: 254–260
27 Burkus JK, Sandhu HS, Gornet MF (2006) Influence of rhBMP-2 on the healing patterns associated with allograft interbody constructs in comparison with autograft. *Spine* 31: 775–781
28 Barnes B, Boden SB, Louis-Ugbo J, Tomak PR, Park J, Park M, Minamide A (2005) Lower dose of rhBMP-2 achieves spine fusion when combined with an osteoconductive bulking agent in non-human primates. *Spine* 30: 1127–1133
29 Glassman SD, Carreon LY, Djurasovic M, Campbell MJ, Puno RM, Johnson JR, Dimar JR (2007) Posterolateral lumbar spine fusion with INFUSE bone graft. *Spine J* 7: 44–49
30 Epstein NE (2006) A preliminary study of the efficacy of beta tricalcium phosphate as a bone expander for instrumented posterolateral lumbar fusions. *J Spinal Disord Tech* 19: 424–429

Bone morphogenetic protein signaling is fine-tuned on multiple levels

Christina Sieber, Gerburg K. Schwaerzer and Petra Knaus

Freie Universität Berlin, Institute for Chemistry/Biochemistry, Thielallee 63, 14195 Berlin, Germany

Signal transduction from extra- to intracellular compartments

The receptors for ligands of the bone morphogenetic protein (BMP) family translate signals from the outside to the inside of the cell. In the cytoplasm, stimuli are received by signal transducer molecules. This process passes through numerous stages and checkpoints, which are depicted in the following paragraphs.

The ligands

BMPs are secreted ligands, which are expressed as large precursor proteins containing an N-terminal signal peptide, a prodomain and the C-terminal mature peptide. The monomers typically form covalent homo- or heterodimers in the endoplasmic reticulum (ER) [1, 2]. The mature protein is released from the proprotein following cleavage by a serine endoprotease within cellular membranes, such as the ER or Golgi. For example, furin, a protease predominantly localized in the Golgi membrane, was shown to cleave off the prodomain of BMP-4 [2]. Growth differentiation factor 2 (GDF-2, BMP-9) and GDF-8 (myostatin) are exceptions: the proregion remains associated with the mature protein after cleavage and even after secretion. Outside the cell, the non-covalently attached proregion may inhibit binding of the ligand to its receptor. However, little is known about the regulation or physiological relevance and activity of these complexes [1].

Mature BMP and GDF ligands are highly homologous. In particular, a set of seven cysteine residues is highly conserved. Dimers are covalently linked by a single intermolecular disulfide bridge, mediated through one of the conserved cysteines in each monomer [3]. While the cysteine bridge is believed to be critical for dimerization and biological activity for most members of the BMP family [4], it appears to be dispensable for some, such as GDF-9 [5], GDF-9B (BMP-15) [6], and GDF-3 [5]. These proteins are biologically active and were shown to form non-covalent homo- and heterodimers *in vivo*. Moreover, a point mutation in GDF-5, removing the critical cysteine and resulting in monomeric protein, was shown to have equal biological activity as its dimeric counterpart [7]. Interestingly, heterodimers of BMP-2/5, BMP-2/6, BMP-2/7 and BMP-4/7 form naturally *in vivo* and *in vitro* when co-expressed and show enhanced activity compared to the corresponding homodimers [8].

Six of the conserved cysteine residues serve to stabilize the monomers [3], forming a structure known as a cystine knot. In BMP-2, two intramolecular disulfide bonds (Cys43/Cys111 and Cys47/Cys113) shape a ring-like topology wide enough for the third cysteine bridge (Cys14/Cys79) to pass through [9].

The crystal structures of BMP-2 [9], BMP-7 [10, 11], BMP-9 [12], and GDF-5 [13] revealed great similarities between these and other members of the TGF-β superfamily ligands. The typical TGF-β fold is often described as a double left-hand, with each monomer containing a wrist epitope, four fingers, and a knuckle epitope. The wrist epitope (concave) comprises residues from both monomers. One part is located on a long α helix, from which two anti-parallel β sheets, the fingers, project away like butterfly wings. The second part of the wrist epitope is made up of residues on the inner side of the β sheets. The knuckle epitope (convex) is located on the outer slope of the β strands (Fig. 1) [14].

Due to the central role of BMPs in development, naturally occurring mutations in all proteins along the BMP pathway have severe effects. In humans, mutations in GDF-5 give rise to various skeletal defects such as abnormally short and deformed limbs (acrosomelic dysplasia), joint fusions (symphalangism), shortened phalanges in fingers and toes (brachydactyly type C and type A2) or a combination of these and other defects (multiple synostosis syndrome) [15–18]. Several naturally occurring mutations in the receptor binding site of GDF-5 have been described that alter affinity of GDF-5 to its type I receptors. One mutation prevents binding of GDF-5 to its high-affinity receptor BRIb and results in Mohr-Wriedt Brachydactyly type A2, a condition typically ascribed to mutations in the receptor itself [16]. Furthermore, a different mutation was found to enhance affinity of GDF-5 to BRIa leading to symphalangism, a phenotype previously described for mutations in the GDF-5 antagonist Noggin [18].

The receptors

BMPs signal through two types of transmembrane serine/threonine kinase receptors (BRI and BRII). Both receptors carry an extracellular ligand binding domain, followed by a transmembrane domain and an intracellular serine/threonine kinase domain. Additionally, BRI contains two distinct motives that are important for receptor activation and signal transduction. First, a glycine/serine (SGSGSG)-rich region, the so-called GS-box, is located at the juxtamembrane region preceding the kinase. Phosphorylation of the GS-box by the constitutively active type II receptor activates type I receptor kinase activity [1]. Secondly, within the kinase domain, a

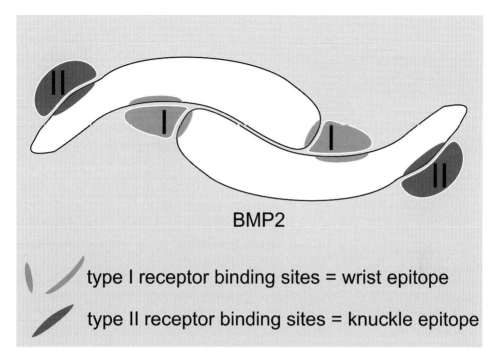

Figure 1
Bone morphogenetic protein (BMP) ligand receptor complex
BMP monomers assume a left-hand shape, forming a double left-hand structure in the dimeric ligand. Binding sites for the receptors are located on the concave and convex sides of the protein. The type I receptor binds to the wrist epitope (concave) where it is in contact with both monomers. The type II receptor on the other hand interacts with the knuckle epitope (convex), making contact with only one monomer.

short region of 8 amino acids termed L45 loop, determines isoform-specific activation of Smads, the intracellular signal transducer molecules of Smad-dependent BMP signaling [19]. Receptor specificity of Smads is conferred by their L3 loop, a 17-amino acid region that protrudes from the C-terminal domain of Smads [20].

Three type I receptors serve the BMP pathway: BMP receptor type Ia (BRIa or Alk3) [21], BMP receptor type Ib (BRIb or Alk6) [22] and activin receptor type Ia (ActRIa or Alk2) [23]. They are highly homologous, yet they have very specific preferences for their respective ligands. Both receptors can bind BMP-2; however, BRIa binds BMP-2 with higher affinity, while BRIb is the high-affinity receptor for GDF-5 (Tab. 1). Both receptors form heteromeric complexes with BMP receptor type II (BRII) or activin receptor type II (ActRII) and both can initiate Smad-dependent as well as Smad-independent pathways [1]. Nevertheless, they play different

Table 1 - Ligands, receptors and respective affinities.

	BRII	ActRII	ActRIIb	BRIa	BRIb
BMP-2	+	+	+	+++	++
BMP-4	+	+	+	+++	++
BMP-6	+	+++	+++	+	++
BMP-7	++	+++	+++	++	+++
BMP-15	+	−	−	++	+++
GDF-5	+	+	+	++	+++
GDF-6	+	?	?	+	+
GDF-8	?	+++	+++	−	−
GDF-9	++	?	?	−	−

Key: unknown (?), no affinity (−), low affinity (+), medium affinity (++), high affinity (+++)
Approximate, relative affinities of each ligand to the receptors are listed. Ligands are viewed separately. This table does not provide sufficient information to allow comparison of affinities of different ligands to a single receptor.

roles during embryogenesis and in the adult organism. While BRIb is essential for differentiation of osteoblasts from mesenchymal precursor cells, BRIa triggers adipogenic differentiation in the same cell line [19]. Studies performed on chick limb development revealed that BRIb is required for the initial steps of mesenchyme condensation and cartilage formation, as well as regulating programmed cell death, necessary to form separate digits and joints. BRIa, however, controls the later stages of chondrocyte differentiation [24].

Mutations in BRIa cause juvenile polyposis syndrome (JPS). JPS predisposes to hamartomatous juvenile polyps in the gastrointestinal tract, which may undergo malignant transformation and result in cancer [25]. Besides BRIa, also mutations in Smad4, the common mediator Smad (co-Smad), can cause JPS [26]. As mentioned, brachydactyly type A2 was described to involve mutations in GDF-5, and other cases have revealed mutations in BRIb [16, 17]. Fibrodysplasia ossificans progressiva (FOP) is a rare autosomal dominant disorder of connective tissue, characterized by congenital malformation of the great toes and progressive heterotopic ossification of tendons, ligaments, fasciae, and striated muscles [27]. The genetic locus was recently mapped to the gene of ActRIa. The mutation affects the GS domain of ActRIa and possibly leads to constitutive activation of the receptor [28]. Indeed, overexpression of BMP-4 was observed in FOP patients in accordance with the fact that constitutive activation of ActRIa upregulates *BMP-4*, downregulates BMP antagonist expression and induces ectopic chondrogenesis as well as joint fusions [28].

BRII exists in two alternative splice variants, referred to as the short form (SF) and the long form (LF) [29–32]. The human BRII gene consists of 13 exons, with exon 12 coding for the so-called BRII-tail, an extension at the C-terminal end of the receptor. BRII SF resembles the typical type II receptor of the TGF-β superfamily. The BRII LF tail region has been postulated to serve as a binding site for multiple adaptor proteins to modulate BMP signaling specificity, complexity and intensity. This was demonstrated with proteomics-based approaches applying different cytoplasmic domains of BRII as GST fusion proteins [33].

Pulmonary arterial hypertension (PAH), a disease characterized by remodeling of small pulmonary arteries, is caused in part by mutations in BRII [34]. The phenotype reveals severe mucosal hemorrhage, incomplete cell coverage on vessel walls, and gastrointestinal hyperplasia. Liu et al. [35] described silencing of BRII expression in mice using a BRII-specific short hairpin RNA transgene. Repressed BRII expression and subsequent disruption of BRII signaling causes increased activation of AKT. AKT in turn suppresses canonical Wnt signaling in response to BMP signaling through BRIa. BRII also controls expression of endothelial guidance molecules that promote vascular remodeling. The effects are dosage-dependent in that the phenotype becomes more severe when BRII expression is constitutively attenuated by two copies of the shRNA transgene [35]. Most recently, RACK1 (receptor for activated C-kinase 1) was identified as a BRII interaction partner, negatively regulating pulmonary arterial smooth muscle cell proliferation and downregulated in a PAH rat model. BRII mutants leading to PAH show reduced interaction with RACK1, suggesting a role for RACK1 in the pathogenesis of PAH [36].

The ligand-receptor complex

Resolution of crystal structures of ligands and receptor domains has been a major achievement in determining important sites for protein-protein interactions.

The crystal structure of BMP-2 in complex with the extracellular domain (ECD) of BRIa confirmed binding of the high-affinity type I receptor to the wrist epitope (Fig. 1), where it interacts with both BMP-2 monomers [37, 38]. The ternary signaling complex, consisting of the ECDs of BRIa and ActRII and BMP-2 was solved recently [39, 40]. While the structure of the binary complex (ligand bound to its high-affinity receptor) gave important hints on binding specificity, the ternary complex revealed that BMP-2 does not undergo significant conformational changes upon binding to its receptors [39]. The ActRII-ECD adopts a three-finger (six β-strands) toxin fold, which creates the complementary binding surface for attaching to the concave face (knuckle epitope) of one BMP-2 monomer. This mostly hydrophobic interface consists of 12 residues from BMP-2 and 10 residues from ActRII. Three of these residues from ActRII are indispensable for ligand binding, forming a hydrophobic core contacting five residues at the BMP-2 knuckle epitope [41]. Even though the five amino acids involved in shaping this core are identical in BMP-2, BMP-6 and BMP-7, with the exception of one residue in BMP-7, ActRII does not have the same affinity to all three ligands (Tab. 1). It is thus speculated that non-conserved residues outside the hydrophobic core account for ligand specificity and affinity. Confirmation comes from studies performed on BMP-2, in which a silent H-bond was activated through mutation of one or two amino acids to transform BMP-2 into a high-affinity ligand for ActRIIB [40]. In contrast to ActRII, BRIa interacts with both monomers in the wrist epitope. For the greatest part, the convex face of one BMP-2 monomer contributes to the BRIa binding interface, while the second monomer provides mostly hydrophobic contacts between its concave side and the receptor ECD (Fig. 1). The interface of BMP-2 with BRIa-ECD has been described as a "knob-into-hole" motif, typical for all type I receptor-binding sites [37].

Analysis of BMP-2 heteromeric muteins, in which the binding epitope for either type I or type II receptor were depleted on one monomer, demonstrated that two functional type II receptors are required for biological activity, while depletion of a single type I receptor only affects Smad-independent signaling. Complete loss of biological activity is achieved in homomeric muteins missing the wrist epitope on both monomers [42].

Different modes of BMP receptor oligomerization

TGF-β and BMP signaling, even though both pathways use their respective type I and type II receptors, follows a very distinct and different mode of receptor activation. TGF-β signaling is initiated by binding of the ligand to its high-affinity type II receptor, whereupon the type I receptor is recruited into the complex. The activated complex eventually translocates to clathrin-coated pits (CCPs) from where it proceeds to early endosomes and initiates signal transduction. This internalization route was also proposed to result in recycling or degradation of the receptors [43]. If, however, the complex is not sequestered from detergent-resistant membrane regions (DRMs), it will shuttle into caveolae, resulting in endocytosis to caveosomes and subsequent degradation [44].

As for BMPs, the binding mode of the ligand to its receptors can trigger Smad-dependent or Smad-independent signaling [45–47]. BMP-2 first binds to its high-affinity receptor (BRI) upon which the low affinity receptor (BRII) is recruited into a ternary complex. This binding mode leads formation to the BMP-induced signaling complex (BISC) and was shown to initiate Smad-independent signaling. BRI receptors are predominantly localized in DRMs, whereas BRII is found in all membrane domains. When BMP-2 binds to BRI, BRII is recruited into BISC, which moves to caveolae and internalizes into caveosomes. This pathway initiates Smad-independent signaling (Fig. 2). In contrast to TGF-β, BMP was shown to bind preformed complexes (PFC) of type I and type II receptors, leading to activation of Smad-dependent signaling. Smad1, 5 and 8 associate with BRI in PFCs at the plasma membrane. Following stimulation with BMP-2, Smad1, 5 and 8 are phosphorylated before the complex internalizes from CCPs to early endosomes. Dissociation of activated Smads from BRI occurs after endocytosis (Fig. 2) [47].

BMP-induced signaling cascades

Smad pathway

As described in the previous chapter, the mode of receptor oligomerization determines the onset of different BMP signaling pathways that do or do not depend on Smads [46].

Smads are proteins that mediate intracellular signals and regulate transcription of distinct BMP target genes. Based on their function, they are divided into receptor-associated Smads (R-Smads), common-mediator Smads (Co-Smad) and inhibitory Smads (I-Smads). They consist of the N-terminal domain Mad homology 1 (MH1), a proline-rich linker region and the C-terminal domain MH2. The MH1 domain represses MH2 activity in the cytoplasm and binds to specific DNA sequences, while MH2 domains interact with BRI and are involved in homomeric or heteromeric complex formation (e.g., with Smad4) [48].

After ligand binding to PFC consisting of BRI and BRII homodimers, the constitutive active kinase of BRII phosphorylates multiple residues in the GS-box of BRI, which allows recruitment of BMP Smad1, 5 and 8 (R-Smads) to BRI and their phosphorylation by BRI kinase [49]. Similar to SARA, endofin acts as a Smad1 anchor for receptor activation and enhances Smad1 phosphorylation and translocation to the nucleus after BMP stimulation [50]. Subsequently, BMP receptors and associated Smads are internalized by CCP-dependent endocytosis [47]. In endosomes, R-Smads dissociate from the complex to form oligomers with Smad4 and to translocate into the nucleus [47]. While activation and therefore nuclear translocation are achieved by C-terminal phosphorylation of Smads, phosphorylation of serines in the linker region, catalyzed by members of the family of ERK kinases, causes inhibition of this process [51]. In the nucleus, Smad oligomers in cooperation with co-repressors or co-activators bind to DNA at the Smad-binding element (SBE) (Tab. 2) and regulate the transcription of *BMP* target genes [52–55]. Phosphorylated Smads accumulate in the nucleus and are dephosphorylated on their C terminus by small C-terminal domain phosphatases (SCP) [56], pyruvate dehydrogenase phosphatase (PDP) [57] or PPM1A [58], and in the linker region by SCP [59]. Export of dephosphorylated Smads, allowing them to be activated once more, relies on mechanisms that to date are not well understood. I-Smads (Smad6 and 7) belong to the immediate early genes following BMP stimulation and participate in negative feedback. While

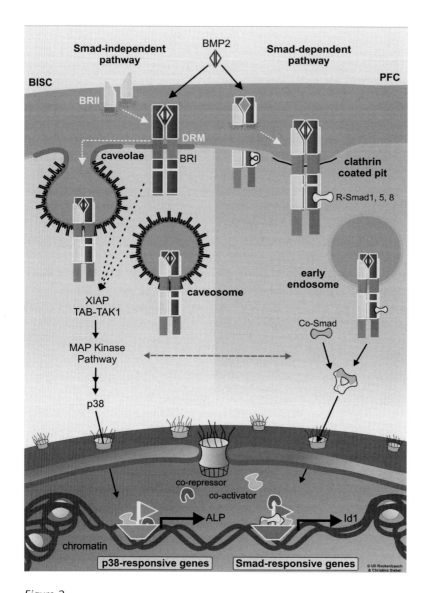

Figure 2
BMP signaling routes from cell surface to nucleus
BMP-2 can bind to its receptors in two different modes. It can either induce a signaling complex (BISC) by first binding its high-affinity type I receptor (BRI), upon which the type two receptor (BRII) is recruited or it can attach to preformed complexes (PFC) consisting of type I and type II receptors. Depending on the binding mode, different pathways are switched on. BRI predominantly resides in detergent-resistant membranes (DRM). Binding of the ligand to DRM-located BRI followed by recruiting of BRII results in transfer of the complex into caveolae and caveosomes. Subsequently, Smad-independent pathways are initiated, here

Smad6 inhibits BMP signaling only, Smad7 inhibits both BMP and TGF-β signaling. There are different reports on the action of I-Smads, such as competition of R-Smads for binding to activated BRI, interference in R- and Co-Smads interactions [60], binding of Smad7 to PP1 (protein phosphatase 1) to cause BRI dephosphorylation [61], and association of I-Smads with HECT type E3 ligases Smurf 1 and 2 [62, 63]. Smurfs induce nuclear export of I-Smads to enhance interaction with BRI and consequently inhibition of BMP signaling. Furthermore, Smurfs induce polyubiquitination and degradation of R-Smads, BMP receptors and the BMP stimulation product Runx2 [64, 65]. Besides Smurf 1 and 2, there are other E3 ligases inhibiting TGF-β/ BMP-dependent signaling, such as CHIP, Jab1, SCFβ-TrCP1, Sumo-1/Ubc9, WWP1 [66–71]. In contrast, binding of Smad6 to AMSH (associated molecule with SH3 domain of STAM) and binding of Smad7 to Jab1 prevents interaction between R- and I-Smads and decreases the inhibitory potential of I-Smads [66, 72].

Smad-independent pathways

Initiation of Smad-independent signaling is mediated by binding of BMP-2 to its high-affinity receptor BRI. This leads to recruitment of BRII into the BISC, which is localized in distinct plasma membrane regions (DRMs) [47, 73]. X-chromosomal linked inhibitor of apoptosis (XIAP), a member of the multifunctional IAP proteins, serves as a bridge between BRI and the MAP kinase signaling pathway, initiated by TAK1. XIAP contains three BIR domains and a Ring domain and can thus regulate caspase activities and apoptosis [74–78]. NF-κB and MAPK kinase (MKK) are activated *via* TAK1 [79–85]. Activated MKK3/6 phosphorylates the stress kinase p38, while MKK 4 activates c-Jun-N-terminal kinases (JNK) [81, 86]. After translocation of p38 and JNK into the nucleus, they activate ATF-2, c-Jun and c-Fos to regulate distinct BMP target genes [87, 88]. BMP-2-dependent phosphorylation of p38 induces expression of specific target genes (e.g., type I collagen, fibronectin, osteopontin, osteocalcin and ALP) [88–90] and regulates apoptosis [91, 92]. Smad6 inhibits the TAK1-MKK3/6–p38 pathway, while binding of Smad7 to MKK3 and p38 promotes TAK1-MKK3-mediated signaling [92, 93].

---

represented by the MAP kinase pathway, resulting in expression of p38-responsive genes like alkaline phosphatase (ALP). The Smad-dependent pathway is targeted when BMP-2 binds to a PFC. After ligand binding R-Smads are phosphorylated by BRI at the cell surface, indicated by structural change of the Smad protein. The complex moves into clathrin-coated pits and is internalized into early endosomes. Here phosphorylated R-Smads detach from BRI to form complexes with Co-Smads, which then translocate into the nucleus to trigger expression of Smad-responsive genes, such as Id1. Gene expression can be further regulated by presence of co-repressors or co-activators.

Table 2 - BMP responsive transcription factors.

Transcriptional factors	Function	Binding partners	References
Cbfa1/Runx2/ PEBP2αA	Transcriptional activator	R-Smads	[199–204]
c-Jun/c-Fos; CREB; ATF-2	Transcriptional activator	R-Smads, Smad4	[87, 205, 206]
EHZF/Evi3	Transcriptional activator	Smad1 and 4	[207]
OAZ	Transcriptional activator	R-Smads	[208]
p300; CBP; p/CAF	Transcriptional activator	Smad1, 4 or 5; Runx2	[209, 210]
GCN5	Transcriptional activator	Smads	[211]
SMIF	Transcriptional activator	Smad4 and p300/CBP	[212]
Xvent-2B	Transcriptional activator	Smad1	[213]
ZEB-1/deltaEF1	Transcriptional activator	p300/CBP	[214–216]
Menin	Transcriptional activator/repressor	R-Smads; Runx2; Smad3	[217]
Hoxc-8	Transcriptional repressor	Smad1 and 6	[218, 219]
mZNf8	Transcriptional repressor	Smad1	[220]
Nkx3.2	Transcriptional repressor	Smad1 and 4	[221, 222]
YY1	Transcriptional repressor	MH1 of Smad1 and 4	[223, 224]
Ski	Transcriptional repressor	MH2 domain of Smad1, 4 or 5	[225–227]
SnoN	Transcriptional repressor	MH2 domain of Smad1, 4 or 5	[228]
DACH1	Transcriptional repressor	Smad 4	[229]
Tob	Transcriptional repressor	I-Smads and BRI; R-Smads	[230, 231]
ZEB-2/SIP1	Transcriptional repressor	ALP promotor	[214–216, 232, 233]
CtBP	Transcriptional repressor	DNA-binding proteins	[234]
SNIP1	Transcriptional repressor	Smad1 and 4 and p300/CBP	[235, 236]
Cdk6	Transcriptional repressor	Smads	[237]
SANE	Transcriptional repressor	Smad1 and 5 and BRI	[238]
CIZ	Transcriptional repressor	Smad1 and 5	[239]

Furthermore, BMP stimulation activates the PI3K-Akt/PKB signaling pathway [94–96] to induce cell survival by inhibition of proapoptotic proteins [97]. More recently it was suggested that the PI3K pathway is activated in several tissues where

Table 3 - BMP receptor associated proteins.

Associated protein	Binding receptor	Function	References
BRAM	BRI	Links BRI and TAK1	[240]
XIAP	BRI	Links BRI and TAK1	[82]
Dullard	BRI and BRII	Dephosphorylates BRI and BRII	[241]
c-Src	BRII	Inhibits BMP signaling if it is dephosphorylated by BRII	[100]
Lim kinase	BRII	Modulates actin dynamics	[102, 103]
Tctex-1	BRII	Is a light chain of Dynein moving along microtubules and interacting with downstream BMP effectors (p38; ALP)	[101]
Trb3	BRII	Triggers degradation of Smurf1 through the ubiquitin-proteasome pathway	[242]

BRIa has been inactivated, including intestine and Mx1-positive skin tumors and the hair follicle stem cell niche [98]. BMP also induces the Ras-Raf-MEK1/2-Erk-pathway [89, 99], which results in ATF-2 and Elk1/2 phosphorylation affecting transcription of distinct target genes (fibronectin and osteopontin) [88]. Furthermore, Erk-dependent phosphorylation of R-Smads at their linker region results in inhibition of nuclear Smad accumulation [51].

Mutations in BRII, which cause PAH [100] or primary pulmonary hypertension (PPH) [101–103], have revealed that BRII-interacting proteins are essential to connect BMP-dependent pathways with other signaling pathways regulating cell proliferation, differentiation and apoptosis [33]. Until now, only some of these proteins are characterized (Tab. 3).

Fine-tuning

The extracellular fine-tuning of BMP signaling is achieved by interaction of BMPs with secreted proteins (e.g., antagonists) or co-receptors. At the receptor level, the mode of receptor oligomerization and the amount of cell surface receptors, as well as the ratio of BRI to BRII determine the activation of different BMP-signaling pathways [35, 73]. The amount and ratio of cell surface receptors are regulated by receptor synthesis, endocytosis and degradation. Intracellular BMP signaling is controlled by cytoplasmic inhibitors such as I-Smads, intracellular binding proteins, signaling cross-talk, and nuclear inhibitors such as co-repressors.

Antagonists

About 15 secreted proteins are known to bind BMPs and prevent interaction with their specific BMP receptors or inhibit BMP signaling indirectly [104–106]. Similar to the proteins they antagonize, BMP antagonists consist of a cysteine-rich (CR) domain, which allows formation of the cystine knot structure and protein–protein interaction [9, 107, 108]. Based on the number of involved cysteines, BMP antagonists are grouped into Dan family (eight-membered ring), in twisted gastrulation (nine-membered ring) and the Noggin and Chordin subgroup (ten-membered ring).

Members of the Dan (differential screening-selected gene aberrative in neuroblastoma) family are Dan, cerberus, coco, PRDC, gremlin, dante, caronte, USAG-1 and sclerostin [104, 105, 109]. Most Dan family members are expressed during embryogenesis, only few adopt a role during skeletogenesis. Dan/NO3 is a 19 kDa glycoprotein, which is downregulated in *v-src*-transformed rat fibroblasts, where it demonstrates tumor-suppressor activity [110, 111]. It binds BMP-2 and -4, and more efficiently GDF-5 [112]. During embryogenesis it is expressed in anterior neural tissues and in the myotome [112, 113]. Cerberus is a secreted glycoprotein of 31 kDa, which forms homodimers; it is expressed in anterior endoderm and known as head-inducing factor [114, 115]. Cerberus binds to BMP, Wnt and Nodal by separate binding sites and inhibits their activities in *Xenopus* [115]. Murine protein related to cerberus (mCer1) binds to BMP and Nodal only and is also expressed in anterior endoderm, in presomitic mesoderm and in the somites, but mCer1 does not seem to be essential for development [116–119]. Another cerberus-like protein, Caronte, regulates the left-right asymmetry in vertebrates during embryogenesis [120, 121]. As for cerberus, coco was identified in *Xenopus* and is only expressed in pre-gastrula embryos as a BMP-4, TGF-β and Wnt inhibitor [122]. During embryogenesis, dante is expressed in a multitude of tissues including cartilage. Its role in skeletogenesis is unknown [120, 123]. PRDC (protein related to Dan and cerberus) shares a degree of homology with gremlin and is expressed in ovary, brain, and spleen, where it inhibits BMP-2 and BMP-4 activity [124].

Gremlin/IHG2 (induced high glucose-2) is a homodimer of 28 kDa that was identified in the neural crest of *Xenopus*. It binds to BMP-2, -4 and -7 with high affinity [125, 126]. During embryogenesis, gremlin is required to establish the apical ectodermal ridge and epithelial-mesenchymal feedback signaling, and is decisive for early limb outgrowth, patterning and morphogenesis in kidney and lung rudiments [127–129]. Gremlin induced by BMPs limits its effects on osteoblastic replication, differentiation and function. Overexpression causes decreased osteoblast numbers, disorganized collagen fibrils at endosteal cortical surfaces and reduced bone mineral density leading to osteopenia and bone fractures [130, 131]. Overexpression of *DRM* (downregulated by v-mos) gremlin homolog in rats inhibits tumorigenesis in tumor-derived cell lines [132–135].

The monomer uterine sensitization associated gene-1 (USAG-1)/Wise/Ectodin is a newly identified BMP antagonist of 28–30 kDa that binds BMP-2, -4, -6, and -7 with high affinity [136–138]. Human USAG-1 is expressed in embryonic tissues, kidney, skin, liver, mammary gland, aorta, and vein [109]. USAG-1 reduces BMP-induced alkaline phosphatase (ALP) activity in C2C12 and MC3T3-E1 cells [136, 137] and affects Wnt signaling as activator as well as inhibitor depending on the microenvironment [139–141].

Sclerostin is a secreted glycosylated monomer, mutations in the *SOST* gene cause sclerosteosis [142–145] or van Buchem disease [146, 147]. Sclerostin is a controversially discussed antagonist of BMP. Contrary to former publications [148], Sclerostin is only expressed in osteocytes in mineralized matrix and in human hypertrophic chondrocytes [106, 149, 150]. Sclerostin decreases activity of ALP and induces apoptosis by activation of different caspases in human mesenchymal stem cells [151]. Current data published on binding affinities of Sclerostin to BMPs are not clear [148, 149]. Winkler et al. [149] show that Sclerostin antagonizes BMP signaling directly by competing with BRI and BRII for binding BMP, which subsequently decreases Smad-phosphorylation and ALP activity, while van Bezooijen et al. [152, 153] did not observe this effect and suggest indirect inhibition of BMP and Wnt signaling. It is known that expression of Sclerostin is induced by BMP-2, -4 and -6 and the transcription factors osterix and Runx-2 as part of a protective mechanism preventing exposure of skeleton to BMPs [149, 151, 153]. Sclerostin suppresses differentiation of osteoblastic cells and mineralization of osteogenic cells [154, 155]. The BMP antagonists Sclerostin and Noggin form a complex, which does not inhibit BMP signaling [156].

Twisted gastrulation (Tsg) is required for dorso-ventral axis formation in *Drosophila* [157]. It is a small glycoprotein of 24 kDa and has two CR domains. The CR domain near the N terminus is similar to the domains of Chordin and is necessary for binding BMP-2 and -4 or BMP-2/4-Chordin complexes. The C-terminal region shows no significant similarity with known domains and interacts with Chordin [158, 159]. It appears that the concentration of Tsg determines whether Tsg acts as promotor or inhibitor of BMP signaling. Binding of Tsg to BMP-2 and -4 inhibits their interaction with BMP receptors. Alternatively, interaction with the BMP-2/4-Chordin complex serves as stabilization and enhances inhibition of BMP signaling. However, Tsg also enhances the cleavage of Chordin by the metalloproteases BMP-1/Tolloid and raises the inhibitory effect of Chordin and its proteolytic products (Fig. 3) [159–161]. Tsg is expressed in marrow stromal cells, in osteoblasts and during T cell differentiation. Overexpression prevents differentiation of stromal cells and decreases potential of differentiated osteoblasts [156, 162, 163].

Noggin is a glycosylated homodimer of 64 kDa. It consists of an acidic N-terminal region and a CR region forming the cystine knot at the C terminus. Noggin binds BMP-2, -4 and -7 and GDF-5 and -6 with different affinities and inhibits their signaling [164, 165]. The crystal structure of BMP-7 together with Noggin shows

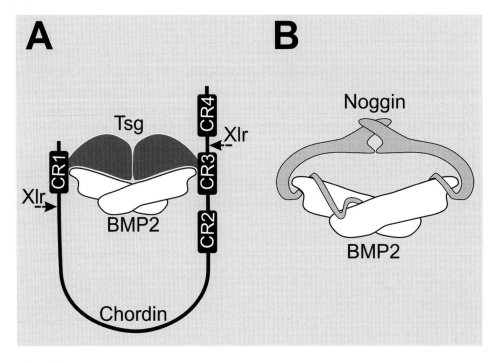

Figure 3
BMP ligand antagonist complex
(A) The cysteine-rich (CR) domains 1 and 3 of chordin bind to BMP-2 and inhibit its activity. Twisted gastrulation (Tsg) can abolish this effect by binding to the BMP-Chordin complex and facilitating the cleavage of Chordin by the zinc metalloproteases Xolloid-related/tolloid (Xlr). Consequently, BMP-2 is released.
(B) Noggin blocks the receptor binding sites of BMP-2, BMP-4 and BMP-7 in a clamp-like fashion, blocking the BMP-binding epitopes and thus inhibiting BMP signaling.

that the homodimer of Noggin binds BMP-7 in a clamp-like fashion and blocks binding epitopes of BRI and BRII [166]. Noggin is secreted during embryogenesis where it induces the formation of neural tissues and dorsalation of mesoderm cells [167–169]. In the adult organism, it is essential for bone and joint formation [164, 170, 171]. In osteoblasts, expression of Noggin is induced after stimulation with BMP-2, -4 or -7 to negatively regulate BMP signaling [172, 173]. Overexpression of Noggin inhibits the differentiation of marrow stromal cells and reduces the function of differentiated osteoblasts [172, 173]. A similar negative feedback inhibition is also observed in chondrocytes, where the expression of Noggin and Chordin increases after inducing endochondral ossification by Indian hedgehog (Ihh) [174]. Heterozygous mutations of human *noggin* decrease Noggin function and cause proximal symphalangism and multiple synostosis syndrome [175, 176].

Chordin/Sog (short gastrulation) is expressed in the Speman organizer as a 120-kDa glycosylated homodimer and plays a role in morphogenesis, formation of the three body axes and the formation of neural tissue [177–179]. It includes four CR domains of about 70 amino acids, each forming a ten-membered cystine knot. The first and the third CR domain determine the function of Chordin and the ability to bind and to inhibit BMP-2, -4 and -7 [180].

As opposed to the established BMP nomenclature, BMP-1/Tolloid do not act as growth factors. They are zinc metalloproteases resulting from alternative splicing of the same gene [181]. They are expressed in chondrocytes, inhibiting their differentiation, and in osteoblasts [172, 182]. Secreted BMP-1/Tolloid recognizes and cleaves Chordin bound to BMP growth factors to release BMPs for receptor binding [183]. This effect is enhanced by Tsg. In contrast, the inhibitory effect of Chordin is enhanced, if the secreted frizzled related protein Sizzled/Ogon blocks the Chordin binding site of BMP-1/Tolloid or BMP-4 and Chordin bind to an extracellular proteoglycan called Biglycan (Fig. 3) [184].

Not only Chordin but also a number of extracellular proteins have CR domains that form a ten-membered ring and influence BMP signaling. This group of proteins consists of Chordin-like-1/Ventroptin/Neuralin-1, Chordin-like-2, procollagen type II, Kielin, Amnioless, Nell, Crossveinless-2, Crim-1, members of the CCN family (i.e., CTGF and WISP) and the novel identified antagonist Brorin [105, 185]. Most of them act as BMP antagonists, with the exception of Crossveinless-2 and WISP, which promote BMP signaling [186, 187].

Co-receptors

Several transmembrane and membrane attached proteins were shown to interact with BMP receptors and modulate signaling outcomes.

One of the first co-receptors to be identified was the pseudoreceptor BMP and activin membrane-bound protein (BAMBI). BAMBI is a transmembrane protein with a similar extracellular domain structure as the TGF-β superfamily type I receptors, but it lacks an intracellular kinase domain. BAMBI was shown to antagonize dorso-anterior structures promoted by activin and BMP-4 during embryogenesis. This negative regulation of TGF-β family signaling is achieved through ligand-independent association of BAMBI with BRIa and BRIb [188].

Recently, several members of the glycosylphosphatidylinositol-anchored repulsive guidance molecule (RGM) family were identified as BMP co-receptors [189]. Dragon (RGMb) is expressed in the developing nervous system, where it mediates homophilic and heterophilic adhesion of neurons. It binds to BMP-2 and BMP-4, as well as to BRI and BRII and enhances BMP signaling [189, 190]. Like Dragon, RGMa is expressed in the central nervous system where it mediates repulsive axonal guidance and neural tube closure. RGMa interacts with BMP-2, BMP-4 and

BRI, and was shown to enhance BMP-mediated Smad signaling [191]. Likewise, hemojuvelin (RGMc) binds to BMP-2 and BMP-4 and to BRIb (in the presence of BMP-2) to enhance Smad signaling [192]. Hemojuvelin is mainly expressed in liver, heart, and skeletal muscle. *HAMP* and *HFE2*, the genes encoding for hepcidin and hemojuvelin, respectively, are loci for mutations causing juvenile hemochromatosis, a disorder of iron overload. Hepcidin was described as soluble mediator of iron homeostasis; however, little was known about the role of hemojuvelin in iron metabolism. Recent findings suggest that hemojuvelin as well as BMP-2 promote expression of hepcidin. The presence of hemojuvelin as co-receptor for BMP-2 is crucial in this respect. The model is further supported by liver-specific conditional Smad4 knockouts in mice, which demonstrate reduced hepcidin expression and iron overload [192, 193]. Knockdown of RGMa, Dragon or hemojuvelin in C2C12 cells resulted in significant reduction of Smad-dependent and Smad-independent BMP signaling. In conclusion, every single RGM BMP co-receptor is required for solid BMP signaling and members of the family cannot compensate for one another [194].

Ror2 is a tyrosine kinase receptor that is involved in the development of cartilage-derived skeleton and was suggested to promote differentiation of growth plate chondrocytes. In humans, mutations in Ror2 cause Robinow syndrome or brachydactyly type B (BDB), genetic diseases featuring skeletal malformations [195, 196]. The striking phenotypic overlaps of BDB and other types of brachydactyly, resulting from mutations in GDF-5 and BRIb, led to studies characterizing the role of Ror2 in BMP signaling. Sammar et al. [197] reported interaction of Ror2 with BRIb, which leads to negative regulation of Smad signaling. However, both receptors alone induce chondrogenesis in response to GDF-5 stimulation.

The tyrosine kinase stem cell factor receptor (c-kit), also a tyrosine kinase receptor, modulates BMP signaling. The proto-oncogene c-kit was isolated in a proteomics-based approach as a BRII tail-associated protein. Effective complex formation requires presence of either BMP-2 or BMP-2 and SCF, the ligand for c-kit, suggesting that the interaction site only becomes available in activated BRII. Smad phosphorylation was enhanced in response to treatment with both BMP-2 and SCF, while Erk activation was delayed but more stable [198]. Interestingly, kit ligand stimulates expression of BMP-15 and mutations in kit ligand and c-kit both cause infertility in female mice [1].

Perspectives

As illustrated, BMP signaling specificity, intensity and duration are regulated on multiple levels, ranging from interaction of the ligand with secreted antagonists, modes of receptor oligomerization and internalization, to cross-talk with multiple other signaling cascades. To interfere with, to alter and to manipulate distinct BMP-induced signaling cascades, the molecular mechanism of receptor activation for these specific pathways needs to be addressed very carefully. Much research is still required for a better understanding of cell type-specific effects of these growth factors to strengthen their great importance in tissue regeneration and their potential as therapeutics for various diseases.

Acknowledgments

We thank Luiza Bengtsson and Jan Boergermann for helpful suggestions on the manuscript, and Uli Rockenbauch for providing us with his cartoon of the nucleus.

References

1. Shimasaki S, Moore RK, Otsuka F, Erickson GF (2004) The bone morphogenetic protein system in mammalian reproduction. *Endocr Rev* 25: 72–101
2. Molloy SS, Bresnahan PA, Leppla SH, Klimpel KR, Thomas G (1992) Human furin is a calcium-dependent serine endoprotease that recognizes the sequence Arg-X-X-Arg and efficiently cleaves anthrax toxin protective antigen. *J Biol Chem* 267: 16396–16402
3. Schlunegger MP, Grutter MG (1992) An unusual feature revealed by the crystal structure at 2.2 Å resolution of human transforming growth factor-beta 2. *Nature* 358: 430–434
4. Kingsley DM (1994) The TGF-beta superfamily: New members, new receptors, and new genetic tests of function in different organisms. *Genes Dev* 8: 133–146
5. McPherron AC, Lee SJ (1993) GDF-3 and GDF-9: Two new members of the transforming growth factor-beta superfamily containing a novel pattern of cysteines. *J Biol Chem* 268: 3444–3449
6. Liao WX, Moore RK, Otsuka F, Shimasaki S (2003) Effect of intracellular interactions on the processing and secretion of bone morphogenetic protein-15 (BMP-15) and growth and differentiation factor-9. Implication of the aberrant ovarian phenotype of BMP-15 mutant sheep. *J Biol Chem* 278: 3713–3719
7. Sieber C, Ploger F, Schwappacher R, Bechtold R, Hanke M, Kawai S, Muraki Y, Katsuura M, Kimura M, Rechtman MM et al (2006) Monomeric and dimeric GDF-5 show equal type I receptor binding and oligomerization capability and have the same biological activity. *Biol Chem* 387: 451–460
8. Israel DI, Nove J, Kerns KM, Kaufman RJ, Rosen V, Cox KA, Wozney JM (1996) Heterodimeric bone morphogenetic proteins show enhanced activity *in vitro* and *in vivo*. *Growth Factors* 13: 291–300
9. Scheufler C, Sebald W, Hulsmeyer M (1999) Crystal structure of human bone morphogenetic protein-2 at 2.7 Å resolution. *J Mol Biol* 287: 103–115
10. Griffith DL, Keck PC, Sampath TK, Rueger DC, Carlson WD (1996) Three-dimensional

structure of recombinant human osteogenic protein 1: Structural paradigm for the transforming growth factor beta superfamily. *Proc Natl Acad Sci USA* 93: 878–883

11 Greenwald J, Groppe J, Gray P, Wiater E, Kwiatkowski W, Vale W, Choe S (2003) The BMP-7/ActRII extracellular domain complex provides new insights into the cooperative nature of receptor assembly. *Mol Cell* 11: 605–617

12 Brown MA, Zhao Q, Baker KA, Naik C, Chen C, Pukac L, Singh M, Tsareva T, Parice Y, Mahoney A et al (2005) Crystal structure of BMP-9 and functional interactions with pro-region and receptors. *J Biol Chem* 280: 25111–2518

13 Schreuder H, Liesum A, Pohl J, Kruse M, Koyama M (2005) Crystal structure of recombinant human growth and differentiation factor 5: Evidence for interaction of the type I and type II receptor-binding sites. *Biochem Biophys Res Commun* 329: 1076–1086

14 Lin SJ, Lerch TF, Cook RW, Jardetzky TS, Woodruff TK (2006) The structural basis of TGF-beta, bone morphogenetic protein, and activin ligand binding. *Reproduction* 132: 179–190

15 Lehmann K, Seemann P, Boergermann J, Morin G, Reif S, Knaus P, Mundlos S (2006) A novel R486Q mutation in BMPR1B resulting in either a brachydactyly type C/symphalangism-like phenotype or brachydactyly type A2. *Eur J Hum Genet* 14: 1248–1254

16 Kjaer KW, Eiberg H, Hansen L, van der Hagen CB, Rosendahl K, Tommerup N, Mundlos S (2006) A mutation in the receptor binding site of GDF-5 causes Mohr-Wriedt brachydactyly type A2. *J Med Genet* 43: 225–231

17 Lehmann K, Seemann P, Stricker S, Sammar M, Meyer B, Suring K, Majewski F, Tinschert S, Grzeschik KH, Muller D et al (2003) Mutations in bone morphogenetic protein receptor 1B cause brachydactyly type A2. *Proc Natl Acad Sci USA* 100: 12277–12282

18 Seemann P, Schwappacher R, Kjaer KW, Krakow D, Lehmann K, Dawson K, Stricker S, Pohl J, Ploger F, Staub E et al (2005) Activating and deactivating mutations in the receptor interaction site of GDF-5 cause symphalangism or brachydactyly type A2. *J Clin Invest* 115: 2373–2381

19 Chen YG, Hata A, Lo RS, Wotton D, Shi Y, Pavletich N, Massague J (1998) Determinants of specificity in TGF-beta signal transduction. *Genes Dev* 12: 2144–2152

20 Lo RS, Chen YG, Shi Y, Pavletich NP, Massague J (1998) The L3 loop: A structural motif determining specific interactions between SMAD proteins and TGF-beta receptors. *EMBO J* 17: 996–1005

21 ten Dijke P, Ichijo H, Franzen P, Schulz P, Saras J, Toyoshima H, Heldin CH, Miyazono K (1993) Activin receptor-like kinases: a novel subclass of cell-surface receptors with predicted serine/threonine kinase activity. *Oncogene* 8: 2879–2887

22 ten Dijke P, Yamashita H, Sampath TK, Reddi AH, Estevez M, Riddle DL, Ichijo H, Heldin CH, Miyazono K (1994) Identification of type I receptors for osteogenic protein-1 and bone morphogenetic protein-4. *J Biol Chem* 269: 16985–16988

23 Zimmerman CM, Mathews LS (1996) Activin receptors: Cellular signalling by receptor serine kinases. *Biochem Soc Symp* 62: 25–38

24 Zou H, Wieser R, Massague J, Niswander L (1997) Distinct roles of type I bone mor-

phogenetic protein receptors in the formation and differentiation of cartilage. *Genes Dev* 11: 2191–2203

25 Howe JR, Bair JL, Sayed MG, Anderson ME, Mitros FA, Petersen GM, Velculescu VE, Traverso G, Vogelstein B (2001) Germline mutations of the gene encoding bone morphogenetic protein receptor 1A in juvenile polyposis. *Nat Genet* 28: 184–187

26 Sayed MG, Ahmed AF, Ringold JR, Anderson ME, Bair JL, Mitros FA, Lynch HT, Tinley ST, Petersen GM, Giardiello FM et al (2002) Germline SMAD4 or BMPR1A mutations and phenotype of juvenile polyposis. *Ann Surg Oncol* 9: 901–906

27 Virdi AS, Shore EM, Oreffo RO, Li M, Connor JM, Smith R, Kaplan FS, Triffitt JT (1999) Phenotypic and molecular heterogeneity in fibrodysplasia ossificans progressiva. *Calcif Tissue Int* 65: 250–255

28 Shore EM, Xu M, Feldman GJ, Fenstermacher DA, Cho TJ, Choi IH, Connor JM, Delai P, Glaser DL, LeMerrer M et al (2006) A recurrent mutation in the BMP type I receptor ACVR1 causes inherited and sporadic fibrodysplasia ossificans progressiva. *Nat Genet* 38: 525–527

29 Kawabata M, Chytil A, Moses HL (1995) Cloning of a novel type II serine/threonine kinase receptor through interaction with the type I transforming growth factor-beta receptor. *J Biol Chem* 270: 5625–5630

30 Rosenzweig BL, Imamura T, Okadome T, Cox GN, Yamashita H, ten Dijke P, Heldin CH, Miyazono K (1995) Cloning and characterization of a human type II receptor for bone morphogenetic proteins. *Proc Natl Acad Sci USA* 92: 7632–7636

31 Liu F, Ventura F, Doody J, Massague J (1995) Human type II receptor for bone morphogenic proteins (BMPs): Extension of the two-kinase receptor model to the BMPs. *Mol Cell Biol* 15: 3479–3486

32 Nohno T, Ishikawa T, Saito T, Hosokawa K, Noji S, Wolsing DH, Rosenbaum JS (1995) Identification of a human type II receptor for bone morphogenetic protein-4 that forms differential heteromeric complexes with bone morphogenetic protein type I receptors. *J Biol Chem* 270: 22522–22526

33 Hassel S, Eichner A, Yakymovych M, Hellman U, Knaus P, Souchelnytskyi S (2004) Proteins associated with type II bone morphogenetic protein receptor (BMPR-II) and identified by two-dimensional gel electrophoresis and mass spectrometry. *Proteomics* 4: 1346–1358

34 Morrell NW (2006) Pulmonary hypertension due to BMPR2 mutation: A new paradigm for tissue remodeling? *Proc Am Thorac Soc* 3: 680–686

35 Liu D, Wang J, Kinzel B, Mueller M, Mao X, Valdez R, Liu Y, Li E (2007) Dosage-dependent requirement of BMP type II receptor for maintenance of vascular integrity. *Blood* 110: 1502–1510

36 Zakrzewicz A, Hecker M, Marsh LM, Kwapiszewska G, Nejman B, Long L, Seeger W, Schermuly RT, Morrell NW, Morty RE et al (2007) Receptor for activated C-kinase 1, a novel interaction partner of type II bone morphogenetic protein receptor, regulates smooth muscle cell proliferation in pulmonary arterial hypertension. *Circulation* 115: 2957–2968

37 Kirsch T, Sebald W, Dreyer MK (2000) Crystal structure of the BMP-2–BRIA ectodomain complex. *Nat Struct Biol* 7: 492–496
38 Kirsch T, Nickel J, Sebald W (2000) Isolation of recombinant BMP receptor IA ectodomain and its 2:1 complex with BMP-2. *FEBS Lett* 468: 215–219
39 Allendorph GP, Vale WW, Choe S (2006) Structure of the ternary signaling complex of a TGF-beta superfamily member. *Proc Natl Acad Sci USA* 103: 7643–7648
40 Weber D, Kotzsch A, Nickel J, Harth S, Seher A, Mueller U, Sebald W, Mueller TD (2007) A silent H-bond can be mutationally activated for high-affinity interaction of BMP-2 and activin type IIB receptor. *BMC Struct Biol* 7: 6
41 Gray PC, Greenwald J, Blount AL, Kunitake KS, Donaldson CJ, Choe S, Vale W (2000) Identification of a binding site on the type II activin receptor for activin and inhibin. *J Biol Chem* 275: 3206–3212
42 Knaus P, Sebald W (2001) Cooperativity of binding epitopes and receptor chains in the BMP/TGFbeta superfamily. *Biol Chem* 382: 1189–1195
43 Mitchell H, Choudhury A, Pagano RE, Leof EB (2004) Ligand-dependent and -independent transforming growth factor-beta receptor recycling regulated by clathrin-mediated endocytosis and Rab11. *Mol Biol Cell* 15: 4166–4178
44 Di Guglielmo GM, Le Roy C, Goodfellow AF, Wrana JL (2003) Distinct endocytic pathways regulate TGF-beta receptor signalling and turnover. *Nat Cell Biol* 5: 410–421
45 Gilboa L, Nohe A, Geissendorfer T, Sebald W, Henis YI, Knaus P (2000) Bone morphogenetic protein receptor complexes on the surface of live cells: A new oligomerization mode for serine/threonine kinase receptors. *Mol Biol Cell* 11: 1023–1035
46 Nohe A, Hassel S, Ehrlich M, Neubauer F, Sebald W, Henis YI, Knaus P (2002) The mode of bone morphogenetic protein (BMP) receptor oligomerization determines different BMP-2 signaling pathways. *J Biol Chem* 277: 5330–5338
47 Hartung A, Bitton-Worms K, Rechtman MM, Wenzel V, Boergermann JH, Hassel S, Henis YI, Knaus P (2006) Different routes of bone morphogenic protein (BMP) receptor endocytosis influence BMP signaling. *Mol Cell Biol* 26: 7791–805
48 Zhang Y, Feng X, We R, Derynck R (1996) Receptor-associated Mad homologues synergize as effectors of the TGF-beta response. *Nature* 383: 168–172
49 Wrana JL, Attisano L, Wieser R, Ventura F, Massague J (1994) Mechanism of activation of the TGF-beta receptor. *Nature* 370: 341–347
50 Shi W, Chang C, Nie S, Xie S, Wan M, Cao X (2007) Endofin acts as a Smad anchor for receptor activation in BMP signaling. *J Cell Sci* 120: 1216–1224
51 Kretzschmar M, Doody J, Massague J (1997) Opposing BMP and EGF signalling pathways converge on the TGF-beta family mediator Smad1. *Nature* 389: 618–622
52 Korchynskyi O, ten Dijke P (2002) Identification and functional characterization of distinct critically important bone morphogenetic protein-specific response elements in the Id1 promoter. *J Biol Chem* 277: 4883–4891
53 Kusanagi K, Inoue H, Ishidou Y, Mishima HK, Kawabata M, Miyazono K (2000) Characterization of a bone morphogenetic protein-responsive Smad-binding element. *Mol Biol Cell* 11: 555–565

54 Korchynskyi O, Dechering KJ, Sijbers AM, Olijve W, ten Dijke P (2003) Gene array analysis of bone morphogenetic protein type I receptor-induced osteoblast differentiation. *J Bone Miner Res* 18: 1177–1185
55 de Jong DS, Vaes BL, Dechering KJ, Feijen A, Hendriks JM, Wehrens R, Mummery CL, van Zoelen EJ, Olijve W, Steegenga WT (2004) Identification of novel regulators associated with early-phase osteoblast differentiation. *J Bone Miner Res* 19: 947–958
56 Knockaert M, Sapkota G, Alarcon C, Massague J, Brivanlou AH (2006) Unique players in the BMP pathway: Small C-terminal domain phosphatases dephosphorylate Smad1 to attenuate BMP signaling. *Proc Natl Acad Sci USA* 103: 11940–11945
57 Chen HB, Shen J, Ip YT, Xu L (2006) Identification of phosphatases for Smad in the BMP/DPP pathway. *Genes Dev* 20: 648–653
58 Duan X, Liang YY, Feng XH, Lin X (2006) Protein serine/threonine phosphatase PPM1A dephosphorylates Smad1 in the bone morphogenetic protein signaling pathway. *J Biol Chem* 281: 36526–36532
59 Sapkota G, Knockaert M, Alarcon C, Montalvo E, Brivanlou AH, Massague J (2006) Dephosphorylation of the linker regions of Smad1 and Smad2/3 by small C-terminal domain phosphatases has distinct outcomes for bone morphogenetic protein and transforming growth factor-beta pathways. *J Biol Chem* 281: 40412–40419
60 Hata A, Lagna G, Massague J, Hemmati-Brivanlou A (1998) Smad6 inhibits BMP/Smad1 signaling by specifically competing with the Smad4 tumor suppressor. *Genes Dev* 12: 186–197
61 Bennett D, Alphey L (2002) PP1 binds Sara and negatively regulates Dpp signaling in *Drosophila melanogaster*. *Nat Genet* 31: 419–423
62 Murakami G, Watabe T, Takaoka K, Miyazono K, Imamura T (2003) Cooperative inhibition of bone morphogenetic protein signaling by Smurf1 and inhibitory Smads. *Mol Biol Cell* 14: 2809–2817
63 Zhang Y, Chang C, Gehling DJ, Hemmati-Brivanlou A, Derynck R (2001) Regulation of Smad degradation and activity by Smurf2, an E3 ubiquitin ligase. *Proc Natl Acad Sci USA* 98: 974–979
64 Zhu H, Kavsak P, Abdollah S, Wrana JL, Thomsen GH (1999) A SMAD ubiquitin ligase targets the BMP pathway and affects embryonic pattern formation. *Nature* 400: 687–693
65 Shen R, Chen M, Wang YJ, Kaneki H, Xing L, O'Keefe R J, Chen D (2006) Smad6 interacts with Runx2 and mediates Smad ubiquitin regulatory factor 1–induced Runx2 degradation. *J Biol Chem* 281: 3569–3576
66 Kim BC, Lee HJ, Park SH, Lee SR, Karpova TS, McNally JG, Felici A, Lee DK, Kim SJ (2004) Jab1/CSN5, a component of the COP9 signalosome, regulates transforming growth factor beta signaling by binding to Smad7 and promoting its degradation. *Mol Cell Biol* 24: 2251–2262
67 Komuro A, Imamura T, Saitoh M, Yoshida Y, Yamori T, Miyazono K, Miyazawa K (2004) Negative regulation of transforming growth factor-beta (TGF-beta) signaling by WW domain-containing protein 1 (WWP1). *Oncogene* 23: 6914–6923

68 Lin X, Liang M, Liang YY, Brunicardi FC, Feng XH (2003) SUMO-1/Ubc9 promotes nuclear accumulation and metabolic stability of tumor suppressor Smad4. *J Biol Chem* 278: 31043–31048

69 Lee PS, Chang C, Liu D, Derynck R (2003) Sumoylation of Smad4, the common Smad mediator of transforming growth factor-beta family signaling. *J Biol Chem* 278: 27853–27863

70 Li L, Xin H, Xu X, Huang M, Zhang X, Chen Y, Zhang S, Fu XY, Chang Z (2004) CHIP mediates degradation of Smad proteins and potentially regulates Smad-induced transcription. *Mol Cell Biol* 24: 856–864

71 Wan M, Tang Y, Tytler EM, Lu C, Jin B, Vickers SM, Yang L, Shi X, Cao X (2004) Smad4 protein stability is regulated by ubiquitin ligase SCF beta-TrCP1. *J Biol Chem* 279: 14484–14487

72 Itoh F, Asao H, Sugamura K, Heldin CH, ten Dijke P, Itoh S (2001) Promoting bone morphogenetic protein signaling through negative regulation of inhibitory Smads. *EMBO J* 20: 4132–4142

73 Nohe A, Keating E, Knaus P, Petersen NO (2004) Signal transduction of bone morphogenetic protein receptors. *Cell Signal* 16: 291–299

74 Lu M, Lin SC, Huang Y, Kang YJ, Rich R, Lo YC, Myszka D, Han J, Wu H (2007) XIAP induces NF-kappaB activation *via* the BIR1/TAB1 interaction and BIR1 dimerization. *Mol Cell* 26: 689–702

75 Chai J, Shiozaki E, Srinivasula SM, Wu Q, Datta P, Alnemri ES, Shi Y (2001) Structural basis of caspase-7 inhibition by XIAP. *Cell* 104: 769–780

76 Huang Y, Park YC, Rich RL, Segal D, Myszka DG, Wu H (2001) Structural basis of caspase inhibition by XIAP: Differential roles of the linker *versus* the BIR domain. *Cell* 104: 781–790

77 Riedl SJ, Renatus M, Schwarzenbacher R, Zhou Q, Sun C, Fesik SW, Liddington RC, Salvesen GS (2001) Structural basis for the inhibition of caspase-3 by XIAP. *Cell* 104: 791–800

78 Shiozaki EN, Chai J, Rigotti DJ, Riedl SJ, Li P, Srinivasula SM, Alnemri ES, Fairman R, Shi Y (2003) Mechanism of XIAP-mediated inhibition of caspase-9. *Mol Cell* 11: 519–527

79 Shibuya H, Yamaguchi K, Shirakabe K, Tonegawa A, Gotoh Y, Ueno N, Irie K, Nishida E, Matsumoto K (1996) TAB1: An activator of the TAK1 MAPKKK in TGF-beta signal transduction. *Science* 272: 1179–1182

80 Wang C, Deng L, Hong M, Akkaraju GR, Inoue J, Chen ZJ (2001) TAK1 is a ubiquitin-dependent kinase of MKK and IKK. *Nature* 412: 346–351

81 Shirakabe K, Yamaguchi K, Shibuya H, Irie K, Matsuda S, Moriguchi T, Gotoh Y, Matsumoto K, Nishida E (1997) TAK1 mediates the ceramide signaling to stress-activated protein kinase/c-Jun N-terminal kinase. *J Biol Chem* 272: 8141–8144

82 Yamaguchi K, Nagai S, Ninomiya-Tsuji J, Nishita M, Tamai K, Irie K, Ueno N, Nishida E, Shibuya H, Matsumoto K (1999) XIAP, a cellular member of the inhibitor of apop-

tosis protein family, links the receptors to TAB1–TAK1 in the BMP signaling pathway. *EMBO J* 18: 179–187

83   Yamaguchi K, Shirakabe K, Shibuya H, Irie K, Oishi I, Ueno N, Taniguchi T, Nishida E, Matsumoto K (1995) Identification of a member of the MAPKKK family as a potential mediator of TGF-beta signal transduction. *Science* 270: 2008–2011

84   Lewis J, Burstein E, Reffey SB, Bratton SB, Roberts AB, Duckett CS (2004) Uncoupling of the signaling and caspase-inhibitory properties of X-linked inhibitor of apoptosis. *J Biol Chem* 279: 9023–9029

85   Sanna MG, Duckett CS, Richter BW, Thompson CB, Ulevitch RJ (1998) Selective activation of JNK1 is necessary for the anti-apoptotic activity of hILP. *Proc Natl Acad Sci USA* 95: 6015–6020

86   Shibuya H, Iwata H, Masuyama N, Gotoh Y, Yamaguchi K, Irie K, Matsumoto K, Nishida E, Ueno N (1998) Role of TAK1 and TAB1 in BMP signaling in early *Xenopus* development. *EMBO J* 17: 1019–1028

87   Sano Y, Harada J, Tashiro S, Gotoh-Mandeville R, Maekawa T, Ishii S (1999) ATF-2 is a common nuclear target of Smad and TAK1 pathways in transforming growth factor-beta signaling. *J Biol Chem* 274: 8949–8957

88   Lai CF, Cheng SL (2002) Signal transductions induced by bone morphogenetic protein-2 and transforming growth factor-beta in normal human osteoblastic cells. *J Biol Chem* 277: 15514–15522

89   Gallea S, Lallemand F, Atfi A, Rawadi G, Ramez V, Spinella-Jaegle S, Kawai S, Faucheu C, Huet L, Baron R et al (2001) Activation of mitogen-activated protein kinase cascades is involved in regulation of bone morphogenetic protein-2–induced osteoblast differentiation in pluripotent C2C12 cells. *Bone* 28: 491–498

90   Guicheux J, Lemonnier J, Ghayor C, Suzuki A, Palmer G, Caverzasio J (2003) Activation of p38 mitogen-activated protein kinase and c-Jun-NH2–terminal kinase by BMP-2 and their implication in the stimulation of osteoblastic cell differentiation. *J Bone Miner Res* 18: 2060–2068

91   Zuzarte-Luis V, Montero JA, Rodriguez-Leon J, Merino R, Rodriguez-Rey JC, Hurle JM (2004) A new role for BMP5 during limb development acting through the synergic activation of Smad and MAPK pathways. *Dev Biol* 272: 39–52

92   Kimura N, Matsuo R, Shibuya H, Nakashima K, Taga T (2000) BMP-2–induced apoptosis is mediated by activation of the TAK1-p38 kinase pathway that is negatively regulated by Smad6. *J Biol Chem* 275: 17647–17652

93   Edlund S, Bu S, Schuster N, Aspenstrom P, Heuchel R, Heldin NE, ten Dijke P, Heldin CH, Landstrom M (2003) Transforming growth factor-beta1 (TGF-beta)-induced apoptosis of prostate cancer cells involves Smad7-dependent activation of p38 by TGF-beta-activated kinase 1 and mitogen-activated protein kinase kinase 3. *Mol Biol Cell* 14: 529–544

94   Ghosh-Choudhury N, Abboud SL, Nishimura R, Celeste A, Mahimainathan L, Choudhury GG (2002) Requirement of BMP-2–induced phosphatidylinositol 3-kinase and Akt

serine/threonine kinase in osteoblast differentiation and Smad-dependent BMP-2 gene transcription. *J Biol Chem* 277: 33361–33368

95 Vinals F, Lopez-Rovira T, Rosa JL, Ventura F (2002) Inhibition of PI3K/p70 S6K and p38 MAPK cascades increases osteoblastic differentiation induced by BMP-2. *FEBS Lett* 510: 99–104

96 Osyczka AM, Leboy PS (2005) Bone morphogenetic protein regulation of early osteoblast genes in human marrow stromal cells is mediated by extracellular signal-regulated kinase and phosphatidylinositol 3-kinase signaling. *Endocrinology* 146: 3428–3437

97 Manning BD, Cantley LC (2007) AKT/PKB signaling: Navigating downstream. *Cell* 129: 1261–1274

98 Kobielak K, Stokes N, de la Cruz J, Polak L, Fuchs E (2007) Loss of a quiescent niche but not follicle stem cells in the absence of bone morphogenetic protein signaling. *Proc Natl Acad Sci USA* 104: 10063–10068

99 Palcy S, Bolivar I, Goltzman D (2000) Role of activator protein 1 transcriptional activity in the regulation of gene expression by transforming growth factor beta1 and bone morphogenetic protein 2 in ROS 17/2.8 osteoblast-like cells. *J Bone Miner Res* 15: 2352–2361

100 Wong WK, Knowles JA, Morse JH (2005) Bone morphogenetic protein receptor type II C-terminus interacts with c-Src: Implication for a role in pulmonary arterial hypertension. *Am J Respir Cell Mol Biol* 33: 438–446

101 Machado RD, Rudarakanchana N, Atkinson C, Flanagan JA, Harrison R, Morrell NW, Trembath RC (2003) Functional interaction between BMPR-II and Tctex-1, a light chain of Dynein, is isoform-specific and disrupted by mutations underlying primary pulmonary hypertension. *Hum Mol Genet* 12: 3277–3286

102 Foletta VC, Moussi N, Sarmiere PD, Bamburg JR, Bernard O (2004) LIM kinase 1, a key regulator of actin dynamics, is widely expressed in embryonic and adult tissues. *Exp Cell Res* 294: 392–405

103 Lee-Hoeflich ST, Causing CG, Podkowa M, Zhao X, Wrana JL, Attisano L (2004) Activation of LIMK1 by binding to the BMP receptor, BMPRII, regulates BMP-dependent dendritogenesis. *EMBO J* 23: 4792–4801

104 Yanagita M (2005) BMP antagonists: their roles in development and involvement in pathophysiology. *Cytokine Growth Factor Rev* 16: 309–317

105 Gazzerro E, Canalis E (2006) Bone morphogenetic proteins and their antagonists. *Rev Endocr Metab Disord* 7: 51–65

106 van Bezooijen RL, ten Dijke P, Papapoulos SE, Lowik CW (2005) SOST/sclerostin, an osteocyte-derived negative regulator of bone formation. *Cytokine Growth Factor Rev* 16: 319–327

107 Tamaoki H, Miura R, Kusunoki M, Kyogoku Y, Kobayashi Y, Moroder L (1998) Folding motifs induced and stabilized by distinct cystine frameworks. *Protein Eng* 11: 649–659

108 Vitt UA, Hsu SY, Hsueh AJ (2001) Evolution and classification of cystine knot-con-

taining hormones and related extracellular signaling molecules. *Mol Endocrinol* 15: 681–694

109 Avsian-Kretchmer O, Hsueh AJ (2004) Comparative genomic analysis of the eight-membered ring cystine knot-containing bone morphogenetic protein antagonists. *Mol Endocrinol* 18: 1–12

110 Ozaki T, Sakiyama S (1993) Molecular cloning of rat calpactin I heavy-chain cDNA whose expression is induced in v-src-transformed rat culture cell lines. *Oncogene* 8: 1707–1710

111 Ozaki T, Sakiyama S (1994) Tumor-suppressive activity of N03 gene product in v-src-transformed rat 3Y1 fibroblasts. *Cancer Res* 54: 646–648

112 Dionne MS, Skarnes WC, Harland RM (2001) Mutation and analysis of Dan, the founding member of the Dan family of transforming growth factor beta antagonists. *Mol Cell Biol* 21: 636–643

113 Stanley E, Biben C, Kotecha S, Fabri L, Tajbakhsh S, Wang CC, Hatzistavrou T, Roberts B, Drinkwater C, Lah M et al (1998) DAN is a secreted glycoprotein related to *Xenopus* cerberus. *Mech Dev* 77: 173–184

114 Bouwmeester T, Kim S, Sasai Y, Lu B, De Robertis EM (1996) Cerberus is a head-inducing secreted factor expressed in the anterior endoderm of Spemann's organizer. *Nature* 382: 595–601

115 Piccolo S, Agius E, Leyns L, Bhattacharyya S, Grunz H, Bouwmeester T, De Robertis EM (1999) The head inducer Cerberus is a multifunctional antagonist of Nodal, BMP and Wnt signals. *Nature* 397: 707–710

116 Belo JA, Bachiller D, Agius E, Kemp C, Borges AC, Marques S, Piccolo S, De Robertis EM (2000) Cerberus-like is a secreted BMP and nodal antagonist not essential for mouse development. *Genesis* 26: 265–270

117 Biben C, Stanley E, Fabri L, Kotecha S, Rhinn M, Drinkwater C, Lah M, Wang CC, Nash A, Hilton D et al (1998) Murine cerberus homologue mCer-1: a candidate anterior patterning molecule. *Dev Biol* 194: 135–151

118 Shawlot W, Min Deng J, Wakamiya M, Behringer RR (2000) The cerberus-related gene, Cerr1, is not essential for mouse head formation. *Genesis* 26: 253–258

119 Shawlot W, Deng JM, Behringer RR (1998) Expression of the mouse cerberus-related gene, Cerr1, suggests a role in anterior neural induction and somitogenesis. *Proc Natl Acad Sci USA* 95: 6198–6203

120 Yokouchi Y, Vogan KJ, Pearse RV, 2nd, Tabin CJ (1999) Antagonistic signaling by Caronte, a novel Cerberus-related gene, establishes left-right asymmetric gene expression. *Cell* 98: 573–583

121 Rodriguez Esteban C, Capdevila J, Economides AN, Pascual J, Ortiz A, Izpisua Belmonte JC (1999) The novel Cer-like protein Caronte mediates the establishment of embryonic left-right asymmetry. *Nature* 401: 243–251

122 Bell E, Munoz-Sanjuan I, Altmann CR, Vonica A, Brivanlou AH (2003) Cell fate specification and competence by Coco, a maternal BMP, TGFbeta and Wnt inhibitor. *Development* 130: 1381–1389

123 Pearce JJ, Penny G, Rossant J (1999) A mouse cerberus/Dan-related gene family. *Dev Biol* 209: 98–110

124 Minabe-Saegusa C, Saegusa H, Tsukahara M, Noguchi S (1998) Sequence and expression of a novel mouse gene PRDC (protein related to DAN and cerberus) identified by a gene trap approach. *Dev Growth Differ* 40: 343–353

125 Hsu DR, Economides AN, Wang X, Eimon PM, Harland RM (1998) The *Xenopus* dorsalizing factor Gremlin identifies a novel family of secreted proteins that antagonize BMP activities. *Mol Cell* 1: 673–683

126 McMahon R, Murphy M, Clarkson M, Taal M, Mackenzie HS, Godson C, Martin F, Brady HR (2000) IHG-2, a mesangial cell gene induced by high glucose, is human gremlin. Regulation by extracellular glucose concentration, cyclic mechanical strain, and transforming growth factor-beta1. *J Biol Chem* 275: 9901–9904

127 Khokha MK, Hsu D, Brunet LJ, Dionne MS, Harland RM (2003) Gremlin is the BMP antagonist required for maintenance of Shh and Fgf signals during limb patterning. *Nat Genet* 34: 303–307

128 Michos O, Goncalves A, Lopez-Rios J, Tiecke E, Naillat F, Beier K, Galli A, Vainio S, Zeller R (2007) Reduction of BMP-4 activity by gremlin 1 enables ureteric bud outgrowth and GDNF/WNT11 feedback signalling during kidney branching morphogenesis. *Development* 134: 2397–2405

129 Michos O, Panman L, Vintersten K, Beier K, Zeller R, Zuniga A (2004) Gremlin-mediated BMP antagonism induces the epithelial-mesenchymal feedback signaling controlling metanephric kidney and limb organogenesis. *Development* 131: 3401–3410

130 Pereira RC, Economides AN, Canalis E (2000) Bone morphogenetic proteins induce gremlin, a protein that limits their activity in osteoblasts. *Endocrinology* 141: 4558–4563

131 Gazzerro E, Pereira RC, Jorgetti V, Olson S, Economides AN, Canalis E (2005) Skeletal overexpression of gremlin impairs bone formation and causes osteopenia. *Endocrinology* 146: 655–665

132 Topol LZ, Bardot B, Zhang Q, Resau J, Huillard E, Marx M, Calothy G, Blair DG (2000) Biosynthesis, post-translation modification, and functional characterization of Drm/Gremlin. *J Biol Chem* 275: 8785–8793

133 Topol LZ, Modi WS, Koochekpour S, Blair DG (2000) DRM/GREMLIN (CKTSF1B1) maps to human chromosome 15 and is highly expressed in adult and fetal brain. *Cytogenet Cell Genet* 89: 79–84

134 Suzuki M, Shigematsu H, Shivapurkar N, Reddy J, Miyajima K, Takahashi T, Gazdar AF, Frenkel EP (2006) Methylation of apoptosis related genes in the pathogenesis and prognosis of prostate cancer. *Cancer Lett* 242: 222–230

135 Chen B, Athanasiou M, Gu Q, Blair DG (2002) Drm/Gremlin transcriptionally activates p21(Cip1) *via* a novel mechanism and inhibits neoplastic transformation. *Biochem Biophys Res Commun* 295: 1135–1141

136 Yanagita M, Oka M, Watabe T, Iguchi H, Niida A, Takahashi S, Akiyama T, Miyazono

K, Yanagisawa M, Sakurai T (2004) USAG-1: A bone morphogenetic protein antagonist abundantly expressed in the kidney. *Biochem Biophys Res Commun* 316: 490–500

137 Laurikkala J, Kassai Y, Pakkasjarvi L, Thesleff I, Itoh N (2003) Identification of a secreted BMP antagonist, ectodin, integrating BMP, FGF, and SHH signals from the tooth enamel knot. *Dev Biol* 264: 91–105

138 Simmons DG, Kennedy TG (2002) Uterine sensitization-associated gene-1: a novel gene induced within the rat endometrium at the time of uterine receptivity/sensitization for the decidual cell reaction. *Biol Reprod* 67: 1638–1645

139 Itasaki N, Jones CM, Mercurio S, Rowe A, Domingos PM, Smith JC, Krumlauf R (2003) Wise, a context-dependent activator and inhibitor of Wnt signalling. *Development* 130: 4295–4305

140 Sokol SY (1996) Analysis of Dishevelled signalling pathways during *Xenopus* development. *Curr Biol* 6: 1456–1467

141 Tamai K, Semenov M, Kato Y, Spokony R, Liu C, Katsuyama Y, Hess F, Saint-Jeannet JP, He X (2000) LDL-receptor-related proteins in Wnt signal transduction. *Nature* 407: 530–535

142 Brunkow ME, Gardner JC, Van Ness J, Paeper BW, Kovacevich BR, Proll S, Skonier JE, Zhao L, Sabo PJ, Fu Y et al (2001) Bone dysplasia sclerosteosis results from loss of the SOST gene product, a novel cystine knot-containing protein. *Am J Hum Genet* 68: 577–589

143 Beighton P (1988) Sclerosteosis. *J Med Genet* 25: 200–203

144 Beighton P, Davidson J, Durr L, Hamersma H (1977) Sclerosteosis – An autosomal recessive disorder. *Clin Genet* 11: 1–7

145 Balemans W, Ebeling M, Patel N, Van Hul E, Olson P, Dioszegi M, Lacza C, Wuyts W, Van Den Ende J, Willems P et al (2001) Increased bone density in sclerosteosis is due to the deficiency of a novel secreted protein (SOST). *Hum Mol Genet* 10: 537–543

146 Jacobs P (1977) Van Buchem disease. *Postgrad Med J* 53: 497–506

147 Van Buchem FS, Hadders HN, Ubbens R (1955) An uncommon familial systemic disease of the skeleton: Hyperostosis corticalis generalisata familiaris. *Acta Radiol* 44: 109–120

148 Kusu N, Laurikkala J, Imanishi M, Usui H, Konishi M, Miyake A, Thesleff I, Itoh N (2003) Sclerostin is a novel secreted osteoclast-derived bone morphogenetic protein antagonist with unique ligand specificity. *J Biol Chem* 278: 24113–24117

149 Winkler DG, Sutherland MK, Geoghegan JC, Yu C, Hayes T, Skonier JE, Shpektor D, Jonas M, Kovacevich BR, Staehling-Hampton K et al (2003) Osteocyte control of bone formation *via* sclerostin, a novel BMP antagonist. *EMBO J* 22: 6267–6276

150 Poole KE, van Bezooijen RL, Loveridge N, Hamersma H, Papapoulos SE, Lowik CW, Reeve J (2005) Sclerostin is a delayed secreted product of osteocytes that inhibits bone formation. *FASEB J* 19: 1842–1844

151 Sutherland MK, Geoghegan JC, Yu C, Turcott E, Skonier JE, Winkler DG, Latham JA (2004) Sclerostin promotes the apoptosis of human osteoblastic cells: A novel regulation of bone formation. *Bone* 35: 828–835

152 van Bezooijen RL, Svensson JP, Eefting D, Visser A, van der Horst G, Karperien M, Quax PH, Vrieling H, Papapoulos SE, ten Dijke P et al (2007) Wnt but not BMP signaling is involved in the inhibitory action of sclerostin on BMP-stimulated bone formation. *J Bone Miner Res* 22: 19–28

153 van Bezooijen RL, Roelen BA, Visser A, van der Wee-Pals L, de Wilt E, Karperien M, Hamersma H, Papapoulos SE, ten Dijke P, Lowik CW (2004) Sclerostin is an osteocyte-expressed negative regulator of bone formation, but not a classical BMP antagonist. *J Exp Med* 199: 805–814

154 Ohyama Y, Nifuji A, Maeda Y, Amagasa T, Noda M (2004) Spaciotemporal association and bone morphogenetic protein regulation of sclerostin and osterix expression during embryonic osteogenesis. *Endocrinology* 145: 4685–4692

155 Sutherland MK, Geoghegan JC, Yu C, Winkler DG, Latham JA (2004) Unique regulation of SOST, the sclerosteosis gene, by BMPs and steroid hormones in human osteoblasts. *Bone* 35: 448–454

156 Winkler DG, Yu C, Geoghegan JC, Ojala EW, Skonier JE, Shpektor D, Sutherland MK, Latham JA (2004) Noggin and sclerostin bone morphogenetic protein antagonists form a mutually inhibitory complex. *J Biol Chem* 279: 36293–36298

157 Mason ED, Konrad KD, Webb CD, Marsh JL (1994) Dorsal midline fate in *Drosophila* embryos requires twisted gastrulation, a gene encoding a secreted protein related to human connective tissue growth factor. *Genes Dev* 8: 1489–1501

158 Oelgeschlager M, Larrain J, Geissert D, De Robertis EM (2000) The evolutionarily conserved BMP-binding protein Twisted gastrulation promotes BMP signalling. *Nature* 405: 757–763

159 Chang C, Holtzman DA, Chau S, Chickering T, Woolf EA, Holmgren LM, Bodorova J, Gearing DP, Holmes WE, Brivanlou AH (2001) Twisted gastrulation can function as a BMP antagonist. *Nature* 410: 483–487

160 Scott IC, Blitz IL, Pappano WN, Maas SA, Cho KW, Greenspan DS (2001) Homologues of Twisted gastrulation are extracellular cofactors in antagonism of BMP signalling. *Nature* 410: 475–478

161 Ross JJ, Shimmi O, Vilmos P, Petryk A, Kim H, Gaudenz K, Hermanson S, Ekker SC, O'Connor MB, Marsh JL (2001) Twisted gastrulation is a conserved extracellular BMP antagonist. *Nature* 410: 479–483

162 Gazzerro E, Deregowski V, Vaira S, Canalis E (2005) Overexpression of twisted gastrulation inhibits bone morphogenetic protein action and prevents osteoblast cell differentiation *in vitro*. *Endocrinology* 146: 3875–3882

163 Petryk A, Shimmi O, Jia X, Carlson AE, Tervonen L, Jarcho MP, O'Connor MB, Gopalakrishnan R (2005) Twisted gastrulation and chordin inhibit differentiation and mineralization in MC3T3-E1 osteoblast-like cells. *Bone* 36: 617–626

164 Aspenberg P, Jeppsson C, Economides AN (2001) The bone morphogenetic proteins antagonist Noggin inhibits membranous ossification. *J Bone Miner Res* 16: 497–500

165 Zimmerman LB, De Jesus-Escobar JM, Harland RM (1996) The Spemann organizer signal noggin binds and inactivates bone morphogenetic protein 4. *Cell* 86: 599–606

166 Groppe J, Greenwald J, Wiater E, Rodriguez-Leon J, Economides AN, Kwiatkowski W, Affolter M, Vale WW, Belmonte JC, Choe S (2002) Structural basis of BMP signalling inhibition by the cystine knot protein Noggin. *Nature* 420: 636–642

167 Anderson RM, Lawrence AR, Stottmann RW, Bachiller D, Klingensmith J (2002) Chordin and noggin promote organizing centers of forebrain development in the mouse. *Development* 129: 4975–4987

168 Smith WC, Harland RM (1992) Expression cloning of noggin, a new dorsalizing factor localized to the Spemann organizer in *Xenopus* embryos. *Cell* 70: 829–840

169 Dionne MS, Brunet LJ, Eimon PM, Harland RM (2002) Noggin is required for correct guidance of dorsal root ganglion axons. *Dev Biol* 251: 283–293

170 Wan DC, Pomerantz JH, Brunet LJ, Kim JB, Chou YF, Wu BM, Harland R, Blau HM, Longaker MT (2007) Noggin suppression enhances *in vitro* osteogenesis and accelerates *in vivo* bone formation. *J Biol Chem* 282: 26450–26459

171 Brunet LJ, McMahon JA, McMahon AP, Harland RM (1998) Noggin, cartilage morphogenesis, and joint formation in the mammalian skeleton. *Science* 280: 1455–1457

172 Gazzerro E, Gangji V, Canalis E (1998) Bone morphogenetic proteins induce the expression of noggin, which limits their activity in cultured rat osteoblasts. *J Clin Invest* 102: 2106–2114

173 Abe E, Yamamoto M, Taguchi Y, Lecka-Czernik B, O'Brien CA, Economides AN, Stahl N, Jilka RL, Manolagas SC (2000) Essential requirement of BMPs-2/4 for both osteoblast and osteoclast formation in murine bone marrow cultures from adult mice: Antagonism by noggin. *J Bone Miner Res* 15: 663–673

174 Pathi S, Rutenberg JB, Johnson RL, Vortkamp A (1999) Interaction of Ihh and BMP/Noggin signaling during cartilage differentiation. *Dev Biol* 209: 239–253

175 Gong Y, Krakow D, Marcelino J, Wilkin D, Chitayat D, Babul-Hirji R, Hudgins L, Cremers CW, Cremers FP, Brunner HG et al (1999) Heterozygous mutations in the gene encoding noggin affect human joint morphogenesis. *Nat Genet* 21: 302–304

176 Marcelino J, Sciortino CM, Romero MF, Ulatowski LM, Ballock RT, Economides AN, Eimon PM, Harland RM, Warman ML (2001) Human disease-causing NOG missense mutations: Effects on noggin secretion, dimer formation, and bone morphogenetic protein binding. *Proc Natl Acad Sci USA* 98: 11353–11358

177 Piccolo S, Sasai Y, Lu B, De Robertis EM (1996) Dorsoventral patterning in Xenopus: Inhibition of ventral signals by direct binding of chordin to BMP-4. *Cell* 86: 589–598

178 Sasai Y, Lu B, Piccolo S, De Robertis EM (1996) Endoderm induction by the organizer-secreted factors chordin and noggin in Xenopus animal caps. *EMBO J* 15: 4547–4555

179 Bachiller D, Klingensmith J, Kemp C, Belo JA, Anderson RM, May SR, McMahon JA, McMahon AP, Harland RM, Rossant J et al (2000) The organizer factors Chordin and Noggin are required for mouse forebrain development. *Nature* 403: 658–661

180 Larrain J, Bachiller D, Lu B, Agius E, Piccolo S, De Robertis EM (2000) BMP-binding modules in chordin: A model for signalling regulation in the extracellular space. *Development* 127: 821–830

181 Scott IC, Blitz IL, Pappano WN, Imamura Y, Clark TG, Steiglitz BM, Thomas CL, Maas

SA, Takahara K, Cho KW et al (1999) Mammalian BMP-1/Tolloid-related metalloproteinases, including novel family member mammalian Tolloid-like 2, have differential enzymatic activities and distributions of expression relevant to patterning and skeletogenesis. *Dev Biol* 213: 283–300

182 Zhang D, Ferguson CM, O'Keefe RJ, Puzas JE, Rosier RN, Reynolds PR (2002) A role for the BMP antagonist chordin in endochondral ossification. *J Bone Miner Res* 17: 293–300

183 Reynolds SD, Zhang D, Puzas JE, O'Keefe RJ, Rosier RN, Reynolds PR (2000) Cloning of the chick BMP1/Tolloid cDNA and expression in skeletal tissues. *Gene* 248: 233–243

184 Moreno M, Munoz R, Aroca F, Labarca M, Brandan E, Larrain J (2005) Biglycan is a new extracellular component of the Chordin-BMP-4 signaling pathway. *EMBO J* 24: 1397–1405

185 Koike N, Kassai Y, Kouta Y, Miwa H, Konishi M, Itoh N (2007) Brorin, a novel secreted bone morphogenetic protein antagonist, promotes neurogenesis in mouse neural precursor cells. *J Biol Chem* 282: 15843–15850

186 Garcia Abreu J, Coffinier C, Larrain J, Oelgeschlager M, De Robertis EM (2002) Chordin-like CR domains and the regulation of evolutionarily conserved extracellular signaling systems. *Gene* 287: 39–47

187 French DM, Kaul RJ, D'Souza AL, Crowley CW, Bao M, Frantz GD, Filvaroff EH, Desnoyers L (2004) WISP-1 is an osteoblastic regulator expressed during skeletal development and fracture repair. *Am J Pathol* 165: 855–867

188 Onichtchouk D, Chen YG, Dosch R, Gawantka V, Delius H, Massague J, Niehrs C (1999) Silencing of TGF-beta signalling by the pseudoreceptor BAMBI. *Nature* 401: 480–485

189 Samad TA, Rebbapragada A, Bell E, Zhang Y, Sidis Y, Jeong SJ, Campagna JA, Perusini S, Fabrizio DA, Schneyer AL et al (2005) DRAGON, a bone morphogenetic protein co-receptor. *J Biol Chem* 280: 14122–14129

190 Samad TA, Srinivasan A, Karchewski LA, Jeong SJ, Campagna JA, Ji RR, Fabrizio DA, Zhang Y, Lin HY, Bell E et al (2004) DRAGON: A member of the repulsive guidance molecule-related family of neuronal- and muscle-expressed membrane proteins is regulated by DRG11 and has neuronal adhesive properties. *J Neurosci* 24: 2027–2036

191 Babitt JL, Zhang Y, Samad TA, Xia Y, Tang J, Campagna JA, Schneyer AL, Woolf CJ, Lin HY (2005) Repulsive guidance molecule (RGMa), a DRAGON homologue, is a bone morphogenetic protein co-receptor. *J Biol Chem* 280: 29820–29827

192 Babitt JL, Huang FW, Wrighting DM, Xia Y, Sidis Y, Samad TA, Campagna JA, Chung RT, Schneyer AL, Woolf CJ et al (2006) Bone morphogenetic protein signaling by hemojuvelin regulates hepcidin expression. *Nat Genet* 38: 531–539

193 Babitt JL, Huang FW, Xia Y, Sidis Y, Andrews NC, Lin HY (2007) Modulation of bone morphogenetic protein signaling *in vivo* regulates systemic iron balance. *J Clin Invest* 117: 1933–1939

194 Halbrooks PJ, Ding R, Wozney JM, Bain G (2007) Role of RGM coreceptors in bone morphogenetic protein signaling. *J Mol Signal* 2: 4
195 Afzal AR, Rajab A, Fenske CD, Oldridge M, Elanko N, Ternes-Pereira E, Tuysuz B, Murday VA, Patton MA, Wilkie AO et al (2000) Recessive Robinow syndrome, allelic to dominant brachydactyly type B, is caused by mutation of ROR2. *Nat Genet* 25: 419–422
196 Schwabe GC, Tinschert S, Buschow C, Meinecke P, Wolff G, Gillessen-Kaesbach G, Oldridge M, Wilkie AO, Komec R, Mundlos S (2000) Distinct mutations in the receptor tyrosine kinase gene ROR2 cause brachydactyly type B. *Am J Hum Genet* 67: 822–831
197 Sammar M, Stricker S, Schwabe GC, Sieber C, Hartung A, Hanke M, Oishi I, Pohl J, Minami Y, Sebald W et al (2004) Modulation of GDF-5/BRI-b signalling through interaction with the tyrosine kinase receptor Ror2. *Genes Cells* 9: 1227–1238
198 Hassel S, Yakymovych M, Hellman U, Ronnstrand L, Knaus P, Souchelnytskyi S (2006) Interaction and functional cooperation between the serine/threonine kinase bone morphogenetic protein type II receptor with the tyrosine kinase stem cell factor receptor. *J Cell Physiol* 206: 457–467
199 Drissi MH, Li X, Sheu TJ, Zuscik MJ, Schwarz EM, Puzas JE, Rosier RN, O'Keefe RJ (2003) Runx2/Cbfa1 stimulation by retinoic acid is potentiated by BMP-2 signaling through interaction with Smad1 on the collagen X promoter in chondrocytes. *J Cell Biochem* 90: 1287–1298
200 Leboy P, Grasso-Knight G, D'Angelo M, Volk SW, Lian JV, Drissi H, Stein GS, Adams SL (2001) Smad-Runx interactions during chondrocyte maturation. *J Bone Joint Surg Am* 83-A Suppl 1: S15–22
201 Nishio Y, Dong Y, Paris M, O'Keefe RJ, Schwarz EM, Drissi H (2006) Runx2-mediated regulation of the zinc finger Osterix/Sp7 gene. *Gene* 372: 62–70
202 Ito Y, Miyazono K (2003) RUNX transcription factors as key targets of TGF-beta superfamily signaling. *Curr Opin Genet Dev* 13: 43–47
203 Lee KS, Kim HJ, Li QL, Chi XZ, Ueta C, Komori T, Wozney JM, Kim EG, Choi JY, Ryoo HM et al (2000) Runx2 is a common target of transforming growth factor beta1 and bone morphogenetic protein 2, and cooperation between Runx2 and Smad5 induces osteoblast-specific gene expression in the pluripotent mesenchymal precursor cell line C2C12. *Mol Cell Biol* 20: 8783–8792
204 Lee MH, Kim YJ, Kim HJ, Park HD, Kang AR, Kyung HM, Sung JH, Wozney JM, Kim HJ, Ryoo HM (2003) BMP-2-induced Runx2 expression is mediated by Dlx5, and TGF-beta 1 opposes the BMP-2-induced osteoblast differentiation by suppression of Dlx5 expression. *J Biol Chem* 278: 34387–34394
205 Akhurst RJ, Derynck R (2001) TGF-beta signaling in cancer – A double-edged sword. *Trends Cell Biol* 11: S44–51
206 Monzen K, Hiroi Y, Kudoh S, Akazawa H, Oka T, Takimoto E, Hayashi D, Hosoda T, Kawabata M, Miyazono K et al (2001) Smads, TAK1, and their common target ATF-2 play a critical role in cardiomyocyte differentiation. *J Cell Biol* 153: 687–698

207 Bond HM, Mesuraca M, Carbone E, Bonelli P, Agosti V, Amodio N, De Rosa G, Di Nicola M, Gianni AM, Moore MA et al (2004) Early hematopoietic zinc finger protein (EHZF), the human homolog to mouse Evi3, is highly expressed in primitive human hematopoietic cells. *Blood* 103: 2062–2070

208 Hata A, Seoane J, Lagna G, Montalvo E, Hemmati-Brivanlou A, Massague J (2000) OAZ uses distinct DNA- and protein-binding zinc fingers in separate BMP-Smad and Olf signaling pathways. *Cell* 100: 229–240

209 de Caestecker MP, Yahata T, Wang D, Parks WT, Huang S, Hill CS, Shioda T, Roberts AB, Lechleider RJ (2000) The Smad4 activation domain (SAD) is a proline-rich, p300–dependent transcriptional activation domain. *J Biol Chem* 275: 2115–2122

210 Pearson KL, Hunter T, Janknecht R (1999) Activation of Smad1-mediated transcription by p300/CBP. *Biochim Biophys Acta* 1489: 354–364

211 Kahata K, Hayashi M, Asaka M, Hellman U, Kitagawa H, Yanagisawa J, Kato S, Imamura T, Miyazono K (2004) Regulation of transforming growth factor-beta and bone morphogenetic protein signalling by transcriptional coactivator GCN5. *Genes Cells* 9: 143–151

212 Bai RY, Koester C, Ouyang T, Hahn SA, Hammerschmidt M, Peschel C, Duyster J (2002) SMIF, a Smad4-interacting protein that functions as a co-activator in TGFbeta signalling. *Nat Cell Biol* 4: 181–190

213 Henningfeld KA, Friedle H, Rastegar S, Knochel W (2002) Autoregulation of Xvent-2B; direct interaction and functional cooperation of Xvent-2 and Smad1. *J Biol Chem* 277: 2097–2103

214 Postigo AA (2003) Opposing functions of ZEB proteins in the regulation of the TGF-beta/BMP signaling pathway. *EMBO J* 22: 2443–2452

215 Postigo AA, Depp JL, Taylor JJ, Kroll KL (2003) Regulation of Smad signaling through a differential recruitment of coactivators and corepressors by ZEB proteins. *EMBO J* 22: 2453–2462

216 van Grunsven LA, Schellens A, Huylebroeck D, Verschueren K (2001) SIP1 (Smad interacting protein 1) and deltaEF1 (delta-crystallin enhancer binding factor) are structurally similar transcriptional repressors. *J Bone Joint Surg Am* 83-A Suppl 1: S40–47

217 Sowa H, Kaji H, Hendy GN, Canaff L, Komori T, Sugimoto T, Chihara K (2004) Menin is required for bone morphogenetic protein 2- and transforming growth factor beta-regulated osteoblastic differentiation through interaction with Smads and Runx2. *J Biol Chem* 279: 40267–40275

218 Bai S, Shi X, Yang X, Cao X (2000) Smad6 as a transcriptional corepressor. *J Biol Chem* 275: 8267–8270

219 Shi X, Yang X, Chen D, Chang Z, Cao X (1999) Smad1 interacts with homeobox DNA-binding proteins in bone morphogenetic protein signaling. *J Biol Chem* 274: 13711–13717

220 Jiao K, Zhou Y, Hogan BL (2002) Identification of mZnf8, a mouse Kruppel-like transcriptional repressor, as a novel nuclear interaction partner of Smad1. *Mol Cell Biol* 22: 7633–7644

221 Provot S, Kempf H, Murtaugh LC, Chung UI, Kim DW, Chyung J, Kronenberg HM, Lassar AB (2006) Nkx3.2/Bapx1 acts as a negative regulator of chondrocyte maturation. *Development* 133: 651–662

222 Kim DW, Kempf H, Chen RE, Lassar AB (2003) Characterization of Nkx3.2 DNA binding specificity and its requirement for somitic chondrogenesis. *J Biol Chem* 278: 27532–27539

223 Kurisaki K, Kurisaki A, Valcourt U, Terentiev AA, Pardali K, Ten Dijke P, Heldin CH, Ericsson J, Moustakas A (2003) Nuclear factor YY1 inhibits transforming growth factor beta- and bone morphogenetic protein-induced cell differentiation. *Mol Cell Biol* 23: 4494–4510

224 Lee KH, Evans S, Ruan TY, Lassar AB (2004) SMAD-mediated modulation of YY1 activity regulates the BMP response and cardiac-specific expression of a GATA4/5/6-dependent chick Nkx2.5 enhancer. *Development* 131: 4709–4723

225 Wu JW, Krawitz AR, Chai J, Li W, Zhang F, Luo K, Shi Y (2002) Structural mechanism of Smad4 recognition by the nuclear oncoprotein Ski: Insights on Ski-mediated repression of TGF-beta signaling. *Cell* 111: 357–367

226 Wang W, Mariani FV, Harland RM, Luo K (2000) Ski represses bone morphogenic protein signaling in *Xenopus* and mammalian cells. *Proc Natl Acad Sci USA* 97: 14394–14399

227 Luo K (2003) Negative regulation of BMP signaling by the ski oncoprotein. *J Bone Joint Surg Am* 85-A Suppl 3: 39–43

228 Luo K (2004) Ski and SnoN: negative regulators of TGF-beta signaling. *Curr Opin Genet Dev* 14: 65–70

229 Wu K, Yang Y, Wang C, Davoli MA, D'Amico M, Li A, Cveklova K, Kozmik Z, Lisanti MP, Russell RG et al (2003) DACH1 inhibits transforming growth factor-beta signaling through binding Smad4. *J Biol Chem* 278: 51673–51684

230 Yoshida Y, Tanaka S, Umemori H, Minowa O, Usui M, Ikematsu N, Hosoda E, Imamura T, Kuno J, Yamashita T et al (2000) Negative regulation of BMP/Smad signaling by Tob in osteoblasts. *Cell* 103: 1085–1097

231 Yoshida Y, von Bubnoff A, Ikematsu N, Blitz IL, Tsuzuku JK, Yoshida EH, Umemori H, Miyazono K, Yamamoto T, Cho KW (2003) Tob proteins enhance inhibitory Smad-receptor interactions to repress BMP signaling. *Mech Dev* 120: 629–637

232 Tylzanowski P, Verschueren K, Huylebroeck D, Luyten FP (2001) Smad-interacting protein 1 is a repressor of liver/bone/kidney alkaline phosphatase transcription in bone morphogenetic protein-induced osteogenic differentiation of C2C12 cells. *J Biol Chem* 276: 40001–40007

233 Verschueren K, Remacle JE, Collart C, Kraft H, Baker BS, Tylzanowski P, Nelles L, Wuytens G, Su MT, Bodmer R et al (1999) SIP1, a novel zinc finger/homeodomain repressor, interacts with Smad proteins and binds to 5'-CACCT sequences in candidate target genes. *J Biol Chem* 274: 20489–20498

234 Lin X, Liang YY, Sun B, Liang M, Shi Y, Brunicardi FC, Shi Y, Feng XH (2003) Smad6

recruits transcription corepressor CtBP to repress bone morphogenetic protein-induced transcription. *Mol Cell Biol* 23: 9081–9093

235 Kim RH, Wang D, Tsang M, Martin J, Huff C, de Caestecker MP, Parks WT, Meng X, Lechleider RJ, Wang T et al (2000) A novel smad nuclear interacting protein, SNIP1, suppresses p300-dependent TGF-beta signal transduction. *Genes Dev* 14: 1605–1616

236 Lin Y, Martin J, Gruendler C, Farley J, Meng X, Li BY, Lechleider R, Huff C, Kim RH, Grasser WA et al (2002) A novel link between the proteasome pathway and the signal transduction pathway of the bone morphogenetic proteins (BMPs). *BMC Cell Biol* 3: 15

237 Ogasawara T, Kawaguchi H, Jinno S, Hoshi K, Itaka K, Takato T, Nakamura K, Okayama H (2004) Bone morphogenetic protein 2-induced osteoblast differentiation requires Smad-mediated down-regulation of Cdk6. *Mol Cell Biol* 24: 6560–6568

238 Raju GP, Dimova N, Klein PS, Huang HC (2003) SANE, a novel LEM domain protein, regulates bone morphogenetic protein signaling through interaction with Smad1. *J Biol Chem* 278: 428–437

239 Shen ZJ, Nakamoto T, Tsuji K, Nifuji A, Miyazono K, Komori T, Hirai H, Noda M (2002) Negative regulation of bone morphogenetic protein/Smad signaling by Cas-interacting zinc finger protein in osteoblasts. *J Biol Chem* 277: 29840–29846

240 Kurozumi K, Nishita M, Yamaguchi K, Fujita T, Ueno N, Shibuya H (1998) BRAM1, a BMP receptor-associated molecule involved in BMP signalling. *Genes Cells* 3: 257–264

241 Satow R, Kurisaki A, Chan TC, Hamazaki TS, Asashima M (2006) Dullard promotes degradation and dephosphorylation of BMP receptors and is required for neural induction. *Dev Cell* 11: 763–774

242 Chan MC, Nguyen PH, Davis BN, Ohoka N, Hayashi H, Du K, Lagna G, Hata A (2007) A novel regulatory mechanism of the bone morphogenetic protein (BMP) signaling pathway Involving the carboxyl-terminal tail domain of BMP type II receptor. *Mol Cell Biol* 27: 5776–5789

Dissection of bone morphogenetic protein signaling using genome-engineering tools

Daniel Graf[1] and Aris N. Economides[2]

[1]Institute of Immunology, Biomedical Sciences Center 'Al. Fleming', 34 Al. Fleming Street, 166 72 Vari, Greece; [2]Genome Engineering Technologies, Regeneron Pharmaceuticals, Inc., 777 Old Saw River Road, Tarrytown, NY 10591, USA

Introduction

Bone morphogenetic proteins (BMPs) encompass a large subgroup of evolutionary conserved, secreted signaling molecules belonging to the TGF-β superfamily. In contrast to that suggested by their name, BMP function is not restricted to the skeleton. Recent studies in several organisms have revealed multiple roles for BMPs during embryogenesis where they are involved in early embryonic patterning, gastrulation, tissue induction and differentiation [1]. BMP signaling activity is regulated at multiple levels (Fig. 1A). In the intracellular space multiple regulatory proteins control BMP signaling after initial BMP receptor activation. In the extracellular space BMP antagonists such as Noggin, Follistatin and related proteins, Gremlin and other members of the Dan family, Chordin and its relatives along with Twisted Gastrulation (Tsg), all regulate the ability of BMPs to engage the BMP receptors. Studies on flies and lower vertebrates have led to the concept that these extracellular interactions result in the formation of BMP activity gradients, which in turn provide positional cues to the cells that encounter them [2–4]. Although direct evidence for the operation of such activity gradients is missing in higher vertebrates, orthologs for all antagonists do exist. In addition, there are data pointing to neighboring or overlapping domains of expression between BMPs and their antagonists as well as evidence for BMP-dependent expression of the latter (for review see [5]). Therefore, when trying to understand BMP function *in vivo*, it is important to examine BMPs in the context of expression of their antagonists as well as other regulatory molecules.

BMPs have roles in multiple tissues – Implications of therapeutic uses

The roles of BMPs in adult organisms are only now starting to be deciphered. Paralleling that observed during development, BMP antagonists are expressed concomitantly or in the vicinity of BMP-expressing cells (for review see [5]). As BMPs

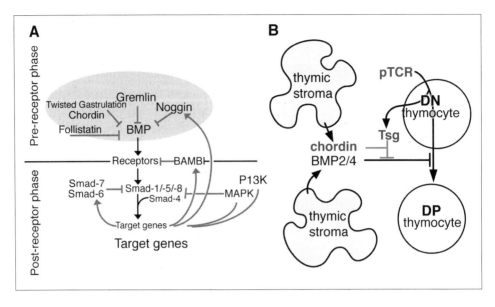

Figure 1
Complex regulation of bone morphogenetic protein (BMP) signaling in the extracellular space to allow various cell types to control biological processes. (A) Balance between BMPs and various BMP antagonists [Noggin, Gremlin, Chordin, Twisted Gastrulation (Tsg), Follistatin] in the extracellular space controls quantity and quality of the BMP signals transduced. BMP-specific, Smad-dependent signal transduction can be visualized by detection of phosphorylated Smad1/5/8. Other signaling cascades also wired to the BMP receptors (MAPK, PI3K/Akt pathways) may also be monitored, but they are not BMP restricted. (B) Dynamic interaction of BMP/BMP antagonists between thymic stroma cells and developing thymocytes regulates the rate of thymocyte maturation. Thymocytes up-regulate Tsg upon preTCR signaling. Tsg in synergy with stroma-derived Chordin blocks BMP-2/4 signals to allow efficient transition from the double-negative (DN) to the double-positive (DP) stage. Adapted from J Exp Med (2002) 196(2); issue cover image. © 2002 The Rockefeller University Press.

have been implicated in the homeostasis of several tissues and have been shown to play central roles in tissue regeneration and repair, it is fair to hypothesize that this coordinated expression may set up an important balance and control local BMP signaling.

The importance of maintaining a fine balance of BMP signaling is illustrated by genetic disorders where BMP activity is altered (for reviews see [6, 7]). These include mutations affecting: (a) BMPs, as for example in genetic mutations that ablate or reduce the activity of *Myostatin* resulting in abnormal muscle mass [8], and the brachypodism and short-ear disorders attributed to *Gdf-5* [9] and *Bmp-5* [9], respectively; (b) BMP receptors, as in fibrodisplasia ossificans progressiva (FOP), a debilitating genetic disorder of heterotopic ossification, which is caused by point mutations in *ACVRI/ALK2* [10], as well as juvenile polyposis syndrome, which is caused by mutations in *BMPRIA* [11]; (c) BMP antagonists, as for example in multiple synostosis syndrome (SYNS1), proximal symphalangism (SYM1) [12], and autosomal dominant stapes ankylosis [13], which are all attributed to mutations in *NOGGIN*; and (d) mutations in other genes that regulate the activity of BMPs or their antagonists, an example of which are mutations of *Fgfr2* in syndromic FGFR-mediated craniosynostoses that regulate the expression of *Noggin* [14]. Interestingly, some of these mutations *per se* have modest effects during embryonic development. In FOP, the mutation does not cause any obvious problems throughout most periods of post-natal life, except during the occasional but catastrophic episodes of bone formation. Although the mechanism by which this mutated ACVRI drives FOP remains unclear, there is evidence that cells of the immune system and local inflammation are involved [15, 16]. The lack of a strong developmental profile and the role of ACVRI in non-skeletal tissues in the adult both point to the importance of study and experimentation in adult organisms.

Two osteogenic BMPs (BMP-2, BMP-7) are currently being used in the clinic, but their use is limited to orthopedic and bone-healing applications [17]. If BMPs are to be developed further for other clinical applications, or if modulation of BMP signaling is to be expanded to other indications as a therapeutic strategy (for examples see [14, 18, 19]), then it will be important to place more emphasis on the identification and functional dissection of the extracellular interactions between BMPs and BMP antagonists in adult animals, and couple this new knowledge with observations in human disease.

Genetically engineered mice for BMPs and BMP antagonists reveal phenotypic complexity and underscore need for new research tools

The importance of extracellular BMP signaling networks is in part illustrated by the severe and complex phenotypes observed in the absence of any single one of participating genes (Tab. 1, as well as [20] for a more comprehensive review). This abbreviated list demonstrates that BMPs and BMP antagonists fulfill critical and unique functions throughout embryonic development and organogenesis. Due to the embryonic/perinatal lethality and multiplicity of phenotypes displayed by mice that are homozygous-null for different BMPs and BMP antagonists, to decipher the roles of BMPs in adult mice and/or in specific tissues or cell types new research tools are needed.

In the adult, despite the fact that BMPs are implicated in a variety of non-bone-associated pathologies, such as pulmonary hypertension and renal failure, as well as various cancers and malignancies, most information on BMP function still comes from bone and cartilage biology. In most cases, the current picture on the physio-

Table 1 - Phenotypes associated with gene deletion of members of the BMP signaling network.

Gene	Phenotype	Reference for null allele	Reference for lacZ reporter
Bmp-2	Embryonic lethal: amnion defects, heart defects	[53]	[39]
Bmp-4	Embryonic lethal: defects in mesoderm formation	[43]	[44]
Bmp-7	Perinatal lethal: kidney agenesis, eye defects, skeletal defects	[59, 60]	[64]
Noggin	Embryonic lethal: skeletal defects, CNS defects	[65, 66]	[65]
Gremlin	Perinatal lethal: kidney, lung and skeletal defects	[36, 67]	[36]
Chordin	Perinatal lethal: pharyngeal defects (DiGeorge syndrome)	[28, 68]	Graf et al, unpublished
Twisted gastrulation	Variable penetrance depending on genetic background: embryonic lethal on C57BL/6 with craniofacial, skeletal, lymphoid, foregut defects	[31, 69]	[70]

logical roles of BMP signaling networks for tissue homeostasis and tissue repair is very incomplete. This is particularly true for malignancies where it is difficult to judge whether BMP or BMP antagonist expression in cancer cells is functionally associated with the progression of the pathology of the tumor, or if it is present only because the tumor cells have a physiological counterpart that normally expresses that particular BMP/BMP antagonist. Studies on primary cells or established cell lines have been very valuable for dissecting the molecular events that occur downstream of the BMP receptors. Recent progress on that front is covered in the preceding two chapters by Knaus et al. and Ghosh-Choudhury et al. However, to dissect and understand the functions of extracellular BMP signaling networks *in vivo*, particularly during late embryogenesis or in adult tissues, this type of approach is not sufficient. A systemic, *in vivo* approach is required where cells are examined in their physiological environment. In this chapter we describe an approach that has been enabled by genome-engineering tools, and we discuss the considerations and methods used. We illustrate our ideas using newly generated conditional alleles for *Bmp-2* and *Bmp-7*, and we examine how design decisions affect the usefulness of genetically engineered alleles.

Non-skeletal role for BMPs – Control of T cell maturation in the thymus

We have previously shown a role for BMP signaling in the thymus and how it affects thymocyte maturation [21]. We proposed a model whereby BMP-2 or BMP-4 signals derived from the thymic stroma control maturation of thymocyte precursors by reducing both their rate of proliferation and differentiation. When individual precursors receive a pre-T cell receptor-dependent competence signal to proceed with their maturation program, they up-regulate Tsg, which in synergy with stroma-derived Chordin alleviates this BMP effect (Fig. 1B). These data were collected using a combination of *in situ* expression profiling, expression studies on isolated cells, and *ex vivo* organ cultures. We wished to obtain proof that this BMP/BMP antagonist interaction operates in the proposed manner *in vivo* and to probe further for other BMP/BMP antagonist interactions in the thymus. Evidence for other BMP interactions derives from our observations that BMP-7 and Gremlin are also expressed in the thymus, although BMP-7 had no effect in the particular setting described above.

As indicated in Table 1, for those BMPs or BMP antagonists that are expressed in the thymus, the corresponding knockout mice die during embryogenesis, shortly after birth, or present a multiplicity of phenotypes. This situation precludes direct functional assessment of the roles of these molecules in thymocyte development or in adult lymphoid tissues, and therefore necessitates a conditional gene ablation approach. As the cellular sources for most BMPs and BMP antagonists in adult tissues are not known, first we had to devise an experimental approach that would allow us to describe BMP/BMP antagonist expression *in situ* to the single cell level, as a prerequisite for successful conditional gene ablation.

Identification of BMP signaling networks *in vivo*

Expression profiling to identify time and site of a gene's expression using a marker gene knocked into the locus encoding for the gene of interest is a widely used approach. In this method, a reporter gene, most often bacterial β-galactosidase (lacZ), is stably introduced by gene targeting in ES cells or bacterial artificial chromosome (BAC) transgenesis into the mouse genome in a manner such that LacZ expression is determined by the promoter (and hopefully also the regulatory elements) of the gene of interest. Staining for lacZ activity is sensitive and allows visualization of expression in whole embryos, or histological sections usually with single cell resolution [22]. Although this strategy has mostly been used in developmental studies, more recently gene reporter mice have found their way into studies examining gene expression in adult tissues, and they are rapidly gaining popularity as they allow the identification and tracking of even small numbers of cells. The decision by EUCOMM and KOMP, the two consortia devoted to knocking out the majority of genes in the mouse genome, to incorporate a reporter gene into their targeting constructs underscores the wide-spread recognition of the usefulness of this technique [23].

As indicated in Table 1, lacZ reporter mice are available for many BMPs and BMP antagonists. Our approach of mapping BMP signaling networks in adult tissues was based on a combination of lacZ reporter mice together with immunohistochemistry for cell specific markers, such as BMP themselves, and the downstream BMP signaling event of phosphorylated Smad1/5/8. In this manner, the identity of the BMP/BMP antagonist-expressing cell, the range of the BMP signal, as well as responding cells can be elucidated. Recording BMP signaling by Smad phosphorylation is convenient as Smads are restricted to mediating BMP signals. In addition, reporter mice for Smad activation have been generated [24]. However, methods restricted to visualizing Smad-dependent signaling overlook other pathways wired to BMP receptors such as MAPK and PI3K [25, 26]. As a result, visualization of Smad-dependent signaling provides us with an incomplete picture of where BMP mediated signals are transmitted.

Visualizing BMPs and BMP antagonists in the thymus as a methodological example

As an illustration of the viability of this approach, we visualized BMP signaling networks in the thymus. It has been proposed by us and others [21, 27] that BMP signals specifically regulate early aspects of thymocyte development, mainly affecting the double-negative (DN) cell population that make up 2–3% of the total thymocytes. There are several indications that BMP signaling is important for correct thymopoiesis, but a direct role of BMPs for thymocyte maturation is less clear. Mice deficient in Chordin are athymic [28], and several Noggin-misexpression studies [29, 30] show severe abnormalities in thymus development. Some of the most compelling data supporting a direct involvement of BMP signals in lymphocyte maturation comes from mice deficient in Tsg, which display a severe reduction of B- and T-lymphopoiesis [31]. The clarification of whether BMP signaling directly affects T cell development and possibly T cell repertoire selection would not be without implications. Altering T cell selection *per se* can lead to the development of autoimmunity in later life. In addition, signaling pathways regulating T cell maturation are commonly also used by mature T cells. Therefore, a direct involvement of BMP signals in the regulation of the mature immune system would potentially have major implications, particularly with respect of inflammatory disorders. At present, however, direct evidence for their involvement is still missing.

Analysis of thymi from *bmp-4 lacZ* mice revealed that BMP-4-expressing cells are sparse and restricted to a few stromal cells in the thymic cortex and the capsu-

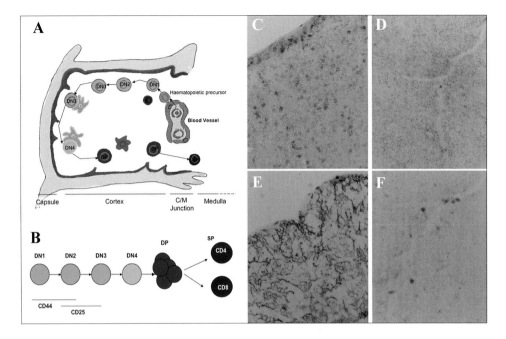

Figure 2
Visualization of BMP expression and signaling in the thymus. (A) Schematic representation of thymocyte maturation steps. Thymocyte precursors enter the thymus from the blood at the cortio-medullary junction. Passing through the cortex they undergo a series of maturation stages (DN1–DN4). Following up-regulation of the TCR (DP) they are selected into CD4 single-positive (SP) or CD8 SP thymocytes and move to the medulla. (B) DN subsets as identified by CD25/CD44 cell surface markers. (C–F) Representative cryostat sections of the subcortical region of the thymus from bmp-4^{LacZ} *reporter mice (C, E) or the cortico-medullary junction (D, F). All sections are stained for lacZ to indicate sites of* bmp-4 *expression. Double staining with CD25 (C, D) to indicate DN2/DN3 thymocytes, NLDC-145 to identify cortical dendritic cells (E), and phosphorylated Smad1/5/8 (F) to map* bmp-4 *activity within thymic compartments. Bmp-4 expression is observed in vicinity of CD25⁺ cells (C, D), is not found in thymic dendritic cells (E). Smad1/5/8 activity is seen in vicinity of BMP-4 lining blood vessels (F).*

lar and subcapsular space (Fig. 2). Localization and cell identity correlate with our findings by immunohistochemistry and RT-PCR; however, the small number of lacZ-positive cells was unexpected. BMP-4 expressing cells are located in regions rich in CD25⁺ cells, which represent the DN2/DN3 thymocytes, a subpopulation of the DN thymocyte precursor population (Fig. 2C, D). Expression is not seen in thymic dendritic cells (Fig. 2E). Nuclear staining for phosphoSmad1/5/8 as an indicator of an actively transduced BMP signal could be detected in mostly CD25⁺ cells located around the *bmp-4-lacZ*-positive cells (Fig. 2F).

LacZ analysis can also be used in combination with flow cytometry using fluorescein di-β-D-galactopyranoside (FDG) as the substrate of β-galactosidase activity. Although this method is somewhat less sensitive, its advantage lies in that it allows for precise identification of particular cell populations through multi-parameter analysis. We used this approach to confirm that DN2/DN3 cells express Tsg (Passa and Graf, unpublished), as predicted by our model. The small thymi occasionally observed in $Tsg^{-/-}$ mice always show either a block in or severely abnormal T cell development ([31], and Passa and Graf, unpublished). Analysis of Tsg reporter mice revealed expression of Tsg apart from thymocyte precursors also in cortical and medullary thymic stroma. Thus, the final functional *in vivo* proof for our model would require conditional deletion of Tsg both in thymocyte precursor populations and in thymic stroma cells to assess the relative functional contribution from the various cell types. For these we plan to use specific Cre-drivers for targeting early thymocytes (Vav-Cre, CD2-Cre) [32] as well as thymic stroma cells [33] allowing us to address the function of Tsg in different compartments of the thymus.

Extending this approach to other BMP pathway genes for which LacZ knock-in mice are available and other tissues, we found evidence for additional BMP/BMP antagonist interactions in the thymus as well as in peripheral lymphoid organs. *BMP-7* and *Gremlin* expression is detected in thymic cortex and medulla (Passa and Graf, unpublished), and several BMP/BMP antagonists are found in mature secondary lymphoid tissues (Tsalavos and Graf, unpublished).

The identification of sites of BMP/BMP antagonist interaction and the expressing cell types is one of the prerequisites for subsequent detailed functional analysis *in vivo*. One of the obvious advantages of having identified and mapped the partners involved in these networks is that multiple gene ablation strategies can be considered. This allows obtaining complementary data, which can give valuable insight on the role of the BMP network.

Considerations for designing gene-targeting constructs

In the following section we aim to provide an overview on 'good practices' with respect to designing and assembling gene targeting constructs, with a particular focus on conditional alleles. Gene targeting, by definition, involves introduction of changes into the genome. Any such process potentially introduces changes that may affect the expression of nearby genes. Although there are not many well-documented examples for the latter, a striking illustration is found in the limb deformity (*ld*) phenotype that was attributed to mutations in the *Formin* gene [34]. Only later was it discovered that the *ld* mutations located within the *Formin* locus serendipitously disrupted a control region required for expression of *Gremlin1*, a gene located approximately 40 kb downstream of *Formin* (in this respect the loci of *Formin* and *Gremlin1* partially overlap). As a result, in these *ld* mutations, *Grem1* expression was specifically abrogated in the limb bud, causing the *ld* phenotype [35, 36]. Albeit rare, such occurrences currently cannot be predicted for most genes, but can complicate the intended studies or lead to erroneous conclusions. Recent advances in our understanding of genome and gene regulation have made us aware that genome organization is far more complex than assumed even only a few years ago. Regulatory elements are not only located in introns, but also in upstream and downstream untranslated and flanking genomic regions, often at substantial distances from the actual locus [37]. With respect to BMP/BMP antagonists such long-range regulatory elements have not only been identified for *Grem1*, but also for *gdf6* [38], *BMP-2* [39], and *BMP-5* [40, 41]. For this reason, when designing knockout constructs, it is important to consider such regulatory elements to avoid altering the characteristics of nearby loci. However, the necessary detailed information is only available for a very small number of genes. For the vast majority of genes, the regulatory elements are not known, which poses a practical problem. As important regulatory elements are usually conserved across species, the best approximation at hand is therefore to identify evolutionarily conserved regions (ECRs) *in silico* [42], and when designing knockout alleles to avoid deleting or disrupting such ECRs.

Similar considerations apply to conditional alleles that are based on the Cre/Lox (or FLP/FRT or other related) systems. Care needs to be taken to avoid accidentally deleting or interrupting evolutionarily conserved regions (ECRs). A good guidance in most cases is the identification of conserved regions by performing careful comparative analysis of the locus of interest between several species, or at least between mouse and human. For the sequences of the human, mouse and other genomes this information is available from a variety of searchable databases (www.ensembl.org, http://genome.ucsc.edu/, http://www.ncbi.nlm.nih.gov/). These considerations are of particular importance for genes that show high degree of cross-species conservation and/or are sensitive to gene dosage.

Upon recombination (by Cre recombinase in the case of lox sites) the flanked part of the gene is permanently removed. Conditional removal of one or more exons does not by definition result in a non-expressed or non-functional protein. Therefore, careful subsequent analysis is required to verify that the observed result corresponds to the designed features of the locus. In addition, exogenous promoter/gene selection cassettes should be removed as they often strongly interfere with the endogenous locus, resulting in hypo/hypermorphic alleles.

The above considerations can be illustrated by the alleles that have been designed for *bmp-4*. Initially a knockout allele was engineered in which the majority of the first protein-encoding exon (exon 3) was replaced by a selection cassette [43] or a lacZ reporter cassette [44]. Mice thus engineered are haplo-insufficient and display a variety of phenotypes, which are particularly strong on the inbred C57BL/6 background [45–47], making it almost impossible to maintain the line in that background. Furthermore, *bmp-4* null mice die early in embryogenesis [43]. To bypass the early embryonic lethality to study the function of this gene at later developmen-

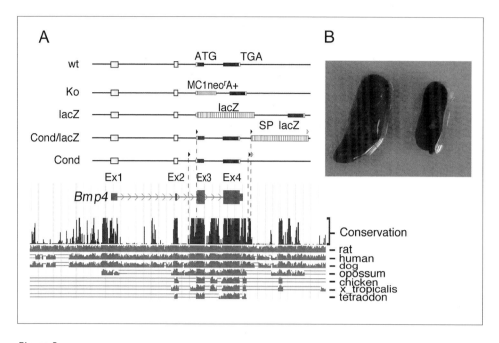

Figure 3
(A) Schematic representation of mouse bmp-4 locus, as well as the design of gene targeting constructs in relation to evolutionarily conserved regions (ECRs), adapted from http://genome.ucsc.edu. Degree of conservation is indicated. In both the bmp-4$^{fl-LacZ}$ allele [48] and the bmp-4^{fl} allele (Hogan et al., unpublished) the loxP elements were inserted directly or very close to ECRs. The conditional/lacZ allele is strongly hypomorph. Aged bmp-4$^{fl/fl}$ mice (6–9 months) develop splenomegaly at high incidence (B).

tal stages or in the adult, a conditional allele was prepared. Following Cre-mediated recombination, this allele turns into a LacZ gene reporter (bmp-$4^{fl-LacZ}$) [48], thereby marking the cells where the floxed region has been deleted. Subsequent analysis revealed that the bmp-$4^{fl-LacZ}$ allele is hypomorphic as bmp-$4^{fl-LacZ/-}$ mice show a significantly stronger phenotype than $bmp^{+/-}$ mice. For example, unlike $bmp^{+/-}$ mice, bmp-$4^{fl-LacZ/-}$ mice cannot be recovered alive at birth [48]. This may be attributed to the presence of the bacteria-derived LacZ element in the 3'UTR, which may alter the expression level and/or expression pattern of bmp-4. In retrospect, it is also likely that regulatory elements were directly affected as the LoxP sites were introduced into and disrupted highly conserved ECRs (Fig. 3A). However, the exact mechanism whereby the bmp-$4^{fl-lacZ/}$ allele gives rise to a phenotype is unknown.

Subsequently, Hogan et al. [48a] prepared a new bmp-4 conditional allele utilizing only loxP sites and lacking a reporter (bmp-4^{fl}). Using this allele for studies on the role of BMP-4 in inflammatory disorders we realized that aged *bmp-4$^{fl/fl}$* mice spontaneously develop splenomegaly (Fig. 3B) (Segklia and Graf unpublished). The underlying reason for this is not clear and requires further investigation. Although this phenotype is irrelevant for most functional studies on BMP-4 and therefore the *bmp-4^{fl}* allele is widely useful, this example illustrates that even alleles with minimal manipulation of the genome can differ from wild type. Analysis of the loxP integration sites reveals that, just as in the *bmp-4$^{fl-LacZ}$* allele, in this new *bmp-4^{fl}* allele both the 5' and the 3' loxP sites have also been introduced directly into, or very close to, ECRs.

In summary, the different alleles created for *bmp-4* illustrate how disrupting ECRs can result in discernable phenotypic alterations. It should be noted that the *bmp-4* alleles described above were designed and assembled at a time when comparative genome sequence information was not readily available. As ECRs are nowadays easily identified, we suggest that in the absence of detailed knowledge regarding their functions, it is best to avoid interrupting them by the insertion of targeting elements.

RedE/T recombineering for precise allele assembly

The generation of targeting vectors for creating conditional alleles in the mouse is a slow and laborious task. The main reason for this is the practical difficulty in assembling targeting constructs with two appropriately placed loxP sites. Classically, two strategies have been pursued: either direct cloning of the flanking genomic sequences using suitable restriction sites in the mouse genome or amplification of the flanking sequences by long distance PCR. The former limits the options on allele design, while the latter restricts the length of homology and central arms for practical considerations.

The advent of bacterial homologous recombination (BHR), also commonly referred to as 'recombineering' [49, 50], provided a major technical breakthrough that revolutionized genome engineering, by enabling the precise assembly of DNA sequences, independent of the presence of restriction sites and the size of the DNA molecule to be modified. It is based on three proteins, RedAlpha, RedBeta and lambda gam that upon expression in bacteria provide a robust recombination machinery that allows homologous recombination to proceed efficiently. Short complementary sequences flanking the integration site on the target DNA are attached to a generic targeting element, which is then electroporated into recombination competent bacteria for homologous recombination to occur (Fig. 4). Successful recombinants are selected using suitable antibiotic resistance genes, followed by diagnostic PCR genotyping. The genomic DNA vectors most often subjected to BHR are BACs. They have the capacity to carry large inserts (100–300 kb), are clonally stable, have a low rate of chimerism, and can be handled with ease. As part of the genome-sequencing project, annotated BACs for virtually all mouse genes are available.

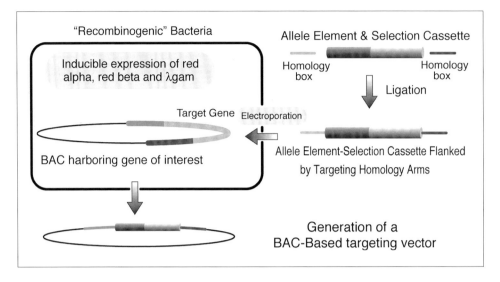

Figure 4
Outline of BHR on BACs. E. coli harboring a BAC that encodes the gene of interest are rendered BHR competent. The allele element and selection cassette are flanked with homology regions that determine the precise integration site into the BAC. Upon electroporation of the flanked allele element, homologous recombination occurs resulting in a modified BAC suitable for direct gene targeting using Velocigene® technology.

The ability to generate targeting vector by modifying BACs, opened the opportunity to utilize these modified BACs directly as targeting vectors. This was made possible by VelociGene®, a process that couples the use of BACs as targeting vectors with a method for automated high throughput qPCR-based genotyping method [51]. We have adapted the two-loxP site strategy for generating conditional alleles to VelociGene®. Apart from the base-pair precision insertion of the loxP sites at the desired sites (empowered by BHR), the other major advantages of this approach is the use of generic modification cassettes for both the 5' and 3'loxP sites, and the ability to use this approach as part of high-throughput genome engineering projects.

The basic steps for BAC targeting vector construction are laid out in Figure 5B, using *bmp-2* as an example. Following careful analysis of the locus to be targeted, the integration sites for the loxP sites are determined. A suitable BAC clone is obtained from one of the repositories and prepared for BHR. In a first round of BHR, the first generic cassette consisting of the 5' loxP site plus an excisable antibiotic resistance gene is introduced. In a second round of BHR, the generic cassette containing the 3'loxP site plus a removable antibiotic resistance gene is introduced. For both rounds of BHR, EM7 promoter-driven antibiotic resistance genes are utilized. Following successful recombineering, the 5' antibiotic resistance gene is

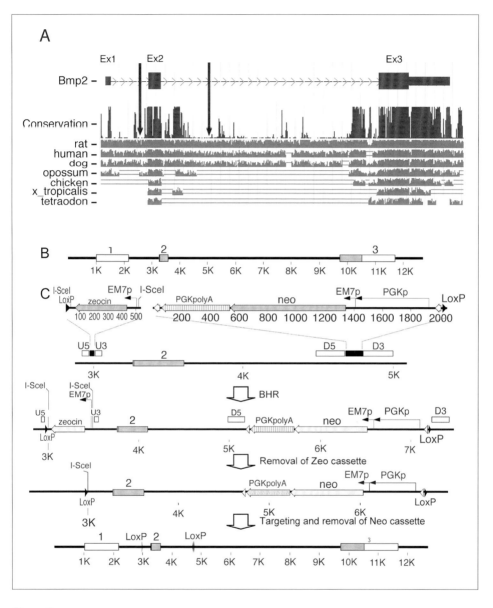

Figure 5
(A) Mouse bmp-2 locus with ECRs adapted from http://genome.ucsc.edu. LoxP integration sites are indicated by arrows. (B) Schematic representation of mouse bmp-2 locus (chromosome 2). Exons 2 and 3 are coding exons. (C) 2-loxP strategy with BHR using generic targeting elements. The 5' element contains a loxP site plus an I-SceI flanked Zeocin selection cassette (removal by digestion/religation). The 3' element contains a loxP site plus a frt-flanked Neomycin selection cassette (removal using the Flp recombinase).

removed using a homing endonuclease* whose recognition site flank the cassette. The BAC targeting vector is then linearized and prepared for introduction into ES cells. Alternatively, the BAC can be trimmed to leave several kilobases of DNA either side of the modified locus. These serve as homology regions for targeting in ES cells, thereby allowing this technology to be used irrespective of automated genotyping capabilities. The whole process can be completed in 4–6 weeks and several targeting constructs can be dealt with in parallel. The generation of conditional alleles for *bmp-2* and *bmp-7* are outlined below. The same strategy has been applied to creating conditional alleles for other *BMP/BMP antagonist* superfamily members, namely *Grem1* [52], *Chordin* (Graf et al., unpublished), *Gdf-3*, as well as other genes not related to BMPs (Economides, unpublished).

Conditional allele for BMP-2

BMP-2 together with BMP-4 are the orthologs of *Drosophila decapentaplegic (dpp)* and form a BMP sub-family. The mature, secreted parts of both molecules are almost identical and both molecules play critical roles in the establishment of the basic embryonic body plan. *Bmp-2*-deficient embryos die early in development as they fail to close the proamniotic canal resulting in a malformed amnion/chorion. They also exhibit a defect in cardiac development, manifested by the abnormal development of the heart in the exocoelomic cavity [53]. *Bmp-2* is expressed both in the extraembryonic mesoderm cells and promyocardium, being compatible with the observed phenotype. The early lethality of *bmp-2* mutant embryos largely prevented further insight into its role in later development stages and organogenesis, although it was possible to show its involvement in neural crest migration [54]. Thus, despite its importance in clinical settings, surprisingly little is known on the *in vivo* roles of *BMP-2*, and almost nothing on its roles in adult tissues. To understand the role of *BMP-2* in lymphopoiesis and function of the mature immune system [21], we generated a conditional mutant for *bmp-2*.

Mouse *bmp-2* is located on chromosome 2, and consists of three exons (two coding exons) that span approximately 10 kb (Fig. 5A). Comparative analysis of mouse and human *BMP-2* loci revealed strong sequence conservation in large parts of the flanking introns of both coding exons, suggesting the presence of conserved functional regulatory elements. In particular, a stretch of several kilobases of conserved sequence can be found downstream of the second coding exon (exon 3, which

---

* Homing endonucleases are rare-cutting enzymes encoded by inteins and introns. They are mobile elements that propagate (home) by a gene conversion process that duplicates the intron or intein. They recognize long, pseudopalindromic homing sites of 14–30 bp in length and cleave their homing site DNA to generate 4nt, 3' extensions. These enzyme characteristics make them suitable for use in molecular biology as extremely rare cutting endonucleases.

encodes most of the mature protein). Targeting the first coding exon in secreted proteins is a preferred strategy as this results in the removal of the signal peptide, which is required for the protein to enter the secretory pathway. In addition, by removing the first coding exon, we excluded the chance that expression of the N-terminal pre-protein could interfere with the processing and maturation of other BMPs. Placement of the LoxP sites was chosen so as not to disrupt ECRs (Fig. 5A). The modification cassettes were prepared and introduced into the BAC as described earlier (Fig. 5B). The modified BAC was used directly as targeting vector. Mice were generated by injection of targeted ES cells into blastocysts, and after germline transmission, the $bmp\text{-}2^{fl(Neo)/+}$ mice were crossed with a germline Flp-deleter mouse to remove *in vivo* the neomycin selection cassette. Subsequent intercross resulted in $bmp\text{-}2^{fl/fl}$ mice that appear healthy and breed normally.

To assess whether the conditionally deleted allele (Δ) phenotypically matches a previously engineered null allele [53], $bmp\text{-}2^{fl/+}$ mice were crossed to a germline Cre-deleter mouse to obtain $bmp\text{-}2^{\Delta/+}$ mice. Intercrosses of those mice did not give any viable $bmp\text{-}2^{\Delta/\Delta}$ offspring, indicating embryonic lethality of the allele. Embryos harvested from timed matings at embryonic days (E) 9.5–10 revealed abnormal development of all $bmp\text{-}2^{\Delta/\Delta}$ embryos (Fig. 6A). All embryos failed to turn and had pronounced defects in headfolds and the heart. This phenotype strongly resembles the phenotype described for $bmp\text{-}2$ homozygous-null embryos [53].

To test the ability of the allele to efficiently recombine *in vivo* in a tissue-specific manner, we crossed the allele to the CD2-Cre mouse [32]. The CD2-Cre line, Cre is active in all lymphoid cells that can conveniently be isolated from lymphoid tissues to assess efficiency of deletion. PCR analysis using a four-primer strategy to simultaneously identify wt, flx, and ko alleles in a single PCR indicated complete deletion in thymocytes, partial deletion in splenocytes and almost no deletion in muscle (Fig. 6B). While this work was in progress, two other groups independently also generated conditional alleles for $bmp\text{-}2$ [55, 56].

Conditional allele for BMP-7

BMP-7 together with BMP-5 and BMP-6 are orthologs of *Drosophila glass bottom boat (Gbb/60A)*. Whereas both *bmp-5-* and *bmp-6*-deficient mice are viable and thus appear to have somewhat more redundant roles for embryonic development [57, 58], $bmp\text{-}7^{-/-}$ mice present with defects in kidney development, lens induction and skeletal formation [59, 60]. Defective kidney organogenesis is most likely responsible for the perinatal death of the $bmp\text{-}7^{-/-}$ mice. Combined $bmp\text{-}5/bmp\text{-}7$ and $bmp\text{-}6/bmp\text{-}7$ deficiency results in more severe, embryonic lethal phenotypes [61, 62], indicating some degree of redundancy amongst members of this subfamily. It appears that BMP-7 is not only important for renal development but also plays a major role in tubulointerstitial fibrosis, a common feature of almost all

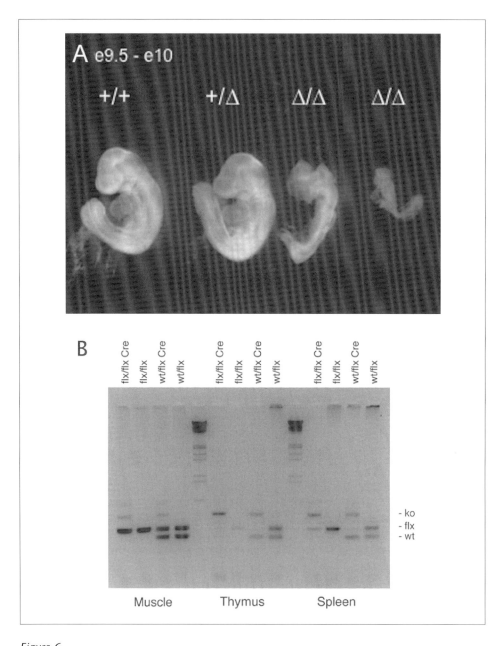

Figure 6
(A) Embryos from timed bmp-$2^{wt/\Delta}$ intercrosses. E9.6–E10 bmp-$2^{\Delta/\Delta}$ embryos show disturbed development and fail to turn. (B) Efficient deletion of bmp-2 in lymphoid cells in vivo using CD2-Cre (pan-lymphoid cell deleter). Genotype was determined using a four-primer PCR strategy for simultaneous detection of wt/flx/ko alleles.

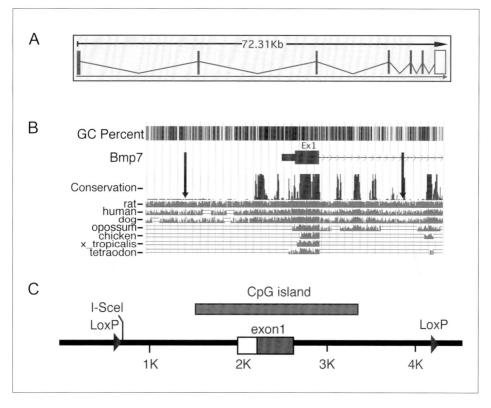

Figure 7
(A) Schematic representation of mouse bmp-7 locus (chromosome 2) with exons 1–7, adapted from http://www.ensembl.org. (B) Exon 1 of mouse bmp-7 locus with ECR, adapted from http://genome.ucsc.edu. Degree of conservation is indicated. LoxP integration sites are indicated. (C) Location of targeting elements flanking exon 1 used to assemble conditional allele.

chronic renal diseases [63]. We have found expression of *bmp-7* in thymic stroma as well as a variety of cell types of the immune system. Although some limited functional studies on thymus and the immune system could be performed using embryonic material, systematic studies on *bmp-7*[−/−] mice are not possible. By generating a conditional allele for *bmp-7*, important *in vivo* studies on its function in adult tissues will be possible, allowing us to address the roles of this molecule in health and disease.

bmp-7 is located towards the end of the long arm on mouse chromosome 2, and contains 7 exons spanning about 70 kb (Fig. 7A). Two strategies have been used to generate the *bmp-7* null allele: removal of the first exon encoding for the signal peptide as well as most of the preBMP-7 protein [59], and removal of exons 6 and 7, which together encode the mature protein [60]. Both $bmp\text{-}7^{-/-}$ lines show largely identical phenotypes, although a few differences with respect to penetrance of the kidney phenotype and the skeletal malformations are noted. This may be attributed to differences in the genetic backgrounds of the cohorts that were phenotyped, although the differences in allele design have not been ruled out. We decided to flank the first coding exon (exon 1) of *bmp-7* such that Cre-mediated recombination would remove the signal peptide and most of preBMP-7, essentially mirroring the strategy used in [59]. Comparative sequence analysis of the *bmp-7* locus revealed the presence of several ECRs, including a well-conserved CpG island stretching either side of exon 1. Both loxP sites were engineered into non-conserved regions (Fig. 7B). Targeting vector construction and all subsequent steps were performed exactly as described for *bmp-2* (Fig. 5C).

To assess whether Cre-mediated deletion would result in a null allele and recapitulate the published phenotypes, mice were crossed to Actin-Cre, and the deletion was then passed through the germline. Resulting $bmp\text{-}7^{+/\Delta}$ mice were intercrossed to yield $bmp\text{-}7^{\Delta/\Delta}$ offspring. No $bmp\text{-}7^{\Delta/\Delta}$ offspring were recovered after birth. $bmp\text{-}7^{\Delta/\Delta}$ embryos recovered at E16.5–E17.5 were analyzed for previously described phenotypes: defective eye and kidney development as well as skeletal malformations [59, 60]. We observed comparable developmental defects as described for $bmp\text{-}7^{-/-}$ embryos (Fig. 8), indicating that, following Cre-mediated recombination, our conditional allele phenotypically mimics the previously described null allele. In addition, $bmp\text{-}7^{fl/fl}$ mice appear healthy and breed normally. As has already been done for the $bmp\text{-}2^{fl/fl}$ mice, $bmp\text{-}7^{fl/fl}$ mice will be subjected to a histopathological analysis to exclude phenotypic changes in the floxed allele.

Concluding remarks

The realization that BMPs are involved in physiology and pathophysiology of many adult tissues offers opportunities but also challenges for future research. The key for success lies clearly in finding ways to trace and accurately manipulate BMP signals *in vivo*. In this chapter we have presented recent progress on two fronts: an experimental approach to identify sites of BMP activity and the generation and characterization of conditional alleles for *bmp-2* and *bmp-7*.

Identification of novel BMP activities in normal adult tissues and during disease processes is of central importance. By combining lacZ reporting with immunohistochemistry it is possible to rapidly screen expression in multiple tissues often with single cell resolution. It also allows sampling the expression dynamics during the course of diseases or tissue repair. The approach presented is not restricted to a particular tissue, and we have been using it to identify sites of BMP action in several other adult tissues. In all cases we found that BMPs and BMP antagonists

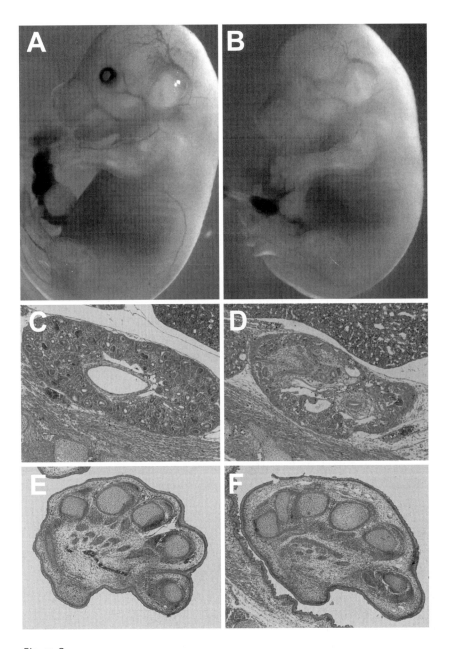

Figure 8
E17.5 bmp-7 embryos from timed bmp-7$^{+/\Delta}$ intercrosses. Wt (bmp-7$^{+/+}$) embryos (A, C, E) in comparison to bmp-7$^{\Delta/\Delta}$ embryos (B, D, F). bmp-7$^{\Delta/\Delta}$ embryos display eye phenotype (B compared to A). H/E-stained paraffin sections reveal disturbed kidney development (D compared to C) and additional digit on fore limbs (F compared to E).

are co-expressed by either the same cell types or by neighboring cells. This clearly underlines the need for describing the extracellular BMP/BMP antagonist interactions.

Many *bmp* loci have evolved complex ways to regulate gene expression that is often conserved across species. For this reason ECRs can be used as an indication for elements involved in gene regulation. ECRs can be found annotated on the publicly accessible genome servers. Interrupting such regions should be duly avoided when assembling gene-targeting vectors. This is possible when BHR on BACs is used for assembly as it allows genome manipulation with base-pair precision. For the generation of conditional alleles for *bmp-2* and *bmp-7* we have adapted the 2-loxP-strategy to the BHR-based Velocigene® platform. This way we could take locus idiosyncrasies into account and avoid disrupting ECRs, an operation that might result in the generation of hypo/hypermorphic alleles. Using the same strategy, we have prepared conditional alleles for several other members of the BMP/BMP antagonist superfamily as well as non-BMP related genes.

By combining identification of extracellular BMP networks with subsequent functional analysis, we will attempt to contribute to the knowledge base required to couple biological processes to observations in human disease. Success on that front will offer new opportunities for therapeutic strategies based on modulating BMP function in clinical settings.

Acknowledgements

Many thanks to Brigid Hogan for the cond. *bmp-4* mice, to Ourania Passa, Katherina Segklia, Vasiliki Zouvelou for their contributions to this work, to David Valenzuela, Andrew Murphy, David Frendewey, Tom Dechiara, William Poueymirou, WojtecAuerbach, Nick Gale, Maria Alexiou, Dimitris Kontoyiannis, Eumorphia Remboutsika and George Kollias for valuable discussions and support, and to the VelociGene production team for ES cell targeting, genotyping and generation of the *bmp-2* and *bmp-7* floxed allele mice. D.G. received support from HFSP, AICR and MUGEN NoE 6FP (LSHG-CT-2005-005203).

References

1   Kishigami S, Mishina Y (2005) BMP signaling and early embryonic patterning. *Cytokine Growth Factor Rev* 16: 265–278
2   Ashe HL, Mannervik M, Levine M (2000) Dpp signaling thresholds in the dorsal ectoderm of the Drosophila embryo. *Development* 127: 3305–3312
3   Mizutani CM, Nie Q, Wan FY, Zhang YT, Vilmos P, Sousa-Neves R, Bier E, Marsh JL,

Lander AD (2005) Formation of the BMP activity gradient in the *Drosophila* embryo. *Dev Cell* 8: 915–924

4 Podos SD, Ferguson EL (1999) Morphogen gradients: New insights from DPP. *Trends Genet* 15: 396–402

5 Canalis E, Economides AN, Gazzerro E (2003) Bone morphogenetic proteins, their antagonists, and the skeleton. *Endocr Rev* 24: 218–235

6 Kornak U, Mundlos S (2003) Genetic disorders of the skeleton: A developmental approach. *Am J Hum Genet* 73: 447–474

7 Hartung A, Sieber C, Knaus P (2006) Yin and Yang in BMP signaling: Impact on the pathology of diseases and potential for tissue regeneration. *Signal Transduction* 6: 314–328

8 Lee SJ (2004) Regulation of muscle mass by myostatin. *Annu Rev Cell Dev Biol* 20: 61–86

9 Storm EE, Huynh TV, Copeland NG, Jenkins NA, Kingsley DM, Lee SJ (1994) Limb alterations in brachypodism mice due to mutations in a new member of the TGF beta-superfamily. *Nature* 368: 639–643

10 Shore EM, Xu M, Feldman GJ, Fenstermacher DA, Cho TJ, Choi IH, Connor JM, Delai P, Glaser DL, LeMerrer M et al (2006) A recurrent mutation in the BMP type I receptor ACVR1 causes inherited and sporadic fibrodysplasia ossificans progressiva. *Nat Genet* 38: 525–527

11 Cao X, Eu KW, Kumarasinghe MP, Li HH, Loi C, Cheah PY (2006) Mapping of hereditary mixed polyposis syndrome (HMPS) to chromosome 10q23 by genome-wide high-density single nucleotide polymorphism (SNP) scan and identification of BMPR1A loss of function. *J Med Genet* 43: e13

12 Gong Y, Krakow D, Marcelino J, Wilkin D, Chitayat D, Babul-Hirji R, Hudgins L, Cremers CW, Cremers FP, Brunner HG et al (1999) Heterozygous mutations in the gene encoding noggin affect human joint morphogenesis. *Nat Genet* 21: 302–304

13 Brown DJ, Kim TB, Petty EM, Downs CA, Martin DM, Strouse PJ, Moroi SE, Gebarski SS, Lesperance MM (2003) Characterization of a stapes ankylosis family with a NOG mutation. *Otol Neurotol* 24: 210–215

14 Warren SM, Brunet LJ, Harland RM, Economides AN, Longaker MT (2003) The BMP antagonist noggin regulates cranial suture fusion. *Nature* 422: 625–629

15 Kaplan FS, Glaser DL, Shore EM, Pignolo RJ, Xu M, Zhang Y, Senitzer D, Forman SJ, Emerson SG (2007) Hematopoietic stem-cell contribution to ectopic skeletogenesis. *J Bone Joint Surg Am* 89: 347–357

16 Kaplan FS, Shore EM, Gupta R, Billings PC, Glaser DL, Pirgnolo RJ, Graf D, Kamoun M (2005) Immunological features of fibrodysplasia ossificans progressiva and the dysregulated BMP-4 pathway. *Clin Rev Bone Miner Metab* 3: 189–193

17 Gautschi OP, Frey SP, Zellweger R (2007) Bone morphogenetic proteins in clinical applications. *ANZ J Surg* 77: 626–631

18 Fuller K, Bayley KE, Chambers TJ (2000) Activin A is an essential cofactor for osteoclast induction. *Biochem Biophys Res Commun* 268: 2–7

19 Lories RJ, Daans M, Derese I, Matthys P, Kasran A, Tylzanowski P, Ceuppens JL, Luyten FP (2006) Noggin haploinsufficiency differentially affects tissue responses in destructive and remodeling arthritis. *Arthritis Rheum* 54: 1736–1746

20 Zhao GQ (2003) Consequences of knocking out BMP signaling in the mouse. *Genesis* 35: 43–56

21 Graf D, Nethisinghe S, Palmer DB, Fisher AG, Merkenschlager M (2002) The developmentally regulated expression of Twisted gastrulation reveals a role for bone morphogenetic proteins in the control of T cell development. *J Exp Med* 196: 163–171

22 Adams NC, Gale NW (2006) High Resolution Gene expression analysis in mice using genetically inserted reporter genes. In: S Pease, C Lois (eds): *Mammalian and Avian Transgenesis – New Approaches*. Springer, Berlin, 131–172

23 Collins FS, Rossant J, Wurst W (2007) A mouse for all reasons. *Cell* 128: 9–13

24 Monteiro RM, de Sousa Lopes SM, Korchynskyi O, ten Dijke P, Mummery CL (2004) Spatio-temporal activation of Smad1 and Smad5 *in vivo*: Monitoring transcriptional activity of Smad proteins. *J Cell Sci* 117: 4653–4663

25 Nohe A, Keating E, Knaus P, Petersen NO (2004) Signal transduction of bone morphogenetic protein receptors. *Cell Signal* 16: 291–299

26 Sugimori K, Matsui K, Motomura H, Tokoro T, Wang J, Higa S, Kimura T, Kitajima I (2005) BMP-2 prevents apoptosis of the N1511 chondrocytic cell line through PI3K/Akt-mediated NF-kappaB activation. *J Bone Miner Metab* 23: 411–419

27 Hager-Theodorides AL, Outram SV, Shah DK, Sacedon R, Shrimpton RE, Vicente A, Varas A, Crompton T (2002) Bone morphogenetic protein 2/4 signaling regulates early thymocyte differentiation. *J Immunol* 169: 5496–5504

28 Bachiller D, Klingensmith J, Shneyder N, Tran U, Anderson R, Rossant J, De Robertis EM (2003) The role of chordin/Bmp signals in mammalian pharyngeal development and DiGeorge syndrome. *Development* 130: 3567–3578

29 Ohnemus S, Kanzler B, Jerome-Majewska LA, Papaioannou VE, Boehm T, Mallo M (2002) Aortic arch and pharyngeal phenotype in the absence of BMP-dependent neural crest in the mouse. *Mech Dev* 119: 127–135

30 Bleul CC, Boehm T (2005) BMP signaling is required for normal thymus development. *J Immunol* 175: 5213–5221

31 Nosaka T, Morita S, Kitamura H, Nakajima H, Shibata F, Morikawa Y, Kataoka Y, Ebihara Y, Kawashima T, Itoh T et al (2003) Mammalian twisted gastrulation is essential for skeleto-lymphogenesis. *Mol Cell Biol* 23: 2969–2980

32 de Boer J, Williams A, Skavdis G, Harker N, Coles M, Tolaini M, Norton T, Williams K, Roderick K, Potocnik AJ et al (2003) Transgenic mice with hematopoietic and lymphoid specific expression of Cre. *Eur J Immunol* 33: 314–325

33 Liston A, Farr AG, Chen Z, Benoist C, Mathis D, Manley NR, Rudensky AY (2007) Lack of Foxp3 function and expression in the thymic epithelium. *J Exp Med* 204: 475–480

34 Mass RL, Zeller R, Woychik RP, Vogt TF, Leder P (1990) Disruption of formin-encoding transcripts in two mutant limb deformity alleles. *Nature* 346: 853–855

35 Zuniga A, Michos O, Spitz F, Haramis AP, Panman L, Galli A, Vintersten K, Klasen C, Mansfield W, Kuc S et al (2004) Mouse limb deformity mutations disrupt a global control region within the large regulatory landscape required for Gremlin expression. *Genes Dev* 18: 1553–1564

36 Khokha MK, Hsu D, Brunet LJ, Dionne MS, Harland RM (2003) Gremlin is the BMP antagonist required for maintenance of Shh and Fgf signals during limb patterning. *Nat Genet* 34: 303–307

37 Long X, Miano JM (2007) Remote control of gene expression. *J Biol Chem* 282: 15941–15945

38 Portnoy ME, McDermott KJ, Antonellis A, Margulies EH, Prasad AB, Kingsley DM, Green ED, Mortlock DP (2005) Detection of potential GDF6 regulatory elements by multispecies sequence comparisons and identification of a skeletal joint enhancer. *Genomics* 86: 295–305

39 Chandler RL, Chandler KJ, McFarland KA, Mortlock DP (2007) BMP-2 transcription in osteoblast progenitors is regulated by a distant 3' enhancer located 156.3 kilobases from the promoter. *Mol Cell Biol* 27: 2934–2951

40 DiLeone RJ, Marcus GA, Johnson MD, Kingsley DM (2000) Efficient studies of long-distance BMP-5 gene regulation using bacterial artificial chromosomes. *Proc Natl Acad Sci USA* 97: 1612–1617

41 DiLeone RJ, Russell LB, Kingsley DM (1998) An extensive 3' regulatory region controls expression of BMP-5 in specific anatomical structures of the mouse embryo. *Genetics* 148: 401–408

42 Ovcharenko I, Nobrega MA, Loots GG, Stubbs L (2004) ECR browser: A tool for visualizing and accessing data from comparisons of multiple vertebrate genomes. *Nucleic Acids Res* 32: W280–286

43 Winnier G, Blessing M, Labosky PA, Hogan BL (1995) Bone morphogenetic protein-4 is required for mesoderm formation and patterning in the mouse. *Genes Dev* 9: 2105–2116

44 Lawson KA, Dunn NR, Roelen BA, Zeinstra LM, Davis AM, Wright CV, Korving JP, Hogan BL (1999) BMP-4 is required for the generation of primordial germ cells in the mouse embryo. *Genes Dev* 13: 424–436

45 Dunn NR, Winnier GE, Hargett LK, Schrick JJ, Fogo AB, Hogan BL (1997) Haploinsufficient phenotypes in BMP-4 heterozygous null mice and modification by mutations in Gli3 and Alx4. *Dev Biol* 188: 235–247

46 Chang B, Smith RS, Peters M, Savinova OV, Hawes NL, Zabaleta A, Nusinowitz S, Martin JE, Davisson ML, Cepko CL et al (2001) Haploinsufficient BMP-4 ocular phenotypes include anterior segment dysgenesis with elevated intraocular pressure. *BMC Genet* 2: 18

47 Frank DB, Abtahi A, Yamaguchi DJ, Manning S, Shyr Y, Pozzi A, Baldwin HS, Johnson JE, de Caestecker MP (2005) Bone morphogenetic protein 4 promotes pulmonary vascular remodeling in hypoxic pulmonary hypertension. *Circ Res* 97: 496–504

48. Kulessa H, Hogan BL (2002) Generation of a loxP flanked BMP-4loxP-lacZ allele marked by conditional lacZ expression. *Genesis* 32: 66–68
48a Chang W, Lin Z, Kulessa H, Hebert J, Hogan BL, Wu DK (2008) *Bmp4* is essential for the formation of the vestibular apparatus that detects angular head movements. *PLOS Genetics, in press*
49. Muyrers JP, Zhang Y, Testa G, Stewart AF (1999) Rapid modification of bacterial artificial chromosomes by ET-recombination. *Nucleic Acids Res* 27: 1555–1557
50. Copeland NG, Jenkins NA, Court DL (2001) Recombineering: A powerful new tool for mouse functional genomics. *Nat Rev Genet* 2: 769–779
51. Valenzuela DM, Murphy AJ, Frendewey D, Gale NW, Economides AN, Auerbach W, Poueymirou WT, Adams NC, Rojas J, Yasenchak J et al (2003) High-throughput engineering of the mouse genome coupled with high-resolution expression analysis. *Nat Biotechnol* 21: 652–659
52. Gazzerro E, Smerdel-Ramoya A, Zanotti S, Stadmeyer L, Durant D, Economides AN, Canalis E (2007) Conditional deletion of gremlin causes a transient increase in bone formation and bone mass. *J Biol Chem* 282: 31549–31557
53. Zhang H, Bradley A (1996) Mice deficient for BMP-2 are nonviable and have defects in amnion/chorion and cardiac development. *Development* 122: 2977–2986
54. Correia AC, Costa M, Moraes F, Bom J, Novoa A, Mallo M (2007) BMP-2 is required for migration but not for induction of neural crest cells in the mouse. *Dev Dyn* 236: 2493–2501
55. Ma L, Martin JF (2005) Generation of a BMP-2 conditional null allele. *Genesis* 42: 203–206
56. Tsuji K, Bandyopadhyay A, Harfe BD, Cox K, Kakar S, Gerstenfeld L, Einhorn T, Tabin CJ, Rosen V (2006) BMP-2 activity, although dispensable for bone formation, is required for the initiation of fracture healing. *Nat Genet* 38: 1424–1429
57. King JA, Marker PC, Seung KJ, Kingsley DM (1994) BMP-5 and the molecular, skeletal, and soft-tissue alterations in short ear mice. *Dev Biol* 166: 112–122
58. Solloway MJ, Dudley AT, Bikoff EK, Lyons KM, Hogan BL, Robertson EJ (1998) Mice lacking BMP-6 function. *Dev Genet* 22: 321–339
59. Dudley AT, Lyons KM, Robertson EJ (1995) A requirement for bone morphogenetic protein-7 during development of the mammalian kidney and eye. *Genes Dev* 9: 2795–2807
60. Luo G, Hofmann C, Bronckers AL, Sohocki M, Bradley A, Karsenty G (1995) BMP-7 is an inducer of nephrogenesis, and is also required for eye development and skeletal patterning. *Genes Dev* 9: 2808–2820
61. Kim RY, Robertson EJ, Solloway MJ (2001) BMP-6 and BMP-7 are required for cushion formation and septation in the developing mouse heart. *Dev Biol* 235: 449–466
62. Solloway MJ, Robertson EJ (1999) Early embryonic lethality in BMP-5; BMP-7 double mutant mice suggests functional redundancy within the 60A subgroup. *Development* 126: 1753–1768

63 Patel SR, Dressler GR (2005) BMP-7 signaling in renal development and disease. *Trends Mol Med* 11: 512–518
64 Godin RE, Takaesu NT, Robertson EJ, Dudley AT (1998) Regulation of BMP-7 expression during kidney development. *Development* 125: 3473–3482
65 Brunet LJ, McMahon JA, McMahon AP, Harland RM (1998) Noggin, cartilage morphogenesis, and joint formation in the mammalian skeleton. *Science* 280: 1455–1457
66 McMahon JA, Takada S, Zimmerman LB, Fan CM, Harland RM, McMahon AP (1998) Noggin-mediated antagonism of BMP signaling is required for growth and patterning of the neural tube and somite. *Genes Dev* 12: 1438–1452
67 Michos O, Panman L, Vintersten K, Beier K, Zeller R, Zuniga A (2004) Gremlin-mediated BMP antagonism induces the epithelial-mesenchymal feedback signaling controlling metanephric kidney and limb organogenesis. *Development* 131: 3401–3410
68 Bachiller D, Klingensmith J, Kemp C, Belo JA, Anderson RM, May SR, McMahon JA, McMahon AP, Harland RM, Rossant J et al (2000) The organizer factors Chordin and Noggin are required for mouse forebrain development. *Nature* 403: 658–661
69 Petryk A, Anderson RM, Jarcho MP, Leaf I, Carlson CS, Klingensmith J, Shawlot W, O'Connor MB (2004) The mammalian twisted gastrulation gene functions in foregut and craniofacial development. *Dev Biol* 267: 374–386
70 Gazzerro E, Deregowski V, Stadmeyer L, Gale NW, Economides AN, Canalis E (2006) Twisted gastrulation, a bone morphogenetic protein agonist/antagonist, is not required for post-natal skeletal function. *Bone* 39: 1252–1260

Alterations of BMP signaling pathway(s) in skeletal diseases

Petra Seemann[1], Stefan Mundlos[1,2] and Katarina Lehmann[2]

[1]Max-Planck-Institut für Molekulare Genetik, Ihnestr. 63–73, 14195 Berlin, Germany;
[2]Institut für Medizinische Genetik, Universitätsmedizin Berlin, Charité, Augustenburger Platz 1, 13353 Berlin, Germany

Introduction

The existence of bone morphogenetic proteins (BMPs) was postulated by Urist already in 1965 based on the observation that dematerialized bone matrix is able to induce bone formation when transplanted into the muscle of rabbits or rats [1]. Because heat-denatured samples did not show this effect, he concluded that the inducing substance had to be a protein and suggested the name ‚bone morphogenetic protein'. It took 20 more years after that pioneering observation to actually prove this theory. The first report on purification of a BMP was in 1988 by Elizabeth Wang and colleagues [2] and soon thereafter a series of ground-breaking discoveries followed. BMPs were either purified or cloned from cartilage and/or bone [3–9]. It turned out that the different BMPs are quite homologous to each other and structurally related to TGF-β proteins [6, 10, 11]. Due to their similarities they were classified as the TGF-β superfamily with a large subgroup of proteins named BMPs and growth and differentiation factors (GDFs). With a few exceptions, BMPs and GDFs are capable of inducing ectopic bone and or cartilage formation *in vitro* and *in vivo*.

Concurrent with the identification of BMPs, a huge scientific field opened and the quest to understand in which way BMPs are able to induce ectopic bone formation was started. It was hypothesized that BMPs are not only responsible for the development of the normal skeleton, but also play a major role in bone regeneration based on the assumption that similar molecular mechanisms are operative during embryogenesis and skeletal homeostasis [12]. Subsequently, many other pivotal roles for BMPs were discovered, establishing the BMP pathway as one of the major signaling events during embryogenesis, homeostasis and ageing.

Mutations in the components of the BMP pathway, including ligands (GDF-5), their receptors and co-receptors (BMPR1B, ACVR1, ROR2), as well as their inhibitors (NOG), can lead to specific skeletal disorders (for review [13, 14]). The following section gives an overview of brachydactylies (BDs) and other more complex skeletal disorders caused by a disturbance of BMP signaling (Fig. 1 and Tab. 1).

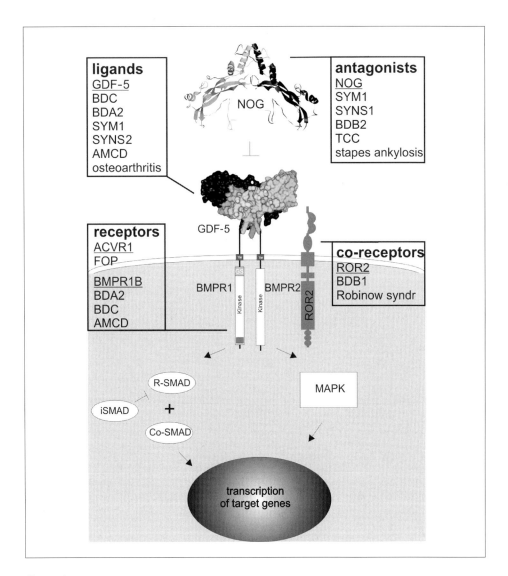

Figure 1
Schematic overview of bone morphogenetic protein (BMP) signaling and associated diseases. BDA2 is caused by heterozygous mutations in BMPR1B, whereas heterozygous NOG mutations lead to SYM1 or SYNS1. Heterozygous mutations in GDF-5 cause BDC, while point mutations in GDF-5 that interfere with specific protein-protein interactions lead to different phenotypes. The GDF-5 mutant L441P does not interact efficiently with BMPR1B and causes BDA2, which cannot be discriminated by phenotype from an inherited BMPR1B mutation. The GDF-5 mutant R438L binds additionally to BMPR1A. This gained function causes a clinical SYM1 phenotype, very similar to the one observed in patients with NOG loss-of-function mutations.

Isolated BDs are a group of hand/foot malformations that are characterized by shortening of the phalanges and/or metacarpal bones. Usually they are inherited in an autosomal dominant manner. The most commonly used clinical classification was introduced by Julia Bell in 1951 based on the anatomical pattern of the abnormal digits distinguishing five different groups (A–E) including three subgroups (A1–A3) [15]. In recent years, the molecular cause has been identified for some of them and pathogenic effects of specific disease causing mutations have been analyzed. After gaining more detailed insights into the different parts of the BMP pathway by investigating the genotype-phenotype association of distinct mutations, the BMP signaling system turned out to be more complex than originally thought. BD types A–C have certain overlapping phenotypic features, suggesting that they are due to defects in different components of the same molecular pathway.

Phenotypic spectrum of mutations in the BMP signaling molecule GDF-5

GDF-5 (also known as CDMP1, BMP-14) was the first member of the TGF-β superfamily reported to be associated with an inherited disease [16]. A homozygous 22-bp tandem duplication was identified, which results in a frameshift in the mature part of GDF-5 causing acromesomelic chondrodysplasia (AMCD) Hunter-Thompson type [Mendelian Inheritance in Men (MIM) 201250], presumably by a loss-of-function (LOF) effect.

Heterozygous mutations in the ligand GDF-5 typically result in BD type C (BDC) (MIM 113100) [17]. The BDC phenotype is characterized by shortened index, middle, and little fingers due to brachymesophalangy (Fig. 2). Ring fingers are usually not affected and appear as the longest fingers. In addition, accessory phalanges, typically located in the index and middle fingers, as well as shortening of the first metacarpal bones can occur. The phenotype of BDC is variable and ranges from severely affected to only mildly affected individuals, and considerable interfamilial as well as intrafamilial variability can be observed. Heterozygous frameshift or nonsense mutations in GDF-5 are causative in most affected BDC patients arguing for an LOF effect of the mutations. In some patients, heterozygous missense mutations in GDF-5 were shown to lead to classic BDC acting presumably also by LOF. In addition, homozygous missense mutations located in the prodomain of GDF-5 can result in BDC in some cases [18]. The underlying pathomechanism of these homozygous mutations is not known so far since the function of the prodomain of GDF-5 is not yet entirely understood.

Missense mutations in GDF-5 can also produce phenotypes different from BDC. Investigating these mutations in more detail showed that they are located at specific sites within the coding sequence of GDF-5. For example, the mutation L441P is located at the receptor interaction site of GDF-5. The phenotype caused by L441P is BD type A2 [BDA2 (MIM 112600)], a condition characterized by short/missing middle phalanges of the index finger and originally shown to be due to mutations in

Table 1 - BMP signaling molecules associated with skeletal malformations in human (Mendelian inheritance in men, MIM).

Genes and synonyms	MIM no.	Inheritance	Human diseases	Mutations: Examples and pathomechanism
ligands				
GDF-5 CDMP1 BMP-14	*601146	Dominant	BDC	Several frameshift, nonsense, missense mutations, LOF predicted
			BDA2	L441P at the receptor binding site decreases binding to BMPR1B [21]
			SYM1	R438L increases binding to BMPR1A leading to a GOF [21, 22]
			SYNS2	E491K is predicted to have a similar effect [76]
			Osteoarthritis	SNP in the promoter region leads to a decreased transcriptional activity [26]
			AMCD, Du Pan type	p.L437del & S439T & H440L on one allele, pathomechanism unclear [24]
			AMCD, Grebe type	C400Y, c.1144delG & C400Y, dominant negative effect [25] c.297insC, homozygous LOF predicted [77] c.206insG, homozygous LOF predicted [78]
		Recessive	AMCD, Hunter-Thompson type	22-bp duplication within the cds resulting in a frameshift [16]
			AMCD Du Pan type	L441P, decreases binding to BMPR1B [21, 23]
			BDC	M173V located in the GDF5 prodomain [18]
receptors				
ACVR1 ALK2 ACVRLK2	*102576	Dominant	FOP	R206H located in the GS-box, pathomechanism unclear [34]
BMPR1B ALK6	*603248	Dominant	BDA2	I200K in the GS-box or R486W in the NANDOR domain [20]
			BDC & SYM1	R486Q in the NANDOR domain [28]
		Recessive	AMCD with genital anomalies	8-bp deletion in the extracellular domain, LOF predicted [29]
ROR2 NTRKR2	*602337	Dominant	BDB1	Several truncating mutations in the intracellular region N-terminal or C-terminal of the kinase domain, dominant effect
		Recessive	Robinow Syndrome	Several frameshift, nonsense, missense mutations in both intra- and extracellular domains, LOF predicted

antagonists	NOG	*602991	Dominant	SYM1	Several frameshift, nonsense, missense mutations, LOF predicted
				SYNS1	
				TCC	
				Stapes ankylosis	
				BDB2	Specific missense mutations affecting codons 35, 36, 48, 167, 187, pathomechanism unclear [52]
	SOST	*605740	Recessive	Sclerosteosis	Nonsense and splice site mutations, LOF [54]
				van Buchem disease	Deletion of a SOST-specific regulatory element [79]
				Bone mineral density	Polymorphisms in the SOST promoter region and in the van Buchem deletion region are predicted to be associated with bone mineral density

AMCD, acromesomelic chondrodysplasia; BD, brachydactyly; SYM1, proximal symphalangism; SYNS, multiple synostoses syndrome; FOP, fibrodysplasia ossificans progressiva; TCC, tarsalcarpal coalition syndrome; LOF, loss of function; GOF, gain of function.

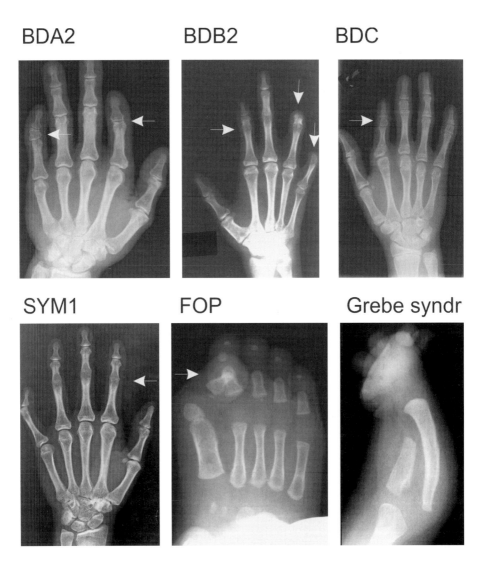

Figure 2
Clinical features associated with mutations in the BMP pathway. Brachydactyly (BD) A2: note hypoplastic or absent middle phalanges of fingers II and V (arrows). BDB2: note hypoplastic or absent distal phalanges (arrows right) and joint fusions of the proximal and middle phalanges (SYM1, arrow left). BDC: note hypoplastic middle phalanges of fingers II, III and V (arrow finger II). The ring finger is typically the least affected. SYM1: note proximal symphalangism (fusion of joints) in fingers II to V (arrow finger II). FOP: note characteristic deviation of a monophalangeal big toe in a 9-month-old child (arrow). Grebe syndrome is the most severe example of the group of acromesomelic chondrodysplasias: note severe shortening and bowing of bones of the forearm and the severely affected hands with rudimentary finger structures in a child.

the BMP receptor 1B (BMPR1B) (Fig. 2) [19–21]. The fact that the same phenotype can be caused by mutations in the ligand as well as the receptor can be explained by the failure of the mutant GDF-5 to form a proper ligand-receptor interaction complex. In contrast, the amino acid change R438L was shown to result in a loss of GDF-5 signaling specificity by a gain-of-function effect [21, 22]. The mutated GDF-5 is altered in its binding affinity and gains BMP-2-like properties by an increased affinity to the BMP receptor 1A (BMPR1A). The clinical picture associated with the R438L mutation is characterized by joint fusions such as proximal symphalangism [SYM1 (MIM 185800)] and multiple synostosis syndrome [SYNS2 (MIM 610017)]. Proximal and middle phalanges are fused in SYM1 (Fig. 2). Multiple joint fusions of the fingers, wrists, ankles, elbows and spine in a variable degree are the characteristics of SYNS2.

Homozygous missense mutations in GDF-5 typically result in rare hereditary skeletal disorders of the acromesomelic chondrodysplasia (AMCD) group. Due to an autosomal recessive inheritance of these disorders, they are more often found in consanguineous families. The main characteristics are severe shortening of the limbs, especially of the mesomelic parts (radii/ulnae and tibiae/fibulae), hand/foot malformations and a normal axial skeleton. The hands and feet show severe brachydactyly with only rudimentary fingers. Different types of AMCD are known: Grebe syndrome (MIM 200700) is the most severe form (Fig. 2), Hunter-Thompson and DuPan syndrome (MIM 228900) are milder forms of the same disease spectrum. A genotype-phenotype correlation seems to exist since the GDF-5 mutation L441P leads in a heterozygous state to the milder BDA2 phenotype [19, 21] and in the homozygous state to DuPan syndrome [23]. In addition, DuPan syndrome can be inherited in an autosomal dominant manner [24]. In this family, three different GDF-5 mutations on one allele were identified. In contrast, heterozygous mutation carriers of the mutation C400Y show variable BDC and individuals who carry the mutation in a homozygous state are affected by the most severe form of AMCD, Grebe syndrome. In addition, compound heterozygosity for C400Y and a frameshift mutation c.1144delG also result in Grebe syndrome [25]. Thomas et al. [25] postulated a dominant negative effect of these mutations since the severe phenotype in Grebe syndrome cannot be explained by an LOF effect alone. Mutations in the mature domain may exert a dominant negative effect by interference with the signaling of other BMP family members through the formation of heterodimers.

More recently, an SNP in the promoter region of *GDF-5* was found to be associated with osteoarthritis (MIM 165720) in the Asian population [26]. The authors claim that this SNP might lead to a decreased expression of GDF-5 and thereby increase the susceptibility of osteoarthritis in these patients. Nevertheless, analysis of this SNP in a Caucasian Greek population did not show association with osteoarthritis [27]. Further analysis is needed to clarify if this SNP is a population-specific risk factor.

Skeletal malformations caused by mutations in BMP receptors

Signaling of the members of the BMP family such as GDF-5 requires binding to cell surface receptors consisting of two kinds of transmembrane serine-threonine kinase receptors classified as types 1 and 2. These receptors form homodimeric and heterodimeric complexes on the cell surface consisting of receptor monomers. BMP signaling can be mediated through three known type 1 receptors: BMPR1A (ALK3), BMPR1B (ALK6), and ACVR1 (ALK2).

Heterozygous missense mutations in BMPR1B typically cause BDA2, a condition characterized by hypoplastic or absent middle phalanges of fingers II and V, resulting in shortened and medially deviated index and little fingers (Fig. 2). Three BDA2-associated missense mutations in BMPR1B are known so far [20]. The mutation I200K is located in a glycine-serine (GS)-rich domain of the receptor that is involved in the phosphorylation and thus activation of BMPR1B. It was shown that this mutation leads to a loss of kinase activity. The second mutation, R486W, lays more C-terminal within a highly conserved region of the kinase domain, called NANDOR box. In contrast to the mutation I200K, the kinase activity is undisturbed in the R486W mutant, arguing for a different pathogenic mechanism, which is not entirely understood to date. Interestingly, another amino acid exchange at the position 486 (R486Q) can either result in BDA2 or in a more severe combination of BDC and SYM1, thus imitating phenotypes caused by GDF-5 or NOG mutations [28].

Similar to the AMCD group caused by GDF-5 mutations, homozygous mutations in BMPR1B have been reported to result in a phenotype of this disease spectrum. A homozygous 8-bp deletion leading to a frameshift in the region coding for the extracellular domain of BMPR1B was described in an individual with disproportionate short stature due to severe acromesomelic limb shortening, severe BD, genital abnormalities, and ovarian dysfunction (MIM 609441) [29]. In contrast to the BDA2-associated mutations, this deletion is likely to result in a complete loss of BMPR1B function. The limb malformations show a considerable overlap to DuPan syndrome such as hypoplasia of the fibulae, knob-like toes and hypoplasia of phalanges resulting in brachydactyly. More specific changes were, however, present in this patient, which are not considered to be part of the GDF-5 spectrum. These include fusion of carpal bones and relatively unaffected metacarpal/metatarsal bones. Inactivation of Bmpr1b in the mouse results in a phenotype very similar to the human disorder [30]. These mice show skeletal malformations that are particularly severe in the distal parts of the skeleton and infertility in female mice due to disturbed ovarian function. The importance of BMP signaling in the regulation of the ovarian function is further emphasized by the identification of mutations in BMPs and BMPR1B in sheep. Specific point mutations affecting interacting partners of that pathway can either lead to a higher ovulation rate resulting in an increased litter size, while other mostly homozygous mutations lead to infertility. For example, Booroola sheep have a mutation (Q249R) in the highly conserved intracellular kinase signaling domain of the BMPR1B receptor that results in an increase in litter size by one to two extra lambs with each copy [31]. As judged by the phenotype and *in vitro* studies, this mutation affects only the ovarian signaling, but not the limbs [20]. Another major locus affecting the ovulation rate in Inverdale sheep has been shown to be due to a mutation in GDF-9b/BMP-15. Those that are heterozygous for the mutation have a higher ovulation rate, but homozygous inactivating mutations result in infertility [32].

Mutations in another BMP type 1 receptor, BMPR1A, are not known to be associated with skeletal dysplasia in humans. However, mutations in BMPR1A predispose to cancer disorders such as juvenile polyposis syndrome (MIM 610069), or Cowden-like syndrome (MIM 158350) [33].

The third BMP type 1 receptor, ACVR1 (also known as ALK2), was recently identified to be involved in the pathogenesis of fibrodysplasia ossificans progressiva (FOP) (MIM 135100) also known as myositis ossificans progressiva [34]. FOP is a very rare genetic disease with an incidence of only about 1 in 2 million. Because of its striking phenotypic appearance, it is well recognized and was already reported in 1692 by a French physician. Patients with FOP are born with a malformed big toe, which is often broadened, shortened, deviated and/or monophalangeal (Fig. 2) [34a]. The disease evolves after birth with a variable onset between age 2 and 15, by rapid formation of heterotopic bone after injury, but also without known triggers. Heterotopic bone is formed in an ordered fashion, affecting the neck and shoulders first and subsequently progressing down the spine and/or affecting muscles, tendons, or ligaments at all major joints. This leads to an ankylosis of the large joints, such as the elbow, hip or knee joint. Because the transformation from soft tissue into bone happens very rapidly, sometimes within days, the position of the locked joint can cause bizarre postures locking the patients in the respective position for the rest of their lives [35]. Unfortunately, the formation of heterotopic bone is often misdiagnosed as cancer and subsequently surgically removed resulting in even more heterotopic bone formation and further deterioration [36]. Marshall Urist's observation that BMPs induce ectopic bone in muscles revealed striking similarities to the pathomechanism of FOP. The BMP signaling components were, of course, the top candidates to find the disease-causing gene and it could be shown that the BMP signaling pathway is deregulated in cells from FOP patients [37]. Finding the disease gene proved to be difficult due the rarity of the condition and the paucity of families large enough for genetic linkage studies. Some years ago, there was a debate on whether the NOG mutation was causative for FOP. The published mutations turned out to be sequencing artifacts [38, 39]. Until now, there are only seven FOP families known worldwide. Five of these were included for a linkage analysis to finally identify the molecular basis for the disorder [34]. In 2006, Shore et al. [34] identified a point mutation in ACVR1, which is associated with FOP and leads to an amino acid exchange of arginine to histidine at position 206 (R206H). This mutation is located in the GS-box of the receptor and is thought to cause an activated isoform of the receptor [40]. Even though FOP has impressed many physicians when they have come into contact with a patient, it is undoubtedly the passion of three people coming together more than 15 years ago from different perspectives has really pushed the knowledge of FOP forward, i.e., the physician Prof. Dr. Frederick Kaplan, the FOP patient Jeannie Peeper and the geneticist Dr. Eileen M. Shore. It is to the credit of these colleagues that an international fibrodysplasia progressiva association was established, providing a platform for patients, relatives, clinicians and scientists to share information and help. Even though there is no treatment until now, the discovery of the FOP gene, ACVR1, gives hope for designing a therapy within the near future [41].

Mutations in the BMP receptor type 2, BMPR2, are not known to be associated with skeletal dysplasia in humans. However, dominant mutations in BMPR2 cause primary pulmonary hypertension progressive disease in which widespread occlusion of the smallest pulmonary arteries leads to increased pulmonary vascular resistance, and subsequently right ventricular failure [42].

Receptors that interact with BMPRs

ROR2 is a receptor tyrosine kinase with a function in skeletal development. It was shown that ROR2 and BMBR1B form a ligand-independent heteromeric complex at the cell surface that is needed for the proper signal transduction into the cell [43]. Intracellular SMAD signaling initiated *via* BMPR1B gets inhibited by an activated tyrosine kinase of ROR2; thus ROR2 can modulate BMP signaling.

Heterozygous truncating mutations in ROR2 [44, 45] lead to BDB (MIM 113000), a condition characterized by hypoplastic or absent distal phalanges imitating an amputation-like phenotype. Associated fusions of interdigital joints can occur and affect typically the distal parts of fingers leading to distal symphalangism. BDB-causing mutations are clustered in regions directly before or after the intracellular tyrosine kinase domain of ROR2. Genotype-phenotype correlations revealed that mutations located at the N-terminus result in a more severe phenotype than mutations that are located before the kinase domain. To date, it is still unknown how these mutations exert their pathogenic effects, but a dominant effect is very likely. The tyrosine domains are necessary for receptor activation and subsequent activation of intracellular messenger molecules. The presence of joint fusions strongly argues for a connection of ROR2 and the BMP antagonist NOGGIN (NOG) as described in the next paragraph.

Homozygous mutations in ROR2 result in autosomal recessive Robinow syndrome (MIM 268310), a complex disorder including short stature, acromesomelic limb shortening, vertebral and rib defects, facial characteristics, and genital abnormalities [46]. Robinow syndrome-causing mutations are scattered throughout the entire coding region either extra- or intracellular of ROR2, indicating an LOF effect.

Heterozygous carriers of Robinow syndrome-causing mutations are not affected by BDB, arguing for two different disease-causing mechanisms.

Skeletal disease due to misregulation of BMP antagonists

NOGGIN (NOG) is an extracellular antagonist of BMPs and GDFs. NOG binds to BMPs and masks the BMP type 1 and 2 receptor-binding sites of certain BMPs and GDFs [47], thus preventing signal transduction *via* the BMP receptors. Different human disorders are related to heterozygous LOF mutations in NOG, including proximal symphalangism (SYM1), multiple synostosis syndrome (SYNS1), tarsal carpal coalition syndrome (fusions of the carpal and tarsal bones) TCC (MIM 186570), and stapes ankylosis with broad thumbs and toes (MIM 184460) [48–50]. These disorders are characterized by various degrees of joint fusions. The mutations are scattered throughout the entire NOG gene and presumably result in an LOF and thus are expected to result in augmented BMP signaling. Mice with inactivated Nog alleles have a severe phenotype consisting of an almost absent spine, open brain, and a massive increase in cartilage in the limbs [51]. Probably due to the overall increase in cartilage, no joints are formed. No human disorder associated with homozygous NOG mutations is known so far. Having in mind the phenotype of the corresponding mouse model, it seems possible that homozygosity of LOF mutations in humans results in early lethality.

In contrast to the above-mentioned LOF mutations, there are specific heterozygous missense mutations in NOG that seem to have another mode of action since the pathogenic effects are quite different. These mutations produce BDB2 (MIM 611377), a novel subtype of BDB characterized by hypoplastic or absent distal phalanges, proximal symphalangism, and carpal fusions (Fig. 2) [52]. Thus, BDB2 is a phenotypic combination of BDB1, which is caused by mutations in ROR2, as well as SYM1 and TCC caused by LOF mutations in NOG. The exact mechanism of these BDB2-causing NOG mutations is not known so far, and a more subtle alteration of BMP signaling than a simple LOF is likely. A partial loss of NOG function resulting in a decreased binding affinity of NOG to the BMPs/GDFs might lead to increased GDF-5 signaling and, consequently, to joint fusions. The clinical phenotype of combined BDB, SYM1, and TCC argues for a functional interaction between the BMP receptors and ROR2.

Sclerosteosis (MIM 269500) was first mentioned in 1967 by Hansen [53] describing an autosomal recessive disorder characterized by bone overgrowth, which leads to high intracranial pressure and subsequent facial palsy and deafness. Cutaneous syndactyly and digit malformations are also associated features. In 2001, a genome-wide linkage analysis revealed a mutation in a novel gene, which they named sclerostin or short SOST [54]. A deletion of the promoter region of SOST is associated with Van Buchem disease (MIM 239100), a similar disorder [55].

SOST is a cystine knot protein, like BMPs or NOG. It is expressed in osteocytes and in the interdigital mesenchyme during limb development. Initially, SOST was thought to act as a BMP inhibitor offering a likely explanation for the bone overgrowth. Indeed, *in vitro* experiments showed that SOST is able to bind to different BMPs [56]. Further studies showed that SOST is, in addition, able to bind to NOG and thereby have a positive effect on BMP signaling [57]. However, SOST was recently shown to primarily act as a Wnt-antagonist by binding to LRP, indicating that not diminished BMP inhibition but rather a lack of Wnt antagonism might explain the phenotype [58, 59].

What do mouse models teach us?

In the 90s, genetic analysis of either spontaneous or radiation/chemically induced mouse mutants revealed the first proof that BMPs are critical for proper bone formation in mammals. In that context, the first analyzed mutant was the short ear mouse [60]. Chromosome walking revealed mutations and/or deletions of the *bmp-5* gene in different short ear mouse lines. *bmp-5* mutant mice exhibit shortened external ears and abnormal formation and repair of skeletal structures. In addition, they show abnormalities in the development of several soft tissues [61]. The next investigated Bmp mouse mutant was the brachypodism (bp) mouse, which was shown to be mutated in the Gdf-5 locus [62]. The bp mouse has shortened long bones, missing phalanges and joint fusions in the autopod. The axial skeleton is not affected. Further analysis of this mutant revealed altered tendon and ligament structures and defective fracture healing [63, 64].

Targeted gene knockout approaches followed to understand the function of more than 20 different *bmps in vivo* (reviewed by [65]). Knockout strategies of many different *bmps* in mice showed early embryonic lethality. This includes *bmp-2* [66], *bmp-4* [67], Acvr1 [68], Bmpr1a [69], and Bmpr2 [70]. This clearly demonstrated that the function of *bmps* is not restricted to the skeleton. Instead, they were shown to have important functions during development within a wide range of organs. The wide spectrum of Bmp function is further exemplified by the inactivation of *bmp-7*, which results in skeletal defects, kidney agenesis and eye defects [71]. Because *bmp* homologues were also identified in species that do not have skeleton (e.g., *Drosophila*), the name 'bone morphogenetic protein' appears somewhat misleading and the term 'body morphogenetic proteins' was suggested as an alternative. The specificity of Bmp receptor interaction is illustrated by the Bmpr1b knockout mouse, which resembles the brachypodism (Gdf-5) phenotype, but additionally has a negative effect on female reproduction [30].

It was a surprise that the genetic inactivation of *bmp-3* led to an increase of bone formation [72] rather than a decrease and identifies Bmp-3 as a Bmp antagonist.

Due to early lethality, more specific approaches were needed to investigate the function of Bmps in certain organs. Conditional alleles of Bmps offered the possibility to inactivate Bmps in a given tissue at a specific time point. As already mentioned, the Bmpr1b knockout model has a relatively mild phenotype in the limbs, affecting primarily the long bones and the distal phalanges. In contrast, the Bmpr1a knockout mouse is embryonic lethal before limbs are developed, but a cartilage-specific conditional knockout under a cartilage-specific Col2-promoter revealed only a mild general skeletal dysplasia. Interestingly, the double inactivation of both receptors showed a nearly complete absence of chondrogenic condensations [73] implying that Bmp signaling is a prerequisite for early skeletal development and that Bmpr1a and Bmpr1b have at least some redundant functions. An unexpected phenotype was observed in the limb-specific conditional knockout of Bmp-2 using the Prx-cre promotor, which restricts cre-activity to the limbs and the skull [74]. There were only very mild effects during limb development and nearly no patterning defect. However, it turned out that Bmp-2 is absolutely necessary for late osteogenesis and bone homeostasis. Adult mice showed microfractures that were unable to heal spontaneously. To investigate this observation in more detail, limb-specific conditional double knockouts of Bmp-2 and Bmp-4 or Bmp-2 and Bmp-7 were analyzed to check whether other Bmps are able to compensate for Bmp-2 in early limb development in prx-cre-Bmp-2 mice [75].

It was shown that compound heterozygous mice for *bmp-2$^{prx-cre+/-}$/bmp-4$^{prx-cre+/-}$* or *bmp-2$^{prx-cre+/-}$/bmp-7$^{+/-}$* as well as *bmp-2$^{prx-cre+/-}$/bmp-7$^{-/-}$* were phenotypically normal. *bmp-4$^{prx-cre-/-}$/bmp-2$^{prx-cre+/-}$* mice displayed the same phenotype as *bmp-4$^{prx-cre-/-}$* mice, which showed variable preaxial and postaxial polydactyly. The opposite approach creating *bmp-2$^{prx-cre-/-}$/bmp-4$^{prx-cre+/-}$* mice showed a different phenotype with an overall thinning of skeletal elements but no obvious patterning defects. *bmp-2$^{prx-cre-/-}$/bmp-7$^{+/-}$* mice show the same phenotype as *bmp-2$^{prx-cre-/-}$* knockouts. Only the limb-specific double-knockout mouse model of *bmp-2$^{prx-cre-/-}$/bmp-4$^{prx-cre-/-}$* showed extreme limb malformations with a shortened stylopod, one single zeugopod element and several joint fusions, e.g., in the fore and hind limbs.

Limb-specific double knockouts for *bmp-2$^{prx-cre-/-}$/bmp-7$^{-/-}$* showed a loss of the distal phalanx of digit III in the limbs with variable penetrance. In addition, shortening of the limbs, especially the fibulae, which were not articulated with the femora were associated features. Overall, it seems that the initial skeletal differentiation in limbs without Bmp-2, 4 or 7 are normal, but Bmp-2 and Bmp-4 are required for late osteogenesis and homeostasis of the bone in mice.

Acknowledgements

We thank Lutz Schomburg for critical remarks on the manuscript.

References

1. Urist MR (1965) Bone: Formation by autoinduction. *Science* 150: 893–899
2. Wang EA, Rosen V, Cordes P, Hewick RM, Kriz MJ, Luxenberg DP, Sibley BS, Wozney JM (1988) Purification and characterization of other distinct bone-inducing factors. *Proc Natl Acad Sci USA* 85: 9484–9488
3. Wozney JM, Rosen V, Celeste AJ, Mitsock LM, Whitters MJ, Kriz RW, Hewick RM, Wang EA (1988) Novel regulators of bone formation: Molecular clones and activities. *Science* 242: 1528–1534
4. Rosen V, Wozney JM, Wang EA, Cordes P, Celeste A, McQuaid D, Kurtzberg L (1989) Purification and molecular cloning of a novel group of BMPs and localization of BMP mRNA in developing bone. *Connect Tissue Res* 20: 313–319
5. Wozney JM, Rosen V, Byrne M, Celeste AJ, Moutsatsos I, Wang EA (1990) Growth factors influencing bone development. *J Cell Sci* Suppl 13: 149–156
6. Celeste AJ, Iannazzi JA, Taylor RC, Hewick RM, Rosen V, Wang EA, Wozney JM (1990) Identification of transforming growth factor beta family members present in bone-inductive protein purified from bovine bone. *Proc Natl Acad Sci USA* 87: 9843–9847
7. Chang SC, Hoang B, Thomas JT, Vukicevic S, Luyten FP, Ryba NJ, Kozak CA, Reddi AH, Moos M Jr (1994) Cartilage-derived morphogenetic proteins. New members of the transforming growth factor-beta superfamily predominantly expressed in long bones during human embryonic development. *J Biol Chem* 269: 28227–28234
8. Vukicevic S, Luyten FP, Reddi AH (1989) Stimulation of the expression of osteogenic and chondrogenic phenotypes *in vitro* by osteogenin. *Proc Natl Acad Sci USA* 86: 8793–8797
9. Luyten FP, Cunningham NS, Ma S, Muthukumaran N, Hammonds RG, Nevins WB, Woods WI, Reddi AH (1989) Purification and partial amino acid sequence of osteogenin, a protein initiating bone differentiation. *J Biol Chem* 264: 13377–13380
10. Sampath TK, Coughlin JE, Whetstone RM, Banach D, Corbett C, Ridge RJ, Ozkaynak E, Oppermann H, Rueger DC (1990) Bovine osteogenic protein is composed of dimers of OP-1 and BMP-2A, two members of the transforming growth factor-beta superfamily. *J Biol Chem* 265: 13198–13205
11. Ozkaynak E, Rueger DC, Drier EA, Corbett C, Ridge RJ, Sampath TK, Oppermann H (1990) OP-1 cDNA encodes an osteogenic protein in the TGF-beta family. *EMBO J* 9: 2085–2093
12. Wall NA, Hogan BL (1994) TGF-beta related genes in development. *Curr Opin Genet Dev* 4: 517–522
13. Schwabe GC, Mundlos S (2004) Genetics of congenital hand anomalies. *Handchir Mikrochir Plast Chir* 36: 85–97
14. Kornak U, Mundlos S (2003) Genetic disorders of the skeleton: A developmental approach. *Am J Hum Genet* 73: 447–474

15 Bell J (1951) *Treasury of Human Inheritance*. Cambridge University Press, London, 1–31
16 Thomas JT, Lin K, Nandedkar M, Camargo M, Cervenka J, Luyten FP (1996) A human chondrodysplasia due to a mutation in a TGF-beta superfamily member. *Nat Genet* 12: 315–317
17 Everman DB, Bartels CF, Yang Y, Yanamandra N, Goodman FR, Mendoza-Londono JR, Savarirayan R, White SM, Graham JM Jr, Gale RP et al (2002) The mutational spectrum of brachydactyly type C. *Am J Med Genet* 112: 291–296
18 Schwabe GC, Turkmen S, Leschik G, Palanduz S, Stover B, Goecke TO, Mundlos S (2004) Brachydactyly type C caused by a homozygous missense mutation in the prodomain of CDMP1. *Am J Med Genet* 124A: 356–363
19 Kjaer KW, Eiberg H, Hansen L, van der Hagen CB, Rosendahl K, Tommerup N, Mundlos S (2005) A mutation in the receptor binding site of GDF-5 causes Mohr-Wriedt brachydactyly type A2. *J Med Genet* 43: 225–231
20 Lehmann K, Seemann P, Stricker S, Sammar M, Meyer B, Suring K, Majewski F, Tinschert S, Grzeschik KH, Muller D et al (2003) Mutations in bone morphogenetic protein receptor 1B cause brachydactyly type A2. *Proc Natl Acad Sci USA* 100: 12277–12282
21 Seemann P, Schwappacher R, Kjaer KW, Krakow D, Lehmann K, Dawson K, Stricker S, Pohl J, Ploger F, Staub E et al (2005) Activating and deactivating mutations in the receptor interaction site of GDF-5 cause symphalangism or brachydactyly type A2. *J Clin Invest* 115: 2373–2381
22 Dawson K, Seeman P, Sebald E, King L, Edwards M, Williams J 3rd, Mundlos S, Krakow D (2006) GDF-5 is a second locus for multiple-synostosis syndrome. *Am J Hum Genet* 78: 708–712
23 Faiyaz-Ul-Haque M, Ahmad W, Zaidi SH, Haque S, Teebi AS, Ahmad M, Cohn DH, Tsui LC (2002) Mutation in the cartilage-derived morphogenetic protein-1 (CDMP1) gene in a kindred affected with fibular hypoplasia and complex brachydactyly (DuPan syndrome). *Clin Genet* 61: 454–458
24 Szczaluba K, Hilbert K, Obersztyn E, Zabel B, Mazurczak T, Kozlowski K (2005) Du Pan syndrome phenotype caused by heterozygous pathogenic mutations in CDMP1 gene. *Am J Med Genet A* 138: 379–383
25 Thomas JT, Kilpatrick MW, Lin K, Erlacher L, Lembessis P, Costa T, Tsipouras P, Luyten FP (1997) Disruption of human limb morphogenesis by a dominant negative mutation in CDMP1. *Nat Genet* 17: 58–64
26 Miyamoto Y, Mabuchi A, Shi D, Kubo T, Takatori Y, Saito S, Fujioka M, Sudo A, Uchida A, Yamamoto S et al (2007) A functional polymorphism in the 5' UTR of GDF-5 is associated with susceptibility to osteoarthritis. *Nat Genet* 39: 529–533
27 Tsezou A, Satra M, Oikonomou P, Bargiotas K, Malizos KN (2007) The growth differentiation factor 5 (GDF-5) core promoter polymorphism is not associated with knee osteoarthritis in the greek population. *J Orthop Res* 26: 136–140
28 Lehmann K, Seemann P, Boergermann J, Morin G, Reif S, Knaus P, Mundlos S (2006) A

novel R486Q mutation in BMPR1B resulting in either a brachydactyly type C/symphalangism-like phenotype or brachydactyly type A2. *Eur J Hum Genet* 14: 1248–1254

29 Demirhan O, Turkmen S, Schwabe GC, Soyupak S, Akgul E, Tastemir D, Karahan D, Mundlos S, Lehmann K (2005) A homozygous BMPR1B mutation causes a new subtype of acromesomelic chondrodysplasia with genital anomalies. *J Med Genet* 42: 314–317

30 Baur ST, Mai JJ, Dymecki SM (2000) Combinatorial signaling through BMP receptor IB and GDF-5: Shaping of the distal mouse limb and the genetics of distal limb diversity. *Development* 127: 605–619

31 Wilson T, Wu XY, Juengel JL, Ross IK, Lumsden JM, Lord EA, Dodds KG, Walling GA, McEwan JC, O'Connell AR et al (2001) Highly prolific Booroola sheep have a mutation in the intracellular kinase domain of bone morphogenetic protein IB receptor (ALK-6) that is expressed in both oocytes and granulosa cells. *Biol Reprod* 64: 1225–1235

32 Galloway SM, McNatty KP, Cambridge LM, Laitinen MP, Juengel JL, Jokiranta TS, McLaren RJ, Luiro K, Dodds KG, Montgomery GW et al (2000) Mutations in an oocyte-derived growth factor gene (BMP15) cause increased ovulation rate and infertility in a dosage-sensitive manner. *Nat Genet* 25: 279–283

33 Zhou XP, Woodford-Richens K, Lehtonen R, Kurose K, Aldred M, Hampel H, Launonen V, Virta S, Pilarski R, Salovaara R et al (2001) Germline mutations in BMPR1A/ALK3 cause a subset of cases of juvenile polyposis syndrome and of Cowden and Bannayan-Riley-Ruvalcaba syndromes. *Am J Hum Genet* 69: 704–711

34 Shore EM, Xu M, Feldman GJ, Fenstermacher DA, Cho TJ, Choi IH, Connor JM, Delai P, Glaser DL, LeMerrer M et al (2006) A recurrent mutation in the BMP type I receptor ACVR1 causes inherited and sporadic fibrodysplasia ossificans progressiva. *Nat Genet* 38: 525–527

34a Kaplan FS (2005) Fibrodysplasia ossificans progressiva – An historical perspective. *Clin Rev Bone Miner Metab* 3: 179–181

35 Mahboubi S, Glaser DL, Shore EM, Kaplan FS (2001) Fibrodysplasia ossificans progressiva. *Pediatr Radiol* 31: 307–314

36 Kitterman JA, Kantanie S, Rocke DM, Kaplan FS (2005) Iatrogenic harm caused by diagnostic errors in fibrodysplasia ossificans progressiva. *Pediatrics* 116: e654–661

37 Fiori JL, Billings PC, de la Pena LS, Kaplan FS, Shore EM (2006) Dysregulation of the BMP-p38 MAPK signaling pathway in cells from patients with fibrodysplasia ossificans progressiva (FOP). *J Bone Miner Res* 21: 902–909

38 Xu MQ, Feldman G, Le Merrer M, Shugart YY, Glaser DL, Urtizberea JA, Fardeau M, Connor JM, Triffitt J, Smith R et al (2000) Linkage exclusion and mutational analysis of the noggin gene in patients with fibrodysplasia ossificans progressiva (FOP). *Clin Genet* 58: 291–298

39 Semonin O, Fontaine K, Daviaud C, Ayuso C, Lucotte G (2001) Identification of three novel mutations of the noggin gene in patients with fibrodysplasia ossificans progressiva. *Am J Med Genet* 102: 314–317

40 Groppe JC, Shore EM, Kaplan FS (2007) Functional modeling of the ACVR1(R206H) mutation in FOP. *Clin Orthop Relat Res* 462: 87–92

41 Kaplan FS, Glaser DL, Pignolo RJ, Shore EM (2007) A new era for fibrodysplasia ossificans progressiva: a druggable target for the second skeleton. *Expert Opin Biol Ther* 7: 705–712

42 Lane KB, Machado RD, Pauciulo MW, Thomson JR, Phillips JA 3rd, Loyd JE, Nichols WC, Trembath RC (2000) Heterozygous germline mutations in BMPR2, encoding a TGF-beta receptor, cause familial primary pulmonary hypertension. The International PPH Consortium. *Nat Genet* 26: 81–84

43 Sammar M, Stricker S, Schwabe GC, Sieber C, Hartung A, Hanke M, Oishi I, Pohl J, Minami Y, Sebald W et al (2004) Modulation of GDF-5/BRI-b signaling through interaction with the tyrosine kinase receptor Ror2. *Genes Cells* 9: 1227–1238

44 Schwabe GC, Tinschert S, Buschow C, Meinecke P, Wolff G, Gillessen-Kaesbach G, Oldridge M, Wilkie AO, Komec R, Mundlos S (2000) Distinct mutations in the receptor tyrosine kinase gene ROR2 cause brachydactyly type B. *Am J Hum Genet* 67: 822–831

45 Oldridge M, Fortuna AM, Maringa M, Propping P, Mansour S, Pollit C, DeChiara TM, Kimble RB, Valenzuela DM, Yancopoulos GD et al (2000) Dominant mutations in ROR2, encoding an orphan receptor tyrosine kinase, cause brachydactyly type B. *Nat Genet* 24: 275–278

46 van Bokhoven H, Celli J, Kayserili H, van Beusekom E, Balci S, Brussel W, Skovby F, Kerr B, Percin EF, Akarsu N et al (2000) Mutation of the gene encoding the ROR2 tyrosine kinase causes autosomal recessive Robinow syndrome. *Nat Genet* 25: 423–426

47 Groppe J, Greenwald J, Wiater E, Rodriguez-Leon J, Economides AN, Kwiatkowski W, Affolter M, Vale WW, Belmonte JC, Choe S (2002) Structural basis of BMP signaling inhibition by the cystine knot protein Noggin. *Nature* 420: 636–642

48 Gong Y, Krakow D, Marcelino J, Wilkin D, Chitayat D, Babul-Hirji R, Hudgins L, Cremers CW, Cremers FP, Brunner HG et al (1999) Heterozygous mutations in the gene encoding noggin affect human joint morphogenesis. *Nat Genet* 21: 302–304

49 Dixon ME, Armstrong P, Stevens DB, Bamshad M (2001) Identical mutations in NOG can cause either tarsal/carpal coalition syndrome or proximal symphalangism. *Genet Med* 3: 349–353

50 Brown DJ, Kim TB, Petty EM, Downs CA, Martin DM, Strouse PJ, Moroi SE, Milunsky JM, Lesperance MM (2002) Autosomal dominant stapes ankylosis with broad thumbs and toes, hyperopia, and skeletal anomalies is caused by heterozygous nonsense and frameshift mutations in NOG, the gene encoding noggin. *Am J Hum Genet* 71: 618–624

51 Brunet LJ, McMahon JA, McMahon AP, Harland RM (1998) Noggin, cartilage morphogenesis, and joint formation in the mammalian skeleton. *Science* 280: 1455–1457

52 Lehmann K, Seemann P, Silan F, Goecke TO, Irgang S, Kjaer KW, Kjaergaard S, Mahoney MJ, Morlot S, Reissner C et al (2007) A new subtype of brachydactyly type B caused by point mutations in the bone morphogenetic protein antagonist NOGGIN. *Am J Hum Genet* 81: 388–396

53 Hansen H (1967) Sklerosteose. In: H Opitz, F Schmid (eds): *Handbuch der Kinderheilkunde*. Springer, Berlin, 351–355

54 Brunkow ME, Gardner JC, Van Ness J, Paeper BW, Kovacevich BR, Proll S, Skonier JE, Zhao L, Sabo PJ, Fu Y et al (2001) Bone dysplasia sclerosteosis results from loss of the SOST gene product, a novel cystine knot-containing protein. *Am J Hum Genet* 68: 577–589

55 Loots GG, Kneissel M, Keller H, Baptist M, Chang J, Collette NM, Ovcharenko D, Plajzer-Frick I, Rubin EM (2005) Genomic deletion of a long-range bone enhancer misregulates sclerostin in Van Buchem disease. *Genome Res* 15: 928–935

56 Winkler DG, Sutherland MK, Geoghegan JC, Yu C, Hayes T, Skonier JE, Shpektor D, Jonas M, Kovacevich BR, Staehling-Hampton K et al (2003) Osteocyte control of bone formation *via* sclerostin, a novel BMP antagonist. *EMBO J* 22: 6267–6276

57 Winkler DG, Yu C, Geoghegan JC, Ojala EW, Skonier JE, Shpektor D, Sutherland MK, Latham JA(2004) Noggin and sclerostin bone morphogenetic protein antagonists form a mutually inhibitory complex. *J Biol Chem* 279: 36293–36298

58 van Bezooijen RL, Roelen BA, Visser A, van der Wee-Pals L, de Wilt E, Karperien M, Hamersma H, Papapoulos SE, ten Dijke P, Lowik CW (2004) Sclerostin is an osteocyte-expressed negative regulator of bone formation, but not a classical BMP antagonist. *J Exp Med* 199: 805–814

59 van Bezooijen RL, Svensson JP, Eefting D, Visser A, van der Horst G, Karperien M, Quax PH, Vrieling H, Papapoulos SE, ten Dijke P et al (2007) Wnt but not BMP signaling is involved in the inhibitory action of sclerostin on BMP-stimulated bone formation. *J Bone Miner Res* 22: 19–28

60 Kingsley DM, Bland AE, Grubber JM, Marker PC, Russell LB, Copeland NG, Jenkins NA (1992) The mouse short ear skeletal morphogenesis locus is associated with defects in a bone morphogenetic member of the TGF beta superfamily. *Cell* 71: 399–410

61 King JA, Marker PC, Seung KJ, Kingsley DM (1994) BMP5 and the molecular, skeletal, and soft-tissue alterations in short ear mice. *Dev Biol* 166: 112–122

62 Storm EE, Huynh TV, Copeland NG, Jenkins NA, Kingsley DM, Lee SJ (1994) Limb alterations in brachypodism mice due to mutations in a new member of the TGF beta-superfamily. *Nature* 368: 639–643

63 Chhabra A, Zijerdi D, Zhang J, Kline A, Balian G, Hurwitz S (2005) BMP-14 deficiency inhibits long bone fracture healing: a biochemical, histologic, and radiographic assessment. *J Orthop Trauma* 19: 629–634

64 Mikic B, Schalet BJ, Clark RT, Gaschen V, Hunziker EB (2001) GDF--5 deficiency in mice alters the ultrastructure, mechanical properties and composition of the Achilles tendon. *J Orthop Res* 19: 365–371

65 Zhao GQ (2003) Consequences of knocking out BMP signaling in the mouse. *Genesis* 35: 43–56

66 Zhang H, Bradley A (1996) Mice deficient for BMP2 are nonviable and have defects in amnion/chorion and cardiac development. *Development* 122: 2977–2986

67 Lawson KA, Dunn NR, Roelen BA, Zeinstra LM, Davis AM, Wright CV, Korving JP,

Hogan BL (1999) Bmp4 is required for the generation of primordial germ cells in the mouse embryo. *Genes Dev* 13: 424–436

68 Gu Z, Reynolds EM, Song J, Lei H, Feijen A, Yu L, He W, MacLaughlin DT, van den Eijnden-van Raaij J, Donahoe PK et al (1999) The type I serine/threonine kinase receptor ActRIA (ALK2) is required for gastrulation of the mouse embryo. *Development* 126: 2551–2561

69 Ahn K, Mishina Y, Hanks MC, Behringer RR, Crenshaw EB 3rd (2001) BMPR IA signaling is required for the formation of the apical ectodermal ridge and dorsal-ventral patterning of the limb. *Development* 128: 4449–4461

70 Beppu H, Kawabata M, Hamamoto T, Chytil A, Minowa O, Noda T, Miyazono K (2000) BMP type II receptor is required for gastrulation and early development of mouse embryos. *Dev Biol* 221: 249–258

71 Jena N, Martin-Seisdedos C, McCue P, Croce CM (1997) BMP7 null mutation in mice: Developmental defects in skeleton, kidney, and eye. *Exp Cell Res* 230: 28–37

72 Daluiski A, Engstrand T, Bahamonde ME, Gamer LW, Agius E, Stevenson SL, Cox K, Rosen V, Lyons KM (2001) Bone morphogenetic protein-3 is a negative regulator of bone density. *Nat Genet* 27: 84–88

73 Yoon BS, Ovchinnikov DA, Yoshii I, Mishina Y, Behringer RR, Lyons KM (2005) Bmpr1a and Bmpr1b have overlapping functions and are essential for chondrogenesis in vivo. *Proc Natl Acad Sci USA* 102: 5062–5067

74 Tsuji K, Bandyopadhyay A, Harfe BD, Cox K, Kakar S, Gerstenfeld L, Einhorn T, Tabin CJ, Rosen V (2006) BMP2 activity, although dispensable for bone formation, is required for the initiation of fracture healing. *Nat Genet* 38: 1424–1429

75 Bandyopadhyay A, Tsuji K, Cox K, Harfe BD, Rosen V, Tabin CJ (2006) Genetic analysis of the roles of BMP2, BMP4, and BMP7 in limb patterning and skeletogenesis. *PLoS Genet* 2: e216

76 Wang X, Xiao F, Yang Q, Liang B, Tang Z, Jiang L, Zhu Q, Chang W, Jiang J, Jiang C et al (2006) A novel mutation in GDF-5 causes autosomal dominant symphalangism in two Chinese families. *Am J Med Genet A* 140: 1846–1853

77 Faiyaz-Ul-Haque M, Ahmad W, Wahab A, Haque S, Azim AC, Zaidi SH, Teebi AS, Ahmad M, Cohn DH, Siddique T et al (2002) Frameshift mutation in the cartilage-derived morphogenetic protein 1 (CDMP1) gene and severe acromesomelic chondrodysplasia resembling Grebe-type chondrodysplasia. *Am J Med Genet* 111: 31–37

78 Stelzer C, Winterpacht A, Spranger J, Zabel B (2003) Grebe dysplasia and the spectrum of CDMP1 mutations. *Pediatr Pathol Mol Med* 22: 77–85

79 Balemans W, Patel N, Ebeling M, Van Hul E, Wuyts W, Lacza C, Dioszegi M, Dikkers FG, Hildering P, Willems PJ et al (2002) Identification of a 52 kb deletion downstream of the SOST gene in patients with van Buchem disease. *J Med Genet* 39: 91–97

Signaling cross-talk by bone morphogenetic proteins

Nandini Ghosh-Choudhury[1] and Goutam Ghosh-Choudhury[2]

South Texas Veterans Health Care System and Departments of Pathology[1] and Medicine[2], University of Texas Health Science Center at San Antonio, 7703 Floyd Curl Drive, San Antonio, TX 78229, USA

Introduction

Bone morphogenetic proteins (BMPs) have diverse effect on different cell types and, for many, BMPs play essential role in their survival and differentiation. BMPs are conserved across the animal kingdom and participate in the development of a wide variety of organisms including vertebrates, arthropods and nematodes. In the growing family of BMPs, more than 30 members have been described so far. In vertebrates, BMPs regulate development of many organs, including heart, kidney and bone, by regulating proliferation, differentiation and apoptotic pathways [1–3]. The wide spectrum of biological responses elicited by BMPs in different cellular set ups are dictated by tightly controlled signaling cross-talk between BMP-induced signaling pathways and interaction of divergent intracellular cofactors. In this review, we describe the signal transduction pathways induced by BMPs and discuss the possible cross-talk between them. We focus mainly on BMP signaling in bone with a brief discussion of other cell types that are regulated by BMPs. Figure 1 summarizes the pathways for BMP-induced signal transduction pathways discussed in this review.

BMP receptors and Smad signaling

BMP signaling propagates by activation of a BMP-specific receptor complex composed of the constitutively active BMPR type II (BMPRII), a serine/threonine kinase that, upon BMP binding, phosphorylates BMPR type I (BMPRI) in the Gly-Ser (GS) domain, resulting in its activation [4–6]. Unlike the TGF-β receptors, BMPRI can bind ligands even in the absence of the BMPRII; however, the binding of BMPs to the BMPRI is enhanced by the presence of BMPRII [7–11]. Recently, it has been reported that the localization of the BMPRI cluster at the cell membrane is influenced by the expression of BMPRII receptor cluster. However, the expression of BMPRI does not affect the positioning of the BMPRII on the cell surface [12].

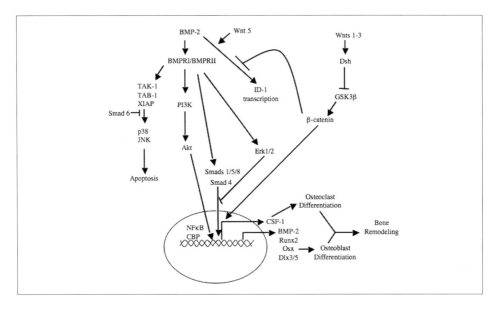

Figure 1
BMP signaling pathways in bone remodeling

Activation of the BMPRI initiates the cascade of activity by phosphorylating Smads 1, 5 and 8 (Fig. 1) These phosphorylated Smads subsequently form complexes with Smad4, a common interacting protein for both the transforming growth factor (TGF)-β and the BMP-regulated Smads. This interaction leads to nuclear transport of the Smad complexes where they execute their role in regulating transcription of specific genes [4, 5, 13, 14]. The BMP-activated Smad signaling is counteracted, or in other words precisely regulated, by the action of two other Smad proteins, Smad6 and -7, that interact with BMPRI or with receptor activated Smads to limit their nuclear localization [15–19]. A recent mechanism of OAZ-induced expression of Smad6 upon BMP-4 treatment highlighted another level of regulation of BMP signaling [20]. OAZ was originally identified in *Xenopus* as a Smad-interacting transcription factor [21]. It contains 30 Kruppel-like zinc fingers and interacts with Smad1–Smad4 complex formed during BMP stimulation to bind to the promoter region and transcriptionally activate Xenopus homeobox gene, *Vent-2* [21]. Upon BMP stimulation OAZ stimulates Smad6 expression and subsequently decreases Smad1 phosphorylation to inhibit BMP signaling [20]. The expression of these inhibitory Smads therefore appears to be a part of a negative feedback loop. The expression pattern of Smads 6 and 7 coincides with those of the Bmps -2 and -4 in *Xenopus* [22–24]. The Smad4–Smad5 complex binds to the Smad6 promoter region upon BMP stimulation to activate its transcription [25]. Therefore, these inhibitory Smads function *via* a feedback loop to regulate the duration of the BMP signaling. The ubiquitin-proteosome pathway also plays a critical role in regulating BMP-induced Smad signaling. Smurf1, an E3 ubiquitin ligase, preferentially binds to the BMP-regulated Smads 1 and 5 and antagonizes BMP signaling in *Xenopus* by degrading these essential downstream effectors of BMPs [26].

Cross-talk of BMP signaling and Erk 1/2 mitogen-activated protein kinase

Signaling cross-talks are the key to ensure proper embryonic development and to maintain tissue homeostasis in adults. It has been reported that the function of the BMPRI-activated Smad1 is negatively regulated by mitogen-activated protein kinase (MAPK) activation by receptor tyrosine kinases for insulin-like growth factor-I (IGF-I), epidermal growth factor (EGF) and fibroblast growth factors (FGFs) [27–30]. The implication of this cross-talk is more related to the neuronal development and perhaps on oncogenic proliferation of cells [29, 31]. Neuronal differentiation requires FGF and IGF signaling and is favored when BMP signaling is compromised [28, 29, 32]. BMP-specific Smad1 is phosphorylated by BMPRIA at the Ser residues in the C-terminal SSVS sequence [33]. Smad1 is phosphorylated even in the absence of BMP signaling in mink lung epithelial cells in the linker region that bridges the highly conserved N- and C-terminal MH1 and MH2 domains, respectively [33]. Four MAPK consensus motifs (PXSP) are present at the linker region of Smad1 that are phosphorylated by EGF and hepatocyte growth factor (HGF) signaling through their tyrosine kinase receptors and evoking strong MAPK activity [27]. It was further reported that this linker-region phosphorylation of Smad1 inhibited its nuclear localization without affecting its interaction with Smad4, implicating a negative regulation of Smad1 transcriptional activity by growth factors activating MAPK signaling [27]. *In vivo* experiments with transgenic mice showed that the Smad1 mutated at two of the linker phosphorylation sites was preferentially retained at the plasma membrane in mouse embryo fibroblasts and in parietal cells, and were not at the expected site, the nucleus [34]. A recent report suggested a regulatory mechanism in which the ubiquitin ligase Smurf1 degrades linker phosphorylated Smad1 and may be important in BMP-induced mouse osteoblast differentiation and in *Xenopus* neural development [35]. However, the disturbing fact in this report is the absence of Erk activation by BMP-2 in HaCaT keratinocytes used for this study, despite the fact that BMP-2 induced Smad1 linker phosphorylation at the MAPK consensus sites [35]. Activation of Erk by BMP-2 has been conclusively documented in human osteoblastic cells and in mouse calvarial osteoblasts [36]. Therefore, the activation of MAPK pathway by BMP-2 can serve as a feedback mechanism for regulation of signal transduction. However, identification of the sites for BMP-mediated phosphorylation of the Smad1 linker region will clarify the importance of the cross-talk between MAPK and Smad signaling pathways.

Smad and p38 cross-talk

BMP-2 binds with high affinity to BMPRI and stimulates its homodimerization and association with BMPRII to preferentially stimulate p38 MAPK signaling [12]. On the other hand, Smad signaling is activated when BMP binds to a preformed BMPRI-RII complex on the plasma membrane [12, 37, 38]. p38 activation by BMP-2 requires association of BMPRI with TAK-1 (TGF-β-activated kinase)-TAB-1 (TGF-β-activated binding protein)-XIAP (X-linked inhibitor of apoptosis) complex [39]. TAK-1, a TGF-β-responsive MAPK kinase kinase (MAPKKK), activates p38, Jun-N terminal kinase and other MAPK [40]. BMP-2 and -4 activate TAK-1, which in turn activates p38 leading to apoptosis (Fig. 1) [41]. Expression of BMP receptor-specific inhibitory Smad6 blocks BMP-2-induced TAK-1 activation and p38 phosphorylation suggesting a complex interaction between Smad and p38 signaling [41]. The *in vivo* relevance of these interactions definitely needs to be examined more closely before assigning significance to cross interactions of these signaling pathways.

BMP and Wnt signaling cross-talk

Wnts are secreted as glycosylated and palmitoylated proteins and play an important role in development and maintenance of many organs and tissues, including bone [42]. Wnt signaling occurs upon binding of Wnt to the seven-domain membrane spanning frizzled receptor and to the low-density lipoprotein receptor-related proteins 5 and 6 (LRP5/6). The relevance of Wnt signaling in bone is emphasized by the bone phenotype of the human mutation in LRP5 gene. Loss-of-function mutations in human LRP5 produce osteoporosis-psuedoglioma characterized by low bone mineral density and skeletal fragility [43]. On the other hand, N-terminal mutations of human LRP5 gene reduce affinity of LRP5 for Dickkopf-1 (Dkk1), an inhibitor of LRP5 activity, resulting in high bone mass phenotype [44–46]. Canonical Wnt proteins (Wnts 1, 2, 3 and 3a) activate cytoplasmic Disheveled (Dsh) protein that in turn inactivates glycogen synthase kinase 3β (GSK3 β) and protects β-catenin from degradation by GSK3 β, axin and adenomatous polyposis coli (APC) protein complex [47–50]. Non-canonical Wnt proteins (Wnts 4, 5a, 5b, 6 and 7a) act independently of β-catenin signaling [51, 52]. While the canonical Wnts induce formation of an active transcription complex consisting of β-catenin and the transcription factors Lef1/Tcf for transcriptional activation of target genes (e.g., c-myc and cyclin D1), the noncanonical Wnts inhibit interaction of DNA with β-catenin-Lef1/Tcf complex by initiating intracellular Ca^{2+} release to activate TAK1 MAPKKK and nemo-like kinase (NLK) by PKC and Ca^{2+}-calmodulin-dependent kinase II (CAM-KII) [51–53]. NLK phosphorylates Lef1/Tcf and inhibits transcriptional activation of β-catenin target genes [54]. Using C2C12 cells stably overexpressing canonical

Wnt3a or non-canonical Wnt 5a, a cross-talk between BMP-2 and Wnt signaling in osteoblast differentiation is documented (Fig. 1) [55]. Activation of β-catenin by Wnt3a inhibited BMP-2-induced expression of Id-1 gene, a negative regulator of muscle differentiation and resulted in suppression of BMP-2 induced inhibition of myotube formation [55]. In contrast, β-catenin-Lef1/Tcf-dependent transcriptional activity was induced by BMP-2 and Smads 1 and 4 [55]. Also, the non-canonical Wnt5a increased BMP-2-induced Id-1 transcription. Taken together, this *in vitro* study suggests a possible role of Wnt 3a signaling in BMP-2-induced mesenchymal cell differentiation. β-Catenin-dependent and -independent activation of Raf-MEK-Erk pathway by Wnt3a is reported in NIH3T3 fibroblasts [56]. Recently, involvement of p38 MAPK was reported in non-canonical Wnt 5a-mediated Ca^{2+} mobilization and activation of NFAT transcription factor in mouse F9 teratocarcinoma cells [57]. Since MAPK signaling pathways are also regulated by BMPs and evidence suggests that BMP and Wnt/β-catenin signaling interact in early *Xenopus* embryogenesis, it is tempting to predict that Wnt signaling acts in parallel to BMPs in bone and embryonic development [58–60]. Further investigation in this interesting area may reveal the mechanism of their integration, involvement in critical regulation and *in vivo* biological significance of such interactions.

Calcium, calmodulin, PKC and Smad signaling

Ca^{2+} signaling is predominantly mediated by its intracellular receptor calmodulin and is part of the classical inositol-phospholipid pathway downstream of certain G-protein-linked receptors. Activated G-proteins (GTP-binding proteins) stimulate phospholipase C (PLC), which in turn hydrolyzes phosphatidylinositol-4,5-bisphosphate (PIP_2) to generate inositol-1,4,5-trisphosphate (IP_3) and diacylglycerol (DAG). DAG activates protein kinase C (PKC) and IP_3 induces Ca^{2+} release from the endoplasmic reticulum, which in turn binds to calmodulin to activate kinases like CAM-KII. The fact that calmodulin can bind Smads 1–4 *in vitro* and in transfected cells in a Ca^{2+}-dependent manner indicates a new input in the BMP signaling pathway [61]. However, the biological outcome of this interaction remains to be proven. *In vivo* experiments in embryonic development suggest that Ca^{2+}-calmodulin might stimulate BMP-Smad signaling [62–64]. PKC activity is induced by BMP-2 in a Smad-independent manner during induction of apoptosis of the immortalized human neonatal calvarial osteoblastic cells [65]. Protein kinase D1 (PKD) is a member of a PKC subgroup characterized by N-terminal cysteine fingers defining the structural basis for lipid-mediated activation and differs from the three major groups of the PKC isozymes (classical, novel and atypical) by the presence of an acidic domain and a plextrin homology (PH) domain [66, 67]. PKD is activated in MC3T3-E1 mouse calvarial cells stimulated with BMP-2, in a PKC-independent manner, as a mechanism to stimulate the activation of stress-activated MAPK, p38 and JNK [68].

These results document the involvement of BMP in engaging a number of critical signaling molecules. It is necessary to follow each of these leads to find a functional consequence *in vivo*. There is a good possibility that all these pathways talk to each other to fine-tune BMP-mediated development and bone remodeling. However, the identification of the master control switch awaits collective inputs from investigators working in this growing field.

The master regulator phosphatidylinositol 3 kinase in BMP-induced bone remodeling

Cell fate and survival is determined by a plethora of signals in which the PI3K/PTEN signaling pathway is considered to be a central integrator of complex signaling networks. It is now established that one of the key signal transduction pathways activated by the cell surface receptors for controlling intracellular events is the class I phosphatidylinositiol 3-kinases (PI3K). The active form of PI3K is composed of a regulatory 85-kDa subunit (p85) and a catalytic subunit of 110 kDa (p110). Activated PI3K catalyzes phosphorylation of the D-3-position of the inositol ring in one or more phosphoinositide species found in mammalian cells [69]. PI3K generates the second messenger lipid phosphatidylinositol-3,4,5-trisphosphate (PIP_3), which in turn recruits phosphatidylinositol-dependent kinase 1 (PDK1) and the critical survival protein kinase B (PKB/Akt) to the membrane *via* its interaction with the PH domains of these two kinases. PDK1 phosphorylates Thr 308 in the catalytic loop, while phosphorylation of Ser 473 in the hydrophobic loop is mediated by the TORC2 containing the mTOR Rictor complex. Both these phosphorylations of Akt are necessary for full activation of Akt. [70–72]. The direct substrates of Akt include p27 (KIP1), the forkhead box transcription factors (FOXO), GSK3, serum and glucocorticoid-induced kinase (SGK1) and tuberous sclerosis complex 2 (TSC2) among many [73, 74]. TSC2/TSC1 heterodimer complex negatively regulates the small GTPase Rheb [75–77]. Rheb in its active GTP-bound form stimulates mammalian target of rapamycin, mTOR in TORC1 complex [76–78].

There are three isoforms of Akt, 1, 2 and 3. Although they perform overlapping functions, Akt1 mediates PI3K signals to promote cell survival and proliferation, while Akt 2 is associated with insulin-mediated metabolic process and Akt 3 controls cell size and number in brain [79–82]. In addition, Akt 1 and Akt 2 are highly expressed in osteoblasts and in osteoclast progenitors [83]. Double knockout mice for Akt 1 and Akt 2 showed severely impaired bone development phenotype and died shortly after birth, indicating a critical role of Akt in bone remodeling [84]. A new dimension for the BMP-induced bone remodeling opened up with our observation that activation of the serine-threonine kinase BMPR stimulated tyrosine phosphorylation of proteins in osteoblasts (Fig. 1) [85]. These results subsequently led to the discovery that BMP-2 strongly stimulates PI3K activity and its downstream target Akt [85]. Activation of these signaling molecules is absolutely necessary for BMP-2-induced osteoblast differentiation [85]. Importantly, this study also suggested cross-talk between PI3K/Akt signaling and Smad signaling in BMP-2-treated osteoblasts [85]. Activation of this signaling pathway is not exclusive to the osteoblast cells. We demonstrated that BMP-2-stimulated cardiomyocyte differentiation is also controlled by this key PI3K signaling pathway [86]. More importantly, we showed that the activation of PI3K was essential for BMP-2-induced cardiomyocyte contractility, thus indicating the physiological significance of this critical signaling pathway in BMP-2-induced differentiation of diverse cell types [87].

The tumor suppressor PTEN (phosphatase with tensin homology deleted in chromosome 10) is a lipid phosphatase that dephosphorylates the second messenger PIP_3 to PIP_2, thereby counteracting PI3K/Akt signaling [88, 89]. As expected, PTEN gene deletion in osteochondroprogenitors resulted in mice with increased bone size and thickness, confirming the central role of PI3K/Akt signaling in regulating skeletal size and bone architecture [90]. This finding is further supported by dramatic increase in bone mineral density in osteoblast specific *Pten* null mice [91]. These recent developments in bone formation put the PI3K/Akt signaling axis at the center of the signaling web in bone remodeling. In fact, the regulation of bone formation by the statins, which act by inducing BMP-2 expression in osteoblasts, activates Ras-mediated PI3K/Akt signaling [92]. Activation of PI3K by statins in turn stimulates Erk MAPK signaling to drive osteoblast differentiation [92]. Therefore, activation of PI3K/Akt signaling by BMP-2 and other bone forming growth factors possibly integrates with Smad and MAPK signaling to control bone formation. BMP-2 plays a critical role in osteoclast formation by inducing expression of one of the key proteins, colony-stimulating factor-1 (CSF-1), for osteoclast formation [93]. Smad signaling is involved in BMP-2-induced expression of CSF-1 and osteoclast formation (Fig. 1) [93]. It will be interesting to investigate the cross-talk of other signaling pathways in BMP-2-induced differentiation of osteoclasts.

BMP signaling in cancer cells

In addition to its role in promoting growth and differentiation of osteoblasts and other cell types, BMP-2 has been reported to inhibit growth of human cancer cells and PDGF- and EGF-induced growth of mesangial cells [94–98]. Work from our laboratory first showed that the treatment of the estrogen-dependent and estrogen-independent human breast cancer cells with BMP-2 arrested the cell cycle progression by inhibiting cyclin-dependent kinase (Cdk) activity, shifting the retinoblastoma tumor suppressor protein to hypophosphorylated active state and by increasing the expression of cyclin-dependent kinase inhibitor, p21/CIP1/WAF1 [96, 97]. The growth of human prostate cancer cells (LNCaP) was also inhibited by recombinant BMP-2 treatment only in the presence of androgen [98]. Incidentally, treatment of LNCaP cells with androgen also increased the expression of BMPRIB mRNA, while the BMPRIA mRNA level remained unchanged [98]. BMP-2 inhibited growth of androgen-sensitive LNCaP cells but had no effect on androgen-insensitive PC3 cells, whereas BMP-7 treatment inhibited growth of PC3 and DU145 cells in an androgen-independent manner [99, 100]. BMP-induced prostate cancer cell growth arrest is carried out by an increase in Smad signaling, increased expression of p21/CIP1/WAF1, decreased activities of cyclin A, Cdk2 and Cdk6 and decreased expression of E2F due to hypophosphorylation of retinoblastoma protein [99–102]. Smad4 was originally identified as a tumor suppressor gene of pancreatic cancer and was named DPC 4 (homogeneously deleted in pancreatic carcinoma, locus 4) [103]. Analysis of human tumor data show that inactivation of Smad4 correlates with the occurrence and metastasis of colorectal cancer and plays a role in acquisition of advanced phenotypes [104]. BMP-7 expression in a human prostate tumor sample was significantly lower than that in normal prostate tissue and the BMP-7 expression in prostate cancer cell lines showed inverse correlation with their tumorigenic and metastatic potential [105]. BMP-7 treatment also significantly reduced the growth of prostate cancer cells in bone [105]. These demonstrate the importance of BMP and Smad signaling pathway in cancer biology and suggest a possible therapeutic intervention.

Transcription factors involved in BMP signaling

A study demonstrating the functional role of PI3K/Akt signaling in BMP-2-induced apoptosis of chondrocytic cells revealed the involvement of this signaling pathway in NF-κB activation [106]. NF-κB transcription factor in turn regulates *BMP-2* gene expression in differentiating chondrocytes [107]. This is a demonstration of fine-tuning of *BMP-2* gene autoregulation, which we reported previously [108]. Functional cooperation of Smad proteins and NF-κB transcription factor in Jun B and in collagen type VII promoters has been reported [109, 110]. Association of Smads and NF-κB with transcriptional coactivators CBP/p300 is essential for DNA binding and subsequent transcriptional activation (Fig. 1) [111–114]. Recently PI3K/Akt signaling was demonstrated to be critical in Smad-CBP interaction, Smad acetylation and transcriptional activation by Smad proteins [115]. Most of these acetylation studies were performed with Smad2 and -3, which are regulated by TGF-β signaling. Similar studies with BMP-regulated Smads may reveal important information in this area. Although *BMP-2* gene expression has been shown to be regulated by Sp1 and Sp3 transcription factors in F9 embryonic carcinoma cells, the role of these transcription factors in BMP-induced differentiation of bone or other cells is yet to be documented [116]. The role of p53 in skeletal development is suggested by the skeletal phenotype of *p53* null mice [117]. BMP-2 restored differentiation potential of p53 null osteoblast cells, demonstrating a role of p53 in BMP expression and subsequent signaling pathways [118]. Runx2 (also known as Cbfa1) is a crucial transcription factor that regulates osteogenesis, and targeted disruption of Runx2 in mice results in complete absence of bone [119]. More recently, osterix (Osx), a novel bone-specific transcription factor with Zn-finger domains, has been shown to be important for terminal differentiation of osteoblasts and *Osx*-null mice also show lack of bone formation [120]. BMP-2 stimulates mRNA expression for both Runx2 and Osx indirectly by inducing Dlx5, a bone inducing homeodomain transcription factor (Fig. 1) [120–123]. BMPs can activate differentiation of central nervous system (CNS) stem cells into a wide variety of dorsal CNS and neural crest cell types. BMP-4 differentiates glial cells through the serine threonine kinase mTOR that associates with STAT3 and activates STAT signaling [124]. Recently, a cross-talk between STAT3 and Smad signaling pathways has been reported in BMP-induced astrogliogenesis, where the leukemia inhibitory factor (LIF) induced BMP expression *via* STAT3 activation, leading to stimulation of Smad signaling to promote astrogliogenic differentiation of neuroepithelial cells [125]. This indicates the complex level of cross-talk between BMP signaling and a variety of transcription factors that are necessary for maintaining the network of differentiation process regulated by BMPs.

Conclusion

The central role of BMPs in growth, differentiation and development of many divergent cell types at various levels of regulation is controlled by a number of signaling pathways and downstream effecter molecules. In this review, an attempt has been made to highlight some of the key players in this complex network. It is evident that the signaling cross-talk activated by BMPs is not limited to Smad- and MAPK-related pathways, but there is a prominent role of the PI3K/Akt signaling, which possibly acts upstream of Smad and MAPK signaling. The contribution of other pathways not yet described for BMP signaling cannot be ruled out, and the possibility of identifying them in near future is very high. This is particularly true since BMP signaling is crucial for embryonic development but needs to be precisely controlled for later growth and differentiation. Therefore, the transient signaling inputs to express BMPs at specific sites need to be closely followed by signals to turn off BMP expression and to counteract signaling initiated by BMPs. This cannot be achieved by the few known straightforward signaling pathways, and necessitates close interaction of the signaling molecules induced by other growth factors that also share the same growth, differentiation and developmental responsibilities with BMPs. With this in mind, we anxiously look forward for cross-talk of many more transcription factors, kinases and phosphatases to be introduced in this exciting network of BMP-related signaling.

Acknowledgements

This work in NGC Laboratory is supported by grants from NIH (RO1 AR52425), Veterans Affairs Medical Research Council (Merit Review) and Morrison Trust. G.G.C. is supported by NIH (RO1 DK50190), Juvenile Diabetes Research Foundation and Veterans Affairs Medical Research Council (Merit Review).

References

1 Hogan BL (1996) Bone morphogenetic proteins in development. *Curr Opin Genet Dev* 6: 432–438
2 Hogan BL (1996) Bone morphogenetic proteins: Multifunctional regulators of vertebrate development. *Genes Dev* 10: 1580–1594
3 Gambaro K, Aberdam E, Virolle T, Aberdam D, Rouleau M (2006) BMP-4 induces a Smad-dependent apoptotic cell death of mouse embryonic stem cell-derived neural precursors. *Cell Death Differ* 13: 1075–1087
4 Kawabata M, Imamura T, Miyazono K (1998) Signal transduction by bone morphogenetic proteins. *Cytokine Growth Factor Rev* 9: 49–61
5 Miyazono K, Maeda S, Imamura T (2005) BMP receptor signaling: Transcriptional targets, regulation of signals, and signaling cross-talk. *Cytokine Growth Factor Rev* 16: 251–263
6 ten Dijke P, Fu J, Schaap P, Roelen BA (2003) Signal transduction of bone morphogenetic proteins in osteoblast differentiation. *J Bone Joint Surg Am* 85-A Suppl 3: 34–38
7 Koenig BB, Cook JS, Wolsing DH, Ting J, Tiesman JP, Correa PE, Olson CA, Pecquet AL, Ventura F, Grant RA et al (1994) Characterization and cloning of a receptor for BMP-2 and BMP-4 from NIH 3T3 cells. *Mol Cell Biol* 14: 5961–5974
8 Nohno T, Ishikawa T, Saito T, Hosokawa K, Noji S, Wolsing DH, Rosenbaum JS (1995) Identification of a human type II receptor for bone morphogenetic protein-4 that forms differential heteromeric complexes with bone morphogenetic protein type I receptors. *J Biol Chem* 270: 22522–22526
9 Rosenzweig BL, Imamura T, Okadome T, Cox GN, Yamashita H, ten Dijke P, Heldin CH, Miyazono K (1995) Cloning and characterization of a human type II receptor for bone morphogenetic proteins. *Proc Natl Acad Sci USA* 92: 7632–7636
10 ten Dijke P, Yamashita H, Sampath TK, Reddi AH, Estevez M, Riddle DL, Ichijo H, Heldin CH, Miyazono K (1994) Identification of type I receptors for osteogenic protein-1 and bone morphogenetic protein-4. *J Biol Chem* 269: 16985–16988
11 Liu F, Ventura F, Doody J, Massague J (1995) Human type II receptor for bone morphogenic proteins (BMPs): Extension of the two-kinase receptor model to the BMPs. *Mol Cell Biol* 15: 3479–3486
12 Nohe A, Hassel S, Ehrlich M, Neubauer F, Sebald W, Henis YI, Knaus P (2002) The

mode of bone morphogenetic protein (BMP) receptor oligomerization determines different BMP-2 signaling pathways. *J Biol Chem* 277: 5330–5338

13 Fujii M, Takeda K, Imamura T, Aoki H, Sampath TK, Enomoto S, Kawabata M, Kato M, Ichijo H, Miyazono K (1999) Roles of bone morphogenetic protein type I receptors and Smad proteins in osteoblast and chondroblast differentiation. *Mol Biol Cell* 10: 3801–3813

14 Miyazono K (1999) Signal transduction by bone morphogenetic protein receptors: Functional roles of Smad proteins. *Bone* 25: 91–93

15 Hayashi H, Abdollah S, Qiu Y, Cai J, Xu YY, Grinnell BW, Richardson MA, Topper JN, Gimbrone MA Jr, Wrana JL et al (1997) The MAD-related protein Smad7 associates with the TGFbeta receptor and functions as an antagonist of TGFbeta signaling. *Cell* 89: 1165–1173

16 Imamura T, Takase M, Nishihara A, Oeda E, Hanai J, Kawabata M, Miyazono K (1997) Smad6 inhibits signalling by the TGF-beta superfamily. *Nature* 389: 622–626

17 Nakao A, Afrakhte M, Moren A, Nakayama T, Christian JL, Heuchel R, Itoh S, Kawabata M, Heldin NE, Heldin CH et al (1997) Identification of Smad7, a TGFbeta-inducible antagonist of TGF-beta signalling. *Nature* 389: 631–635

18 Topper JN, Cai J, Qiu Y, Anderson KR, Xu YY, Deeds JD, Feeley R, Gimeno CJ, Woolf EA, Tayber O et al (1997) Vascular MADs: Two novel MAD-related genes selectively inducible by flow in human vascular endothelium. *Proc Natl Acad Sci USA* 94: 9314–9319

19 Topper JN, Wasserman SM, Anderson KR, Cai J, Falb D, Gimbrone MA Jr (1997) Expression of the bumetanide-sensitive Na-K-Cl cotransporter BSC2 is differentially regulated by fluid mechanical and inflammatory cytokine stimuli in vascular endothelium. *J Clin Invest* 99: 2941–2949

20 Ku M, Howard S, Ni W, Lagna G, Hata A (2006) OAZ regulates bone morphogenetic protein signaling through Smad6 activation. *J Biol Chem* 281: 5277–5287

21 Hata A, Seoane J, Lagna G, Montalvo E, Hemmati-Brivanlou A, Massague J (2000) OAZ uses distinct DNA- and protein-binding zinc fingers in separate BMP-Smad and Olf signaling pathways. *Cell* 100: 229–240

22 Casellas R, Brivanlou AH (1998) *Xenopus* Smad7 inhibits both the activin and BMP pathways and acts as a neural inducer. *Dev Biol* 198: 1–12

23 Nakayama T, Gardner H, Berg LK, Christian JL (1998) Smad6 functions as an intracellular antagonist of some TGF-beta family members during *Xenopus* embryogenesis. *Genes Cells* 3: 387–394

24 Tsuneizumi K, Nakayama T, Kamoshida Y, Kornberg TB, Christian JL, Tabata T (1997) Daughters against dpp modulates dpp organizing activity in *Drosophila* wing development. *Nature* 389: 627–631

25 Ishida W, Hamamoto T, Kusanagi K, Yagi K, Kawabata M, Takehara K, Sampath TK, Kato M, Miyazono K (2000) Smad6 is a Smad1/5-induced smad inhibitor. Characterization of bone morphogenetic protein-responsive element in the mouse Smad6 promoter. *J Biol Chem* 275: 6075–6079

26 Zhu H, Kavsak P, Abdollah S, Wrana JL, Thomsen GH (1999) A SMAD ubiquitin ligase targets the BMP pathway and affects embryonic pattern formation. *Nature* 400: 687–693
27 Kretzschmar M, Doody J, Massague J (1997) Opposing BMP and EGF signalling pathways converge on the TGF-beta family mediator Smad1. *Nature* 389: 618–622
28 Massague J (2003) Integration of Smad and MAPK pathways: A link and a linker revisited. *Genes Dev* 17: 2993–2997
29 Pera EM, Ikeda A, Eivers E, De Robertis EM (2003) Integration of IGF, FGF, and anti-BMP signals *via* Smad1 phosphorylation in neural induction. *Genes Dev* 17: 3023–3028
30 Yue J, Frey RS, Mulder KM (1999) Cross-talk between the Smad1 and Ras/MEK signaling pathways for TGFbeta. *Oncogene* 18: 2033–2037
31 Kretzschmar M, Doody J, Timokhina I, Massague J (1999) A mechanism of repression of TGFbeta/ Smad signaling by oncogenic Ras. *Genes Dev* 13: 804–816
32 De Robertis EM, Larrain J, Oelgeschlager M, Wessely O (2000) The establishment of Spemann's organizer and patterning of the vertebrate embryo. *Nat Rev Genet* 1: 171–181
33 Kretzschmar M, Liu F, Hata A, Doody J, Massague J (1997) The TGF-beta family mediator Smad1 is phosphorylated directly and activated functionally by the BMP receptor kinase. *Genes Dev* 11: 984–995
34 Aubin J, Davy A, Soriano P (2004) *In vivo* convergence of BMP and MAPK signaling pathways: Impact of differential Smad1 phosphorylation on development and homeostasis. *Genes Dev* 18: 1482–1494
35 Sapkota G, Alarcon C, Spagnoli FM, Brivanlou AH, Massague J (2007) Balancing BMP signaling through integrated inputs into the Smad1 linker. *Mol Cell* 25: 441–454
36 Lai CF, Cheng SL (2002) Signal transductions induced by bone morphogenetic protein-2 and transforming growth factor-beta in normal human osteoblastic cells. *J Biol Chem* 277: 15514–15522
37 Gilboa L, Nohe A, Geissendorfer T, Sebald W, Henis YI, Knaus P (2000) Bone morphogenetic protein receptor complexes on the surface of live cells: A new oligomerization mode for serine/threonine kinase receptors. *Mol Biol Cell* 11: 1023–1035
38 Knaus P, Sebald W (2001) Cooperativity of binding epitopes and receptor chains in the BMP/TGFbeta superfamily. *Biol Chem* 382: 1189–1195
39 Yamaguchi K, Nagai S, Ninomiya-Tsuji J, Nishita M, Tamai K, Irie K, Ueno N, Nishida E, Shibuya H, Matsumoto K (1999) XIAP, a cellular member of the inhibitor of apoptosis protein family, links the receptors to TAB1-TAK1 in the BMP signaling pathway. *EMBO J* 18: 179–187
40 Shirakabe K, Yamaguchi K, Shibuya H, Irie K, Matsuda S, Moriguchi T, Gotoh Y, Matsumoto K, Nishida E (1997) TAK1 mediates the ceramide signaling to stress-activated protein kinase/c-Jun N-terminal kinase. *J Biol Chem* 272: 8141–8144
41 Kimura N, Matsuo R, Shibuya H, Nakashima K, Taga T (2000) BMP2–induced apop-

tosis is mediated by activation of the TAK1–p38 kinase pathway that is negatively regulated by Smad6. *J Biol Chem* 275: 17647–17652
42 Cadigan KM, Nusse R (1997) Wnt signaling: A common theme in animal development. *Genes Dev* 11: 3286–3305
43 Gong Y, Slee RB, Fukai N, Rawadi G, Roman-Roman S, Reginato AM, Wang H, Cundy T, Glorieux FH, Lev D et al (2001) LDL receptor-related protein 5 (LRP5) affects bone accrual and eye development. *Cell* 107: 513–523
44 Ai M, Holmen SL, Van Hul W, Williams BO, Warman ML (2005) Reduced affinity to and inhibition by DKK1 form a common mechanism by which high bone mass-associated missense mutations in LRP5 affect canonical Wnt signaling. *Mol Cell Biol* 25: 4946–4955
45 Boyden LM, Mao J, Belsky J, Mitzner L, Farhi A, Mitnick MA, Wu D, Insogna K, Lifton RP (2002) High bone density due to a mutation in LDL-receptor-related protein 5. *N Engl J Med* 346: 1513–1521
46 Van Wesenbeeck L, Cleiren E, Gram J, Beals RK, Benichou O, Scopelliti D, Key L, Renton T, Bartels C, Gong Y et al (2003) Six novel missense mutations in the LDL receptor-related protein 5 (LRP5) gene in different conditions with an increased bone density. *Am J Hum Genet* 72: 763–771
47 Akiyama T (2000) Wnt/beta-catenin signaling. *Cytokine Growth Factor Rev* 11: 273–282
48 Brown JD, Moon RT (1998) Wnt signaling: Why is everything so negative? *Curr Opin Cell Biol* 10: 182–187
49 Ikeda S, Kishida S, Yamamoto H, Murai H, Koyama S, Kikuchi A (1998) Axin, a negative regulator of the Wnt signaling pathway, forms a complex with GSK-3beta and beta-catenin and promotes GSK-3beta-dependent phosphorylation of beta-catenin. *EMBO J* 17: 1371–1384
50 Lee JS, Ishimoto A, Yanagawa S (1999) Characterization of mouse dishevelled (Dvl) proteins in Wnt/Wingless signaling pathway. *J Biol Chem* 274: 21464–21470
51 Huelsken J, Behrens J (2002) The Wnt signalling pathway. *J Cell Sci* 115: 3977–3978
52 Huelsken J, Birchmeier W (2001) New aspects of Wnt signaling pathways in higher vertebrates. *Curr Opin Genet Dev* 11: 547–553
53 Ishitani T, Ninomiya-Tsuji J, Nagai S, Nishita M, Meneghini M, Barker N, Waterman M, Bowerman B, Clevers H, Shibuya H et al (1999) The TAK1-NLK-MAPK-related pathway antagonizes signalling between beta-catenin and transcription factor TCF. *Nature* 399: 798–802
54 Ishitani T, Kishida S, Hyodo-Miura J, Ueno N, Yasuda J, Waterman M, Shibuya H, Moon RT, Ninomiya-Tsuji J, Matsumoto K (2003) The TAK1-NLK mitogen-activated protein kinase cascade functions in the Wnt-5a/Ca(2+) pathway to antagonize Wnt/beta-catenin signaling. *Mol Cell Biol* 23: 131–139
55 Nakashima A, Katagiri T, Tamura M (2005) Cross-talk between Wnt and bone morphogenetic protein 2 (BMP-2) signaling in differentiation pathway of C2C12 myoblasts. *J Biol Chem* 280: 37660–37668

56 Yun MS, Kim SE, Jeon SH, Lee JS, Choi KY (2005) Both ERK and Wnt/beta-catenin pathways are involved in Wnt3a-induced proliferation. *J Cell Sci* 118: 313–322

57 Ma L, Wang HY (2007) Mitogen-activated protein kinase p38 regulates the Wnt/cyclic GMP/Ca^{2+} non-canonical pathway. *J Biol Chem* 282: 28980–28990

58 Baker JC, Beddington RS, Harland RM (1999) Wnt signaling in *Xenopus* embryos inhibits bmp4 expression and activates neural development. *Genes Dev* 13: 3149–3159

59 Hoppler S, Moon RT (1998) BMP-2/-4 and Wnt-8 cooperatively pattern the *Xenopus* mesoderm. *Mech Dev* 71: 119–129

60 Marom K, Fainsod A, Steinbeisser H (1999) Patterning of the mesoderm involves several threshold responses to BMP-4 and Xwnt-8. *Mech Dev* 87: 33–44

61 Zimmerman CM, Kariapper MS, Mathews LS (1998) Smad proteins physically interact with calmodulin. *J Biol Chem* 273: 677–680

62 Creton R, Kreiling JA, Jaffe LF (2000) Presence and roles of calcium gradients along the dorsal-ventral axis in *Drosophila* embryos. *Dev Biol* 217: 375–385

63 Kuhl M, Sheldahl LC, Malbon CC, Moon RT (2000) Ca(2+)/calmodulin-dependent protein kinase II is stimulated by Wnt and Frizzled homologs and promotes ventral cell fates in *Xenopus*. *J Biol Chem* 275: 12701–12711

64 Kume S, Muto A, Inoue T, Suga K, Okano H, Mikoshiba K (1997) Role of inositol 1,4,5-trisphosphate receptor in ventral signaling in *Xenopus* embryos. *Science* 278: 1940–1943

65 Hay E, Lemonnier J, Fromigue O, Marie PJ (2001) Bone morphogenetic protein-2 promotes osteoblast apoptosis through a Smad-independent, protein kinase C-dependent signaling pathway. *J Biol Chem* 276: 29028–29036

66 Gschwendt M, Johannes FJ, Kittstein W, Marks F (1997) Regulation of protein kinase Cmu by basic peptides and heparin. Putative role of an acidic domain in the activation of the kinase. *J Biol Chem* 272: 20742–20746

67 Valverde AM, Sinnett-Smith J, Van Lint J, Rozengurt E (1994) Molecular cloning and characterization of protein kinase D: A target for diacylglycerol and phorbol esters with a distinctive catalytic domain. *Proc Natl Acad Sci USA* 91: 8572–8576

68 Lemonnier J, Ghayor C, Guicheux J, Caverzasio J (2004) Protein kinase C-independent activation of protein kinase D is involved in BMP-2-induced activation of stress mitogen-activated protein kinases JNK and p38 and osteoblastic cell differentiation. *J Biol Chem* 279: 259–264

69 Cantley LC (2002) The phosphoinositide 3-kinase pathway. *Science* 296: 1655–1657

70 Alessi DR, James SR, Downes CP, Holmes AB, Gaffney PR, Reese CB, Cohen P (1997) Characterization of a 3-phosphoinositide-dependent protein kinase which phosphorylates and activates protein kinase Balpha. *Curr Biol* 7: 261–269

71 Sarbassov DD, Guertin DA, Ali SM, Sabatini DM (2005) Phosphorylation and regulation of Akt/PKB by the rictor-mTOR complex. *Science* 307: 1098–1101

72 Stephens L, Anderson K, Stokoe D, Erdjument-Bromage H, Painter GF, Holmes AB, Gaffney PR, Reese CB, McCormick F, Tempst P et al (1998) Protein kinase B kinases

that mediate phosphatidylinositol 3,4,5-trisphosphate-dependent activation of protein kinase B. *Science* 279: 710–714
73   Brazil DP, Park J, Hemmings BA (2002) PKB binding proteins. Getting in on the Akt. *Cell* 111: 293–303
74   Datta SR, Brunet A, Greenberg ME (1999) Cellular survival: A play in three Akts. *Genes Dev* 13: 2905–2927
75   Holland EC, Sonenberg N, Pandolfi PP, Thomas G (2004) Signaling control of mRNA translation in cancer pathogenesis. *Oncogene* 23: 3138–3144
76   Long X, Lin Y, Ortiz-Vega S, Yonezawa K, Avruch J (2005) Rheb binds and regulates the mTOR kinase. *Curr Biol* 15: 702–713
77   Long X, Ortiz-Vega S, Lin Y, Avruch J (2005) Rheb binding to mammalian target of rapamycin (mTOR) is regulated by amino acid sufficiency. *J Biol Chem* 280: 23433–23436
78   Inoki K, Corradetti MN, Guan KL (2005) Dysregulation of the TSC-mTOR pathway in human disease. *Nat Genet* 37: 19–24
79   Bellacosa A, Testa JR, Moore R, Larue L (2004) A portrait of AKT kinases: Human cancer and animal models depict a family with strong individualities. *Cancer Biol Ther* 3: 268–275
80   Garofalo RS, Orena SJ, Rafidi K, Torchia AJ, Stock JL, Hildebrandt AL, Coskran T, Black SC, Brees DJ, Wicks JR et al (2003) Severe diabetes, age-dependent loss of adipose tissue, and mild growth deficiency in mice lacking Akt2/PKB beta. *J Clin Invest* 112: 197–208
81   Stiles B, Gilman V, Khanzenzon N, Lesche R, Li A, Qiao R, Liu X, Wu H (2002) Essential role of AKT-1/protein kinase B alpha in PTEN-controlled tumorigenesis. *Mol Cell Biol* 22: 3842–3851
82   Tschopp O, Yang ZZ, Brodbeck D, Dummler BA, Hemmings-Mieszczak M, Watanabe T, Michaelis T, Frahm J, Hemmings BA (2005) Essential role of protein kinase B gamma (PKB gamma/Akt3) in postnatal brain development but not in glucose homeostasis. *Development* 132: 2943–2954
83   Kawamura N, Kugimiya F, Oshima Y, Ohba S, Ikeda T, Saito T, Shinoda Y, Kawasaki Y, Ogata N, Hoshi K et al (2007) Akt1 in osteoblasts and osteoclasts controls bone remodeling. *PLoS ONE* 2: e1058
84   Peng XD, Xu PZ, Chen ML, Hahn-Windgassen A, Skeen J, Jacobs J, Sundararajan D, Chen WS, Crawford SE, Coleman KG et al (2003) Dwarfism, impaired skin development, skeletal muscle atrophy, delayed bone development, and impeded adipogenesis in mice lacking Akt1 and Akt2. *Genes Dev* 17: 1352–1365
85   Ghosh-Choudhury N, Abboud SL, Nishimura R, Celeste A, Mahimainathan L, Choudhury GG (2002) Requirement of BMP-2-induced phosphatidylinositol 3-kinase and Akt serine/threonine kinase in osteoblast differentiation and Smad-dependent BMP-2 gene transcription. *J Biol Chem* 277: 33361–33368
86   Ghosh-Choudhury N, Abboud SL, Mahimainathan L, Chandrasekar B, Choudhury GG (2003) Phosphatidylinositol 3-kinase regulates bone morphogenetic protein-2 (BMP-2)-

induced myocyte enhancer factor 2A-dependent transcription of BMP-2 gene in cardiomyocyte precursor cells. *J Biol Chem* 278: 21998–22005

87 Ghosh-Choudhury N, Abboud SL, Chandrasekar B, Ghosh Choudhury G (2003) BMP-2 regulates cardiomyocyte contractility in a phosphatidylinositol 3 kinase-dependent manner. *FEBS Lett* 544: 181–184

88 Stambolic V, Suzuki A, de la Pompa JL, Brothers GM, Mirtsos C, Sasaki T, Ruland J, Penninger JM, Siderovski DP, Mak TW (1998) Negative regulation of PKB/Akt-dependent cell survival by the tumor suppressor PTEN. *Cell* 95: 29–39

89 Maehama T, Dixon JE (1998) The tumor suppressor, PTEN/MMAC1, dephosphorylates the lipid second messenger, phosphatidylinositol 3,4,5-trisphosphate. *J Biol Chem* 273: 13375–13378

90 Ford-Hutchinson AF, Ali Z, Lines SE, Hallgrimsson B, Boyd SK, Jirik FR (2007) Inactivation of Pten in osteo-chondroprogenitor cells leads to epiphyseal growth plate abnormalities and skeletal overgrowth. *J Bone Miner Res* 22: 1245–1259

91 Liu X, Bruxvoort KJ, Zylstra CR, Liu J, Cichowski R, Faugere MC, Bouxsein ML, Wan C, Williams BO, Clemens TL (2007) Lifelong accumulation of bone in mice lacking Pten in osteoblasts. *Proc Natl Acad Sci USA* 104: 2259–2264

92 Ghosh-Choudhury N, Mandal CC, Choudhury GG (2007) Statin-induced Ras activation integrates the phosphatidylinositol 3-kinase signal to Akt and MAPK for bone morphogenetic protein-2 expression in osteoblast differentiation. *J Biol Chem* 282: 4983–4993

93 Ghosh-Choudhury N, Singha PK, Woodruff K, St Clair P, Bsoul S, Werner SL, Choudhury GG (2006) Concerted action of Smad and CREB-binding protein regulates bone morphogenetic protein-2-stimulated osteoblastic colony-stimulating factor-1 expression. *J Biol Chem* 281: 20160–20170

94 Ghosh Choudhury G, Jin DC, Kim Y, Celeste A, Ghosh-Choudhury N, Abboud HE (1999) Bone morphogenetic protein-2 inhibits MAPK-dependent Elk-1 transactivation and DNA synthesis induced by EGF in mesangial cells. *Biochem Biophys Res Commun* 258: 490–496

95 Ghosh Choudhury G, Kim YS, Simon M, Wozney J, Harris S, Ghosh-Choudhury N, Abboud HE (1999) Bone morphogenetic protein 2 inhibits platelet-derived growth factor-induced c-fos gene transcription and DNA synthesis in mesangial cells. Involvement of mitogen-activated protein kinase. *J Biol Chem* 274: 10897–10902

96 Ghosh-Choudhury N, Ghosh-Choudhury G, Celeste A, Ghosh PM, Moyer M, Abboud SL, Kreisberg J (2000) Bone morphogenetic protein-2 induces cyclin kinase inhibitor p21 and hypophosphorylation of retinoblastoma protein in estradiol-treated MCF-7 human breast cancer cells. *Biochim Biophys Acta* 1497: 186–196

97 Ghosh-Choudhury N, Woodruff K, Qi W, Celeste A, Abboud SL, Ghosh Choudhury G (2000) Bone morphogenetic protein-2 blocks MDA MB 231 human breast cancer cell proliferation by inhibiting cyclin-dependent kinase-mediated retinoblastoma protein phosphorylation. *Biochem Biophys Res Commun* 272: 705–711

98 Ide H, Yoshida T, Matsumoto N, Aoki K, Osada Y, Sugimura T, Terada M (1997)

Growth regulation of human prostate cancer cells by bone morphogenetic protein-2. *Cancer Res* 57: 5022–5027

99 Kim IY, Lee DH, Lee DK, Ahn HJ, Kim MM, Kim SJ, Morton RA (2004) Loss of expression of bone morphogenetic protein receptor type II in human prostate cancer cells. *Oncogene* 23: 7651–7659

100 Miyazaki H, Watabe T, Kitamura T, Miyazono K (2004) BMP signals inhibit proliferation and *in vivo* tumor growth of androgen-insensitive prostate carcinoma cells. *Oncogene* 23: 9326–9335

101 Brubaker KD, Corey E, Brown LG, Vessella RL (2004) Bone morphogenetic protein signaling in prostate cancer cell lines. *J Cell Biochem* 91: 151–160

102 Tomari K, Kumagai T, Shimizu T, Takeda K (2005) Bone morphogenetic protein-2 induces hypophosphorylation of Rb protein and repression of E2F in androgen-treated LNCaP human prostate cancer cells. *Int J Mol Med* 15: 253–258

103 Hahn SA, Schutte M, Hoque AT, Moskaluk CA, da Costa LT, Rozenblum E, Weinstein CL, Fischer A, Yeo CJ, Hruban RH et al (1996) DPC4, a candidate tumor suppressor gene at human chromosome 18q21.1. *Science* 271: 350–353

104 Miyaki M, Iijima T, Konishi M, Sakai K, Ishii A, Yasuno M, Hishima T, Koike M, Shitara N, Iwama T et al (1999) Higher frequency of Smad4 gene mutation in human colorectal cancer with distant metastasis. *Oncogene* 18: 3098–3103

105 Buijs JT, Rentsch CA, van der Horst G, van Overveld PG, Wetterwald A, Schwaninger R, Henriquez NV, Ten Dijke P, Borovecki F, Markwalder R et al (2007) BMP7, a putative regulator of epithelial homeostasis in the human prostate, is a potent inhibitor of prostate cancer bone metastasis *in vivo*. *Am J Pathol* 171: 1047–1057

106 Sugimori K, Matsui K, Motomura H, Tokoro T, Wang J, Higa S, Kimura T, Kitajima I (2005) BMP-2 prevents apoptosis of the N1511 chondrocytic cell line through PI3K/Akt-mediated NF-kappaB activation. *J Bone Miner Metab* 23: 411–419

107 Feng JQ, Xing L, Zhang JH, Zhao M, Horn D, Chan J, Boyce BF, Harris SE, Mundy GR, Chen D (2003) NF-kappaB specifically activates BMP-2 gene expression in growth plate chondrocytes *in vivo* and in a chondrocyte cell line *in vitro*. *J Biol Chem* 278: 29130–29135

108 Ghosh-Choudhury N, Choudhury GG, Harris MA, Wozney J, Mundy GR, Abboud SL, Harris SE (2001) Autoregulation of mouse BMP-2 gene transcription is directed by the proximal promoter element. *Biochem Biophys Res Commun* 286: 101–108

109 Kon A, Vindevoghel L, Kouba DJ, Fujimura Y, Uitto J, Mauviel A (1999) Cooperation between SMAD and NF-kappaB in growth factor regulated type VII collagen gene expression. *Oncogene* 18: 1837–1844

110 Lopez-Rovira T, Chalaux E, Rosa JL, Bartrons R, Ventura F (2000) Interaction and functional cooperation of NF-kappa B with Smads. Transcriptional regulation of the junB promoter. *J Biol Chem* 275: 28937–28946

111 Chen LF, Greene WC (2003) Regulation of distinct biological activities of the NF-kappaB transcription factor complex by acetylation. *J Mol Med* 81: 549–557

112 Simonsson M, Kanduri M, Gronroos E, Heldin CH, Ericsson J (2006) The DNA bind-

ing activities of Smad2 and Smad3 are regulated by coactivator-mediated acetylation. *J Biol Chem* 281: 39870–39880

113 Inoue Y, Itoh Y, Abe K, Okamoto T, Daitoku H, Fukamizu A, Onozaki K, Hayashi H (2007) Smad3 is acetylated by p300/CBP to regulate its transactivation activity. *Oncogene* 26: 500–508

114 Tu AW, Luo K (2007) Acetylation of Smad2 by the co-activator p300 regulates activin and transforming growth factor beta response. *J Biol Chem* 282: 21187–21196

115 Das F, Ghosh-Choudhury N, Venkatesan B, Li X, Mahimainathan L, Choudhury GG (2007) Akt kinase targets association of CBP with SMAD 3 to regulate TGFbeta-induced expression of plasminogen activator inhibitor-1. *J Cell Physiol* 214: 513–527

116 Xu J, Rogers MB (2007) Modulation of bone morphogenetic protein (BMP) 2 gene expression by Sp1 transcription factors. *Gene* 392: 221–229

117 Ohyama K, Chung CH, Chen E, Gibson CW, Misof K, Fratzl P, Shapiro IM (1997) p53 influences mice skeletal development. *J Craniofac Genet Dev Biol* 17: 161–171

118 Ghosh-Choudhury N, Harris MA, Wozney J, Mundy GR, Harris SE (1997) Clonal osteoblastic cell lines from p53 null mouse calvariae are immortalized and dependent on bone morphogenetic protein 2 for mature osteoblastic phenotype. *Biochem Biophys Res Commun* 231: 196–202

119 Komori T, Yagi H, Nomura S, Yamaguchi A, Sasaki K, Deguchi K, Shimizu Y, Bronson RT, Gao YH, Inada M et al (1997) Targeted disruption of Cbfa1 results in a complete lack of bone formation owing to maturational arrest of osteoblasts. *Cell* 89: 755–764

120 Nakashima K, Zhou X, Kunkel G, Zhang Z, Deng JM, Behringer RR, de Crombrugghe B (2002) The novel zinc finger-containing transcription factor osterix is required for osteoblast differentiation and bone formation. *Cell* 108: 17–29

121 Lee KS, Kim HJ, Li QL, Chi XZ, Ueta C, Komori T, Wozney JM, Kim EG, Choi JY, Ryoo HM et al (2000) Runx2 is a common target of transforming growth factor beta1 and bone morphogenetic protein 2, and cooperation between Runx2 and Smad5 induces osteoblast-specific gene expression in the pluripotent mesenchymal precursor cell line C2C12. *Mol Cell Biol* 20: 8783–8792

122 Lee MH, Javed A, Kim HJ, Shin HI, Gutierrez S, Choi JY, Rosen V, Stein JL, van Wijnen AJ, Stein GS et al (1999) Transient upregulation of CBFA1 in response to bone morphogenetic protein-2 and transforming growth factor beta1 in C2C12 myogenic cells coincides with suppression of the myogenic phenotype but is not sufficient for osteoblast differentiation. *J Cell Biochem* 73: 114–125

123 Ryoo HM, Lee MH, Kim YJ (2006) Critical molecular switches involved in BMP-2-induced osteogenic differentiation of mesenchymal cells. *Gene* 366: 51–57

124 Rajan P, Panchision DM, Newell LF, McKay RD (2003) BMPs signal alternately through a SMAD or FRAP-STAT pathway to regulate fate choice in CNS stem cells. *J Cell Biol* 161: 911–921

125 Fukuda S, Abematsu M, Mori H, Yanagisawa M, Kagawa T, Nakashima K, Yoshimura A, Taga T (2007) Potentiation of astrogliogenesis by STAT3-mediated activation of bone morphogenetic protein-Smad signaling in neural stem cells. *Mol Cell Biol* 27: 4931–4937

The role and mechanisms of bone morphogenetic protein 4 and 2 (BMP-4 and BMP-2) in postnatal skeletal development

Stephen E. Harris[1], Wuchen Yang[1], Jelica Gluhak-Heinrich[1], Dayong Guo[2], Xiao-Dong Chen[1], Marie A. Harris[1], Holger Kulessa[3†], Brigid L. M. Hogan[4], Alex Lichtler[5], Barbara E. Kream[5], Jianhong Zhang[3], Jian Q. Feng[6], Gregory R. Mundy[3], James Edwards[3] and Yuji Mishina[6]

[1]University of Texas Health Science Center at San Antonio, San Antonio, TX 78229, USA; [2]University of Missouri at Kansas City, KC, MO 64108, USA; [3]Vanderbilt University, Nashville, TN 37232, USA; ; [4]Duke University, Durham, NC 27708, USA; [5]University of Connecticut Medical Center, Farmington, CT 06032, USA; [6]National Institute of Environmental Health Science, Research Triangle Park, NC 27709, USA

Dedication

This paper is dedicated to Holger Kulessa who developed the BMP-4 floxed mouse model used in these studies.

Introduction

There is accumulating evidence that BMP-2 and BMP-4 expression patterns have unique and selective domains of expression in a variety of tissues [1]. During osteoblast differentiation *in vitro*, BMP-4 and BMP-2 are expressed at different levels and at different times [2]. The original knockouts of *bmp-2* and *bmp-4* affect different embryonic pathways and tissues. Extensive DNA sequence comparison of the conserved non-coding regions from Fugu fish, chick, and 11 different mammalian species shows islands of high sequence of conservation in the *cis*-regulatory regions, while little homology of the *cis*-regulatory regions between BMP-4 and BMP-2 exists [3].

BMP signaling through different BMP receptors can have a profound effect on the pathways of growth, death and differentiation of not only osteoblasts [4, 5], but also neural stem cells [6]. *In vitro* and *in vivo*, osteoblasts that contain a dominant negative BMP receptor 1B have greatly reduced levels of BMP-2 and BMP-4, suggesting that both BMP-4 and BMP-2 are important for both osteoblast growth and differentiation. Recent evidence in single- and double-knockout studies of BMP receptor 1A and 1B during retinal development has shown both redundant roles and unique roles for these two BMP signaling receptors [7].

Recent experiments in knocking out *bmp-2* or *bmp-4* in early limb bud mesenchyme, using the Prx1-Cre model, have shown that there is little phenotype in removal of either one of these BMPs. However, removal of both BMP-2 and BMP-4 in limb bud mesenchyme with Prx1-Cre in the early limb bud results in severe defects in osteogenesis. Thus, BMP-2 and BMP-4 may be redundant to a certain degree, or have unique interacting roles that depend on the presence of a certain threshold level for both BMP-2 and BMP-4 ligands [8].

The mutated *Bmp-2* gene was identified as a major risk factor for osteoporosis and associated bone fractures in a recent study of osteoporotic patients (both male and female), before and after menopause from 207 pedigrees in Iceland, [9]. Another recent study of related and overlapping markers in another population found no association of BMP-2 with peak bone mass [10]. However, in a Chinese population, an association of several SNP polymorphisms in the *Bmp-2* gene was found associated with population-based BMD changes, when stratified according to BMI, age, and sex, and menopausal state [11]. A recent study has now found that a coding polymorphism in the *Bmp-4* gene is strongly associated with hipbone density in postmenopausal women [12].

Developing animal models with altered BMP pathways and how the BMP pathways are linked to Wnt and Hh signaling will lead to new and effective diagnostic methods to identify osteoporotic patients who would best benefit with a specific drug or treatment that alters key nodes of these BMP/Wnt/Hh pathways.

Targeted deletion of BMP-4 in early osteoblasts

In the normal *bmp-4* knockout model, the embryos do not develop past 8.5 days post conception (dpc) [13]. In the genetic background 129/SvEv × black Swiss, some embryos survive until the early somite stage at 9.5 dpc. The alternative to a traditional Bmp-4ckO was Bmp-4 conditional knockout (cKO). Bmp-4cKO in osteoblasts was created by crossing two mouse models. The first model was *bmp-4* floxed (*bmp-4-fx*) mouse with loxP sites that flank exon 4 of the *bmp-4* gene [14]. The Cre mouse model used for generating Bmp-4cKO was a 3.6 Col1a1-Cre mouse model characterized and described by Liu et al. [15]. The Cre–loxP system is well established and works for a variety of genomic manipulations in mice [16–18]. After crossing *bmp-4-fx* homozygous mice and 3.6 Col1a1-Cre mice that contain at least one *bmp-4* fx allele or one *bmp-4* knockout allele, the resulting Bmp-4cKO mouse model was a 3.6 Col1a1-cre Bmp-4cKO mouse that allowed selective deletion of Bmp-4 in osteoblast and odontoblast lineages. We could now to begin to uncover roles for BMP-4 in osteoblast biology *in vivo* [19]. Bmp-4cKO mice described above were used to show involvement of BMP-4 during the early postnatal skeletal formation and the consequences of deleting BMP-4 in osteoblast up to 1.5 years.

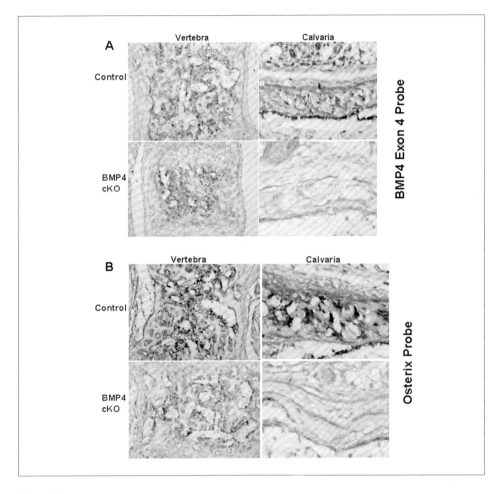

Figure 1
BMP-4 expression in osteoblasts is decreased by >80% using the 3.6 Col1a1-Cre and bmp-4 floxed mouse models (Bmp-4cKO). Osterix expression is also reduced in the bmp-4cKO animals. (A) In situ hybridization of bmp-4 exon 4 probe, newborn animals, vertebrae and calvaria (10×). (B) Osterix expression in control and Bmp-4cKO vertebrae and calvariae, 2-day-old animals (10×).

BMP-4 ablation during early and late postnatal skeletal development

The role and potential mechanisms of BMP-4 action during tooth cytodifferentiation and secretory stages of tooth development were investigated utilizing this 3.6 Col1a1-cre Bmp-4cKO mouse system. Experimentally, in this model, the *bmp-4* gene was abolished after birth. Quantitative *in situ* localization of Bmp-4, using

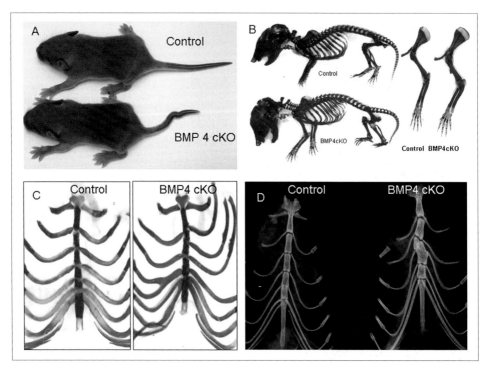

Figure 2
Bmp-4cKO mice are 10–20% smaller with a kinky tail, and an overall normal skeleton at birth with few patterning defects. (A) A 5-day-old Bmp-4cKO mouse clearly shows kinky tail (lower) compared to its control (upper). (B) Alizarin red and alcian blue staining of whole skeleton of control and Bmp-4cKO mice showing an overall normal pattern with delay in mineralization of skull and high frequency of polydactyl. (C) Alizarin red and alcian blue staining also reveals delayed mineralization and disorganized sternum and ribs conjunction at birth (right), as compared to control heterozygotes (left). (D) Distorted sternum is also observed in some 4-month-old mouse (right) versus its control (left).

a exon 4), demonstrated a greater than 90% reduction in Bmp-4 expression in osteoblasts in alveolar bone and in long-bone osteoblastic 18-day-old Bmp-4cKO mice (Fig. 1A). Expression of the BMP-regulated transcription factor, osterix, was also reduced greater than 80% in the osteoblasts of the Bmp-4cKO animals (Fig. 1B).

The Bmp-4cKO animals are 10–20% smaller and also have osteopenia that develops rapidly within the first 2–3 weeks of life. Of the 25% Bmp-4cKO animals expected, only about 18% were born. Thus, some of the Bmp-4cKO animals died prenatally; nevertheless, we had sufficient animals to carry out our postnatal studies.

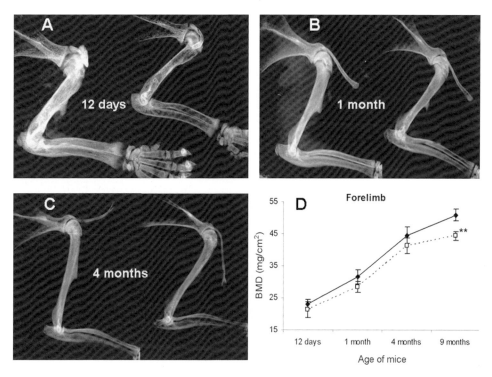

Figure 3
Reduced radio-opacity and BMD in the forearm of Bmp-4cKO mice; 12 day to 9 month study. Left is control heterozygotes and right is Bmp-4cKO. (A) 12 days, (B) 1 month, and (C) 4 months. (D) BMD in forearm from 12 days to 9 months.

As shown in Figure 2A, the Bmp-4cKO mice were slightly smaller and had a kinky tail. Alizarin red/alcian blue analysis of newborn mice showed an overall normal appearing skeleton with delays in mineralization of some of the skull bones and ribs, and a 50% incidence of polydactyly or extra nubbin (Fig. 2B). Figure 2C shows some of the altered rib patterning defects observed in these newborn animals. Figure 2D shows an X-ray of the ribs of a 4-month-old animal with the persistent abnormal patterning of the sternum. Overall, the Bmp-4cKO animals were healthy and have survived up to 1.5 years.

The mice began to develop an osteopenic phenotype very rapidly, within a few days after birth, as observed by x-ray and bone mineral density (BMD) measurements. Figure 3A–C shows the x-ray patterns of the forearm from 12 days to 4 months. Note the overall decrease in radio-opacity and overall decreased diameter of the humerus, radius, and ulna. There was an overall 10–15% decrease in BMD at 12 days that remained low up to 9 months. From the x-ray analysis of the spine

183

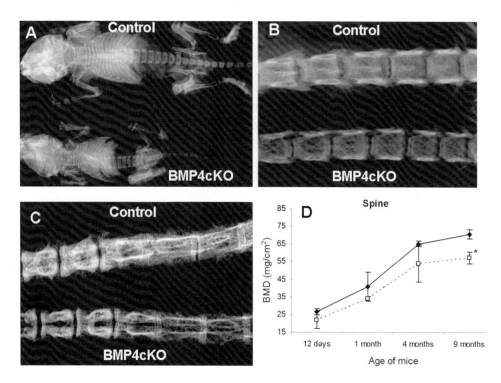

Figure 4
The vertebrae are most sensitive to loss of BMP-4. Left is control heterozygotes and right is Bmp-4cKO; (A) 12 days,. (B) 1 month, and (C) 4 months. (D) BMD in spine and tail region from 12 days to 9 months.

and tail region, these regions appeared to be most susceptible to loss of BMP-4. As noted in Figure 4A, the vertebrae are smaller and show less radio-opacity. Figure 4B and C show x-rays of the tail region with dramatically reduced radio-opacity and vertebral size, at 1 month (B) and at 4 months (C). Figure 4D shows an overall 30% reduction in BMD from 12 days up to 9 months of age. In animals older that 1 month, the analysis was carried out with sex-matched littermates.

Histomorphometric analysis of the distal femur region of 12-day-old animals revealed a 50% decrease in trabecular bone volume (BV)/total volume (TV), and trabecular number per mm, and a 30–40% decrease in cortical thickness, as shown in Figure 5A. Analyses of the radius and ulnae regions showed by ultrasound computed tomography (uCT) indicate an overall 40% reduction in bone volume fraction (BVF), with a 33% reduction in cortical thickness, similar to that observed in the femur by standard histology (Fig. 5B).

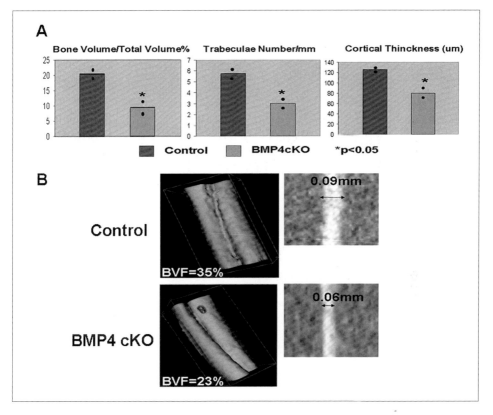

Figure 5
Decrease of approximately 50% in bone volume (BV)/total volume (TV), trabecular number, and cortical thickness in 12-day Bmp-4cKO animal femurs and ultrasound (u) CT analysis of radius-ulna of same animals. (A) Histomorphometric analysis of bones from Bmp-4cKO, left, middle and right panels. (B) MicroCT showed a smaller bone volume fraction (BVF) in 12-day-old Bmp-4cKO animals. BVF of humerus is reduced from 35% in control (upper) to 23% in radius-ulna of cKO (lower). The cortical thickness is reduced 33%.

Using an acid-etch technique with plastic-embedded tibia and fibula from 1-month-old animals, combined with scanning electron microscope analysis, where the acid treatment of the polished surface removes areas that are more mineralized, we observed an overall reduced cortical thickness as shown in Figure 6A, panels A and B, Bmp-4cKO and control, respectively. At higher magnification, it was apparent that the mineral surface of the Bmp-4cKO was altered compared to the control, as seen in Figure 6A, panels C and D. There were pits and areas that were more easily removed by the acid, which gave a rough appearance in the Bmp-4cKO

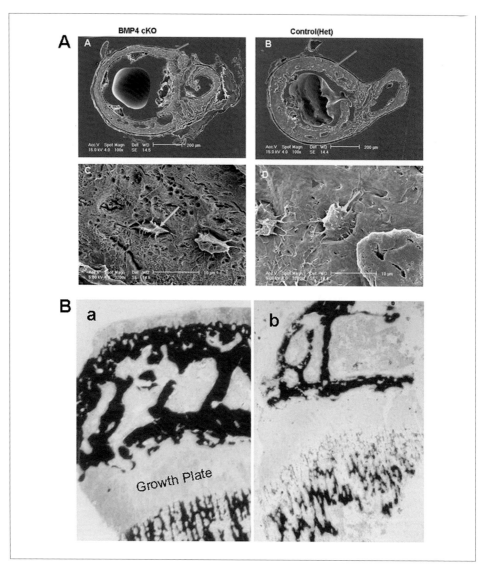

Figure 6
Altered mineral quality in the Bmp-4cKO animals. (A) SEM images reveal different mineral structures of Bmp-4cKO and control bones. After acid etching of plastic-embedded sections, the extrusions in the SEM image represent areas where the plastic has penetrated into the sample, while the embossed areas were previously occupied by mineral contents. Tibia and fibula from Bmp-4cKO mouse (left) and Control heterozygotes (right), shown at low (A, B) and high (C, D) magnification. (B) Alizarin red staining of plastic-embedded 1-mm sections suggests reduced mineral content in the proximal tibia of a 1-month-old Bmp-4cKO mouse compared to its control heterozygous littermate. (a) Control, 10×. (b) Bmp-4cKO, 10×.

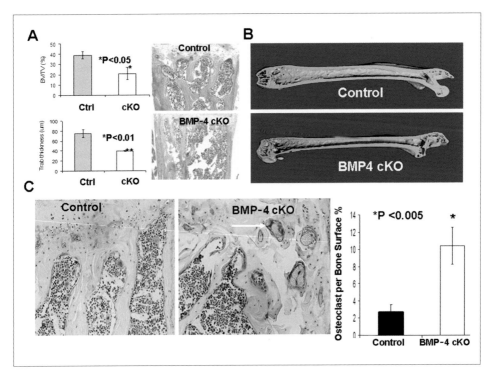

Figure 7
Bmp-4cKO studies at 9 months, n=3. (A) Reduced BV/TV and trabecular thickness in the BMP-4cKO mice compared to control heterozygotes. (B) uCT analysis of femur from 9-month control and Bmp-4cKO animals. Note the overall smaller diameter and reduced cortical thickness. (C) Osteoclast number and activity are increased in the Bmp-4cKO animals compared to control. Left, trap staining of control. Middle, trap staining of Bmp-4cKO. Right, quantitation of the percent of the bone surface covered by osteoclasts.

compared to the heterozygote control bone surface. Using 1-mm plastic-embedded sections stained with alizarin red, we observed a much less intense staining in these proximal tibia sections, suggesting an altered mineral to matrix quality of the bone. This difference in staining in 2-month-old tibia was much less pronounced and suggested that this mineral quality parameter may be transient; this requires further analysis.

One set of animals was allowed to age for 9 months, with three control heterozygotes and three Bmp-4cKO. As shown in Figure 7A, a reduced BV/TV and reduced trabecular number was observed. By uCT analysis of the femur, bones with smaller diameter and thinner cortical thickness were observed (Fig. 7B). When we looked at osteoclast number and area covering the bone surface, we found a fivefold increase

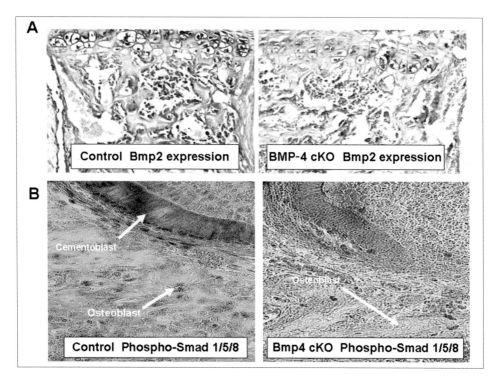

Figure 8
Deletion of Bmp-4 from osteoblasts results in reduced bmp-2 expression and reduced BMP signaling, as assayed by Phospho-Smad1/5/8 immunocytochemistry. (A) bmp-2 in situ hybridization, left is control and right is Bmp-4cKO, 12-day-old tibia. (B) Phospho-Smad1/5/8 immunocytochemistry, red signal, control to the left and Bmp-4cKO to the right.

in osteoclast area per bone surface in these mature skeletons. This suggests that the absence of BMP-4 in osteoblasts in older animals disrupts the remodeling cycle, favoring bone resorption.

To begin to explore the mechanism of BMP-4 loss in early osteoblasts, we explored expression of *bmp-2*. As shown in Figure 8A, deletion of *bmp-4* resulted in a dramatic decrease in *bmp-2* expression, suggesting that BMP-4 action is upstream of *bmp-2* expression. When we analyzed BMP signaling in the 12-day-old animals, sectioning in the mandible using an antibody specific to Phospho-Smad1/5/8, we observed an almost complete shut down of immunoreactivity in the osteoblasts. Similar decreases in BMP signaling were also observed in the long bones (data not shown).

Overall, these results demonstrate that BMP-4 is not redundant for other BMPs, has its own unique role, and maybe linked to expression to other BMPs such as

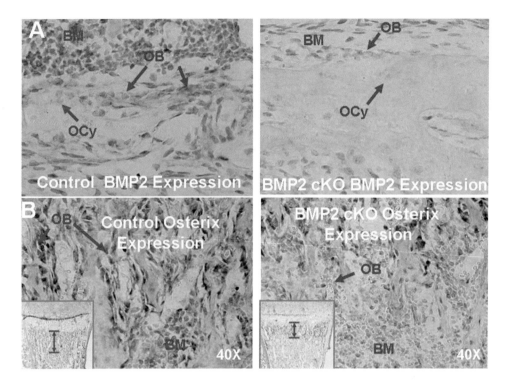

Figure 9
Bmp-2cKO in osteoblasts with the 3.6 Col1a1-Cre results in >90% reduction in bmp-2 expression in osteoblasts lining the bone surface and a 40% reduction in osterix expression, as quantitated using ImageJ in the 5-day-old animal alveolar regions of the jaw. (A) bmp-2 in situ hybridization with an exon 3 probe (region deleted after Cre recombination), left is control heterozygote. (B) Osterix expression in the same tissues as in (A), left is control and right is Bmp-2cKO tissue.

BMP-2 and BMP-6, which are also reduced in the BMP-4cKO animals (data not shown).

Targeted deletion of *bmp-2* using the 3.6 Col1a1-Cre model

We developed a *bmp-2* floxed allele with loxP sites flanking exon 3 of the *bmp-2* gene. By crossing *bmp-2* homozygous floxed mice with the Mox2-Cre line, which activates Cre recombinase in the blastula stage, we demonstrated that the phenotype of the Bmp-2cKO with the Mox2-Cre were identical to the global Bmp-2 KO, with the embryos dying at 8.5–9 dpc and failure in heart development and chorion formation (data not shown).

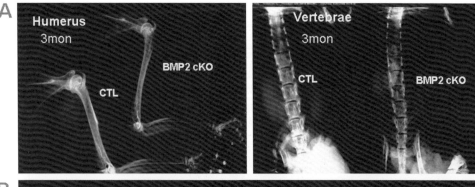
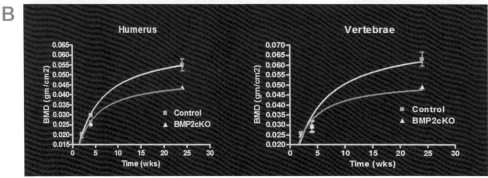

Figure 10
Decreased radio-opacity and BMD in Bmp-2cKO animals. (A) X-ray of humerus and vertebrae region at 3 months. Smaller diameter bone for both humerus and vertebrae can be seen. (B) BMD at 12 days, 1 month, and 6 months of age, in the humerus and vertebrae. Note the 15–20% reduction of BMD in the humerus and over 30% reduction of BMD in the vertebrae region.

We first analyzed exon 3 *bmp-2* expression in alveolar bones of 5-day-old control and BmpP-2cKO mice using the 3.6 Col1a1-Cre model. As shown in Figure 9A, there was a greater than 85% reduction in exon 3 BMP-2 *in situ* hybridization signal from the Bmp-2cKO tissue (left panel) compared to a heterozygote control (right panel). Figure 9B shows that osterix expression was reduced by 40% at this stage, suggesting that BMP signaling was altered.

Overall, the mice were similar in some ways to the Bmp-4cKO, but possibly with a different mechanism. The mice develop a kinky tail within a few days after birth. As shown in Figure 10A, in humerus and vertebrae of 3-month-old mice, there was a marked decrease in diameter of the bone and decreased radio-opacity (left panel, humerus; right panel, vertebrae). We followed the BMD longitudinally in several sites and observed an overall 10–15% decrease in BMD. As shown in

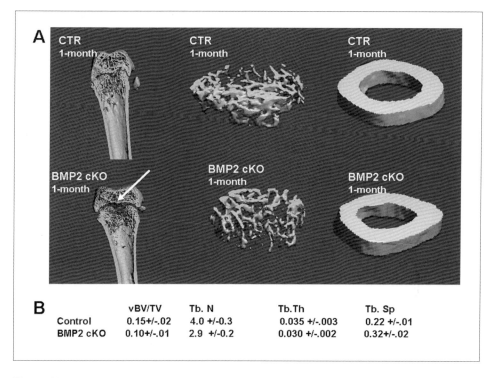

Figure 11
uCT analysis of femur of 1-month-old Bmp-2cKO compared to control. The distal region epiphysis is reduced and there is reduced trabecular number and overall length (insert) in the metaphysis of the Bmp-2cKO compared with control. The cortical thickness is unchanged in the Bmp-2cKO animals compared to control mice. (A) uCT images of the distal femur region, trabecular region and cross-section of cortical area. (B) Quantitation of BV/TV, trabecular number (Tb. N), trabecular thickness (Tb. Th.), and trabecular spacing (Tb. Sp).

Figure 10B, the BMD was reduced by 10–15% in the humerus from 12 days to 6 months (left panel), and the BMD was reduced 30% in the vertebrae by 6 months of age, with less severe reduction at earlier times (12 day and 1 month; right panel of Fig. 10B).

Figure 11 shows an uCT analysis of the femur of 1-month-old female control heterozygote and Bmp-2cKO mice. As shown in the left panel, the distal epiphysis was reduced in size and the metaphysis was reduced in length. There appeared to be fewer trabeculae in the metaphysis. Overall, the bone was smaller with decreased diameter, but the cortical thickness between control and Bmp-2cKO were similar. Quantitation of the trabecular region, as shown in Figure 11B, demonstrated a 35% reduction in BV/TV and trabecular number with increased trabecular separation and less of a plate-like structure in the Bmp-2cKO as compared to control.

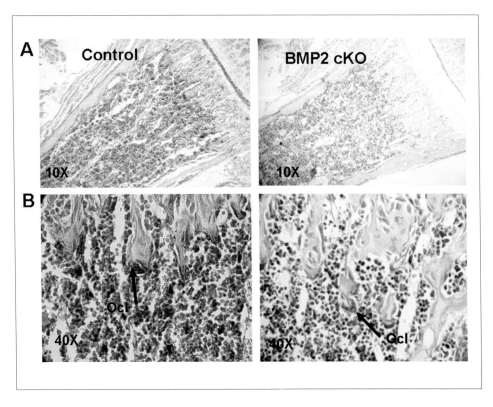

Figure 12
Reduced osteoclast number and activity (TRAP staining) in tibia from Bmp-2cKO compared to control heterozygotes; 12-day-old mice. (A) 10×, left is control and the right is Bmp-2cKO. (B) 40×, left is control and right is the Bmp-2cKO. The red stain represents the TRAP activity. Ocl, osteoclasts.

In our initial exploration of the mechanism of this osteopenia induced by the loss of BMP-2 in osteoblasts, we analyzed the level of osteoclastic activity with TRAP stain for tartrate-resistant alkaline phosphatase activity. As shown in Figure 12, there was greatly reduced TRAP staining in the trabecular region of the tibia, with fewer osteoclasts covering the bone surface (data not shown). Thus, removal of BMP-2 in osteoblast had altered the overall modeling dynamics of the bone. It will be interesting to see if this same dynamic holds true for bone from older animals.

We then explored expression of osteocalcin and collagen type 1a1 in alveolar bone of 12-day-old mice. Figure 13 shows that osteocalcin expression was reduced over 75% and collagen type 1a1 expression was reduced 45%. These results suggest

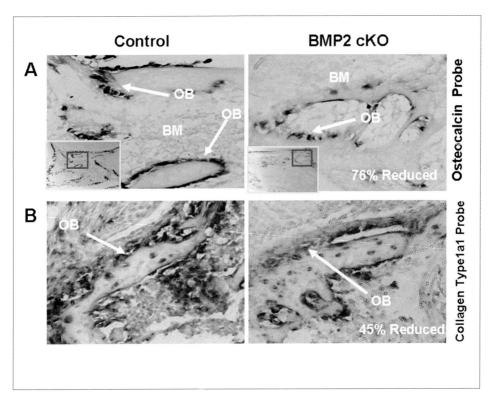

Figure 13
Osteocalcin and collagen type 1a1 expression are reduced in the Bmp-2cKO compared to control heterozygotes in the alveolar bone region of 12-day-old animals. (A) Osteocalcin expression is reduced 76% in Bmp-2cKO animals compared to control. 40×, with inserts showing area at low magnification. Left is control and right is Bmp-2cKO. (B) Collagen type 1a1 expression is reduced 45% in Bmp-2cKO compared to control, 40×. BM, bone marrow; OB, osteoblasts.

that BMP-2 was at least important in progression of osteoblasts to this late stage of differentiation.

Recently, we have started to look at the level of osteoblast precursors or CFU-F *in vitro* colonies, and the capacity of these osteoblast precursors to differentiate into a mineralized matrix *in vitro* [20]. Using bone marrow cells from control heterozygotes, wild type, and Bmp-2cKO, and isolation and culturing under appropriate conditions, we found a slight decrease in the number of CFU-F colonies comparing the control heterozygotes with the wild-type cultures. However, comparing the control heterozygote cultures with the Bmp-2cKO mice, we observed a 30–40% decrease in the number of CFU-F colonies in the bone marrow cultures from the

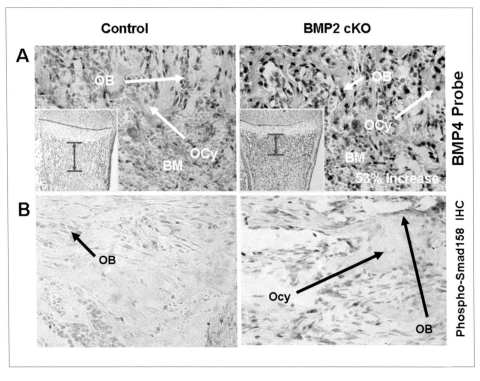

Figure 14
Bmp-4 expression is increased over 50% in Bmp-2cKO compared to control heterozygotes in the tibia of the 12-day-old animals. PhosphoSmad1/5/8 immunocytochemistry reaction is dramatically increased in this same region. (A) bmp-4 in situ hybridization, and quantitation using ImageJ. Insert shows low magnification with the trabecular area denoted by a red bar, and reduced trabecular area in the Bmp-2cKO animals. (B) PhosphoSmad1/5/8 immunocytochemistry, as denoted by the red stain. Note the dramatically increased BMP signaling in the Bmp-2cKO animals compared to control. Osteocytes show little BMP signaling. OB, osteoblasts; Ocy, osteocytes; BM, bone marrow.

Bmp-2cKO animals. We also observed a decrease in the number of colonies that can differentiate into mineralized matrix. These results suggest to us that BMP-2 is multi-functional and plays roles in bone biology at several stages.

In exploration of other mechanisms *in vivo*, we carried out *in situ* hybridization with a BMP-4 probe. To our surprise, and as demonstrated in Figure 14A, *bmp-4* expression was increased over 50% in the Bmp-2cKO. As noted in the insert in Figure 1A, the trabecular region in the metaphysis was greatly reduced, as was the width of the growth plate. We then looked at BMP signaling levels in these same 12-day-old tibias. Figure 1B demonstrates that there were dramatically increased

Phospho-Smad1/5/8 levels in the Bmp-2cKO osteoblasts. Removal of BMP-2 in early osteoblasts has thus led to a compensatory response and activation of the *Bmp-4* gene that leads to increased BMP signaling, and/or the population distribution of early and late stage osteoblast cells has changed.

Model and conclusion

We have shown that removal of BMP-4 in early osteoblasts leads to an osteopenic phenotype with reduced diameter of the bone, reduced trabeculi, and decreased cortical thickness. This phenotype is persistent up to 1.5 years. Late in life, part of this phenotype is due to increased osteoclastic activity, as well as to reduced osteoblast function. Removal of BMP-4 from osteoblasts results in decreased BMP signaling, decreased *bmp-2* and *bmp-6* expression, and reduced late marker expression such as osteocalcin. On the other hand, removal of BMP-2 using the same 3.6 Col1a1-Cre model also results in an osteopenic phenotype, with a similar decrease in overall size and diameter of the bone, decreased trabecular number, but little change in the cortical thickness. Late markers for osteoblast stages, such as osteocalcin and type 1a1 collagen are also reduced, suggesting that BMP-2 is required for late stage differentiation. Bone marrow culture assays, *in vitro*, show that there are fewer CFU-F colonies in the absence of BMP-2, suggesting that BMP-2 is required for early osteoblast commitment. The decrease in CFU-OB colonies in the Bmp-2cKO animals lends support to BMP-2 being required for both these early and late osteoblast functions. Our hypothesis is that early *BMP-4* expression is linked to increased *BMP-2* expression and BMP-2 drives mesenchymal precursor cells to early osteoblast precursors, depleting the mesenchymal stem cell population. These osteoblast precursors are blocked from further differentiation to more mature stages, due to a requirement that is specific to BMP-2 and not BMP-4. The osteoblast precursors pile up with increased levels of BMP-4 and increased BMP signaling. However, they cannot progress to late mineralizing stages in the absence of BMP-2. Further experiments with single and double knockout of *bmp-2* and *bmp-4* in different stages of osteoblast development and detailed expression profiling of cells with and without these various BMP ligands are now required.

Acknowledgements

This study was supported by National Institutes of Health research grants and AR44728 and AR054616.

References

1   Hogan BL (1996) Bone morphogenetic proteins: multifunctional regulators of vertebrate development. *Genes Dev* 10: 1580–1594
2   Harris SE, Bonewald LF, Harris MA, Sabatini M, Dallas S, Feng JQ, Ghosh-Choudhury N, Wozney J, Mundy GR (1994) Effects of transforming growth factor beta on bone nodule formation and expression of bone morphogenetic protein 2, osteocalcin, osteopontin, alkaline phosphatase, and type I collagen mRNA in long-term cultures of fetal rat calvarial osteoblasts. *J Bone Miner Res* 9: 855–863
3   Abrams KL, Xu J, Nativelle-Serpentini C, Dabirshahsahebi S, Rogers MB (2004) An evolutionary and molecular analysis of Bmp2 expression. *J Biol Chem* 279: 15916–15928
4   Chen D, Ji X, Harris MA, Feng JQ, Karsenty G, Celeste AJ, Rosen V, Mundy GR, Harris SE (1998) Differential roles for bone morphogenetic protein (BMP) receptor type IB and IA in differentiation and specification of mesenchymal precursor cells to osteoblast and adipocyte lineages. *J Cell Biol* 142: 295–305
5   Zhao M, Harris SE, Horn D, Geng Z, Nishimura R, Mundy GR, Chen D (2002) Bone morphogenetic protein receptor signaling is necessary for normal murine postnatal bone formation. *J Cell Biol* 157: 1049–1060
6   Panchision DM, Pickel JM, Studer L, Lee SH, Turner PA, Hazel TG, McKay RD (2001) Sequential actions of BMP receptors control neural precursor cell production and fate. *Genes Dev* 15: 2094–2110
7   Murali D, Yoshikawa S, Corrigan RR, Plas DJ, Crair MC, Oliver G, Lyons KM, Mishina Y, Furuta Y (2005) Distinct developmental programs require different levels of Bmp signaling during mouse retinal development. *Development* 132: 913–923
8   Bandyopadhyay A, Tsuji K, Cox K, Harfe BD, Rosen V, Tabin CJ (2006) Genetic analysis of the roles of BMP2,4, and 7 in limb patterning and skeletogenesis. *PLOS* 2: 2116–2130
9   Styrkarsdottir U, Cazier J, Rolfsson O, Larsen H, Bjarnadottir E, Johannsdottir, Sigurdardottir, Jonasson K, Frigge ML, Kong A et al (2003) The bone morphogenetic protein 2 gene contributes to bone density and osteoporotic fractures. *J Bone Miner Res* 18 (Suppl 2): s23
10 Ichikawa S, Koller DL, Johnson ML, Lai D, Johnston CC, Hui SL, Foroud TM, Peacock M, Econs MJ (2004) Polymorphisms in the bone morphogenetic protein 2 (BMP2) gene do not affect peak bone mineral density in men and women. *J Bone Miner Res* 19 (Suppl 1): s249
11 Hsu Y, Zhang Y, Xu X, Feng Y, Terwexdow H, Zhang T, Wu D, Tang G, Li Z, Hong X et al (2004) Genetic variation in ALOX15, BMP2, PPARr and IGF1 gene is associated with BMD. *J Bone Miner Res* 19 (Suppl 1): S386
12 Ramesh Babu L, Wilson SG, Dick IM, Islam FM, Devine A, Prince RL (2005) Bone mass effects of a BMP4 gene polymorphism in postmenopausal women. *Bone* 36: 555–561
13 Winnier G, Blessing M, Labosky PA, Hogan BL (1995) Bone morphogenetic protein-4

is required for mesoderm formation and patterning in the mouse. *Genes Dev* 9: 2105–2116

14 Kulessa H, Hogan BL (2002) Generation of a loxP flanked bmp4loxP-lacZ allele marked by conditional lacZ expression. *Genesis* 32: 66–68

15 Liu F, Woitge HW, Braut A, Kronenberg MS, Lichtler AC, Mina M, Kream BE (2004) Expression and activity of osteoblast-targeted Cre recombinase transgenes in murine skeletal tissues. *Int J Dev Biol* 48: 645–653

16 Nagy A (2000) Cre recombinase: The universal reagent for genome tailoring. *Genesis* 26: 99–109

17 Rossant J, McMahon A (1999) "Cre"-ating mouse mutants-a meeting review on conditional mouse genetics. *Genes Dev* 13: 142–145

18 Kwan KM (2002) Conditional alleles in mice: Practical considerations for tissue-specific knockouts. *Genesis* 32: 49–62

19 Harris SE, Guo D, Yang W, Harris M, Gluhak-Heinrich J, Edwards J, Zhao M, Mundy GR, Anderson HC, Kream B et al (2006) BMP2 and BMP4 action and signaling in bone: Conditional deletion studies in mice. In: *Abstracts of the 6th International Conference on BMPs 2006*, 28–116

20 Chen X-D, Dusevich V, Feng JQ, Manolagas SC, Jilka RL (2007) Extracellular matrix made by bone marrow cells facilitates expansion of marrow-derived mesenchymal progenitor cells and prevents their differentiation into osteoblasts. *J Bone Miner Res* 22: 1943–1956

The role of bone morphogenetic protein 4 (BMP-4) in tooth development

Jelica Gluhak-Heinrich[1], Dayong Guo[2], Wuchen Yang[1], Lilia E. Martinez[1], Marie A. Harris[1], Holger Kulessa[3†], Alexander Lichtler[4], Barbara E. Kream[4], Jianhong Zhang[3], Jian Q. Feng[5] and Stephen E. Harris[1]

[1]The University of Texas Health Science Center at San Antonio, 7703 Floyd Curl Dr., San Antonio, TX 78229-3900, USA; [2]The University of Missouri at Kansas City School of Dentistry, 650 E. 25th St., Kansas City, MO 64109, USA; [3]Vanderbilt University, Nashville, 2201 West End Avenue, Nashville, TN 37240, USA; [4]The University of Connecticut Medical Center, Farmington, 270 Farmington Av., CT 06032, USA; [5]Baylor College of Dentistry, Dallas, TX, P. O. Box 660677, Dallas, TX 75266-0677, USA

Dedication

This paper is dedicated to Holger Kulessa who developed the BMP4 floxed mouse model used in these studies.

Role of BMP-4 in prenatal tooth development

Bone morphogenetic protein 4 (BMP-4) is expressed throughout the main stages of tooth formation: initiation, bud, cap, bell (cytodifferentiation), and secretory stages [1–13]. During these stages, expression of *BMP-4* shifts from dental and oral epithelium and dental mesenchyme to the enamel knot, dental papilla, and dental sac [6].

Tooth formation is initiated at embryonic day 10 (E10) by epithelial-mesenchymal interactions in which secreted fibroblast growth factor 8 (FGF8) from the dental epithelium begins the process by binding to its receptor on the adjacent dental mesenchyme to induce expression of paired box gene 9 (Pax9) [7] and homeobox gene Msx1 [14–16]. Recent studies suggest interaction between Pax9 and Msx1 controlling early tooth development by modulating expression of *BMP-4* [17]. In this initiation stage of tooth development, *BMP-4* expressed in epithelium induces homeobox gene Msx1 and lymphoid enhancer factor (Lef1) as well as itself in the underlying mesenchyme [6, 18]. At E11–E12, the dental lamina is formed as epithelial thickening and *BMP-4* expression is detected transiently in both epithelium and mesenchyme, but before the bud stage shifts completely to dental mesenchyme [1]. In the mesenchyme, BMP-4 plays an inhibitory role by preventing induction of Pax9 [7]. BMP-4 also inhibits expression of the homeobox gene Barx-1 in E10 mouse embryos, restricting it to proximal presumptive molar mesenchyme. The inhibition of BMP-4 signaling during early mandible development causes ectopic homeobox gene Barx-1 expression with exchanged tooth identity to molar from incisor [10]. Thus, BMP-4 directs odontogenic gene expression in mesenchyme cells of the developing mandibular arch and mediates epithelial-mesenchymal interactions during early tooth development.

During the following bud stage (E13), a transition occurs in which Pax9 secreted from dental mesenchyme cells induces *BMP-4* expression within the condensed dental mesenchyme around the epithelial tooth bud with stronger expression at the buccal side of the bud [1, 7]. Mesenchymal BMP-4 binds to its receptor in the overlying dental epithelium to induce expression of Lef1 [5]. Recent work has shown that preventing the BMP-4 mesenchymal-epithelial interactions in this critical bud stage will arrest tooth development by not allowing production of epithelial Lef1, an important transcription factor in the Wnt pathway that interacts with beta catenin. In addition, Lef1 knockout mice have a tooth-arrested phenotype that inhibits progression from the bud stage to the cap stage [3–11, 13, 19, 20], but Lef1 and Wnt signaling can reverse teeth in Lef1 knockout mice [21]. Furthermore, a knockout mouse without the Runx2 gene, which is also part of the BMP signaling pathway during tooth development, showed arrested tooth development at the late bud stage [22].

The cap stage (E14–E15) of tooth development displays the first signs of cellular arrangement in the tooth bud. This stage is identifiable by the enamel knot, the signaling center regulating tooth shape [1]. Expression of *BMP-4* is intense in dental papilla mesenchyme reappearing also in epithelium in the distal part of the enamel knot [23, 24]. This epithelial structure acts as an organizer and secretes the signaling factors including FGF8 and Sonic hedgehog (Shh) required to control tooth identity and shape [4, 25, 26]. *BMP-4* expression in the enamel knot has been shown to be dependent on homeobox gene Msx2 [27]. BMP-4 plays a role in enamel knot progression by inducing p21 CDK inhibitor in the dental epithelium and inducing apoptosis in these epithelial cells after they have specified the positions of the developing cusps of the tooth [4, 28].

The succeeding bell stage (E16–E19) is characterized as the beginning of distinct morphodifferentiation within the tooth germ occurring between E18 and continuing up to postnatal day 18 (P18) in most mouse strains. Images are provided to assist in identifying *BMP-4* expression during different stages of tooth development at http://bite-it.helsinki.fi/BMP-4.HTM.

BMP-4 role during cytodifferentiation and morphodifferentiation of tooth development

The numbers of studies have established a crucial role for BMP-4 in both epithelial and mesenchymal signaling during early tooth development as described in the previous section. However, from about E16 to E21, there is dramatic transformation of the tooth germs into mineralized and erupted teeth with periodontal ligament, and all the other support structures. It is during this phase that there is a little knowledge of the mechanism of how teeth undergo this cytodifferentiation process. The hypothesis that the BMPs, and in particular BMP-4, are important in cytodifferentiation of the teeth has been around for a long time, but could not be tested until the past few years. It was known only that *BMP-4* is highly expressed in both differentiating odontoblasts at several stages, and in ameloblasts [29].

The cytodifferentiation of tooth development is characterized by formation of mineralized tissues: the bone-like dentin, cementum, and epithelial derived enamel. This stage of tooth development ultimately results in amelogenesis from ameloblasts and dentinogenesis from odontoblasts. The supportive evidence of BMP-4 signaling during this cytodifferentiation stage is the high and selective expression of BMPs and their receptors [1, 4, 8, 30]. In particular, *bmp-4* expression disappeared from dental epithelium with the removal of the enamel knot, and is highly expressed in cuspal area of the dental papilla including preodontoblast cell layer, at stage E16, which is the beginning of tooth cytodifferentiation [1]. One day later, when preodontoblasts become polarized, *bmp-2* and *bmp-7* mRNA transcripts appear with the already-expressing *bmp-4* mRNA [1]. Early expression at E18 of *bmp-4* mRNA in preodontoblasts was confirmed by another study in the first mouse molars from stage E18 [31] (Fig. 1A). The quantitative *in situ* hybridization localization showed 43% of preodontoblasts expressing *bmp-4* mRNA. Other potential marker genes like Dlx5 and osterix were expressed in 13% and 41% of preodontoblasts, respectively (unpublished observation) (Fig. 1A–C). *bmp-4* mRNA expression was very abundant in preodontoblasts at E18 (Fig. 1A–C). During the first day after birth, when predentine secretion begins, expression of *bmp-4* is continued, although with lower intensity in the same area, i.e., now secretory odontoblasts [1]. *bmp-4* expression appeared also in differentiating ameloblasts, secretory ameloblasts and terminal differentiation of ameloblasts, but not in pre-ameloblasts [1]. However, in the final secretory stage of tooth development, *bmp-4* expression is found in both ameloblasts and odontoblasts that will form enamel and dentin respectively [4, 6, 10]. During P4, BMP-4, BMP-2 and BMP-7 are all shown to be present in ameloblasts but expression of *bmp-4* was the most intense [1]. At this stage, there are thick layers of dentine and enamel separating odontoblasts from ameloblasts. BMP-4 most likely plays important roles in both amelogenesis and odontogenesis, as well as dental pulp stem cell recruitment in dentogenesis (Fig. 1D).

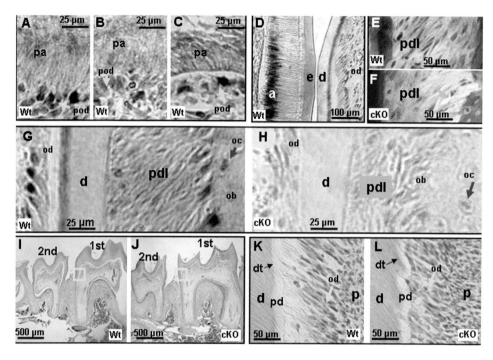

Figure 1
Expression of bmp-4 *(A),* Dlx5 *(B) and Osterix (C), in preodontoblasts in embryo stage E18, localized with quantitative in situ hybridization (signal: dark blue), shows that on average, 43%, 13% and 41% of preodontoblasts express* bmp-4, Dlx5 *and Osterix, respectively. In situ hybridization* bmp-4 *mRNA expression in ameloblasts and odontoblasts of 18-day-old mice incisor (D). Bmp-4cKO mice (18 days old) (F) have decrease of collagen (Van Gieson's pink stain) in periodontium compared to wild type (Wt) mice (E). Bmp-4cKO mice (18 days old) (H) have an 95% reduction in* bmp-4 *mRNA in situ hybridization expression (blue color) in odontoblasts as compared to BMP-4 wild type mice (G). Hematoxylin and eosin staining of first and second molar in 18-day-old Wt (I) and Bmp-4cKO (J) mice. Disorganized pre-dentin, dentin tubules and odontoblasts in Bmp-4cKo mice (L) in comparison to the Wt mice (K). (K) and (L) represent higher magnifications marked with yellow rectangles in (I), (J), respectively. pod, preodontoblasts; pa, preameloblasts; a, ameloblasts; e, enamel; d, dentin; od, odontoblasts (yellow arrow); pdl, periodontium; ob, osteoblasts (green arrow); oc, osteocytes (red arrow); dt, dentin tubules (black arrows); pd, predentin; p, dental pulp.*

The following section describes some of the mechanisms of BMP-4 action in teeth during cytodifferentiation phase using a Bmp-4 conditional knockout (cKO) mouse line that targets gene deletion to both early osteoblasts and to early polarizing odontoblasts.

Targeted deletion of BMP-4 in odontoblasts

In the normal *bmp-4* knockout model, the embryos do not develop past 8.5 days post conception (dpc) [32]. In the genetic background 129/SvEv _ black Swiss, some embryos survive until the early somite stage at 9.5 dpc. The alternative to a traditional Bmp-4-KO was Bmp-4cKO. Bmp-4cKO in odontoblasts was created by crossing two mouse models. The first model was the Bmp-4 floxed (Bmp-4-fx) mouse with loxP sites that flank exon 4 of the *bmp-4* gene [33]. This Bmp-4 floxed mouse model has been used to demonstrate the role of Bmp-4 in heart outflow tract septation and branchial arch remodeling [34]. The other mouse model used for generating Bmp-4cKO was a 3.6 Col1a1-Cre mouse model characterized and described in [34]. The Cre–loxP system is well established and works for a variety of genomic manipulations in mice [35–37]. After crossing Bmp-4-fx homozygote mice and 3.6 Col1a1-Cre mice that contain at least one Bmp-4-fx allele or one *bmp-4* knockout allele, the resulting Bmp-4cKO mouse model was a 3.6 Col1a1-cre Bmp-4cKO mouse that allowed selective deletion of BMP-4 in osteoblast and odontoblast lineages and was used to uncover roles for BMP-4 in odontoblast biology *in vivo* [31, 38]. Bmp-4cKO mice described above were used to show involvement of BMP-4 during the tooth cytodifferentiation and secretory stages in tooth formation, as described in following section.

Consequences of BMP-4 ablation during the tooth cytodifferentiation

The role and potential mechanisms of BMP-4 during tooth cytodifferentiation and secretory stages of tooth development were investigated utilizing the 3.6 Col1a1-cre Bmp-4cKO mouse. Experimentally, in this model, the *bmp-4* gene was abolished after birth. Quantitative *in situ* localization of BMP-4, using a *bmp-4* exon 4 probe (exon 4 was deleted after Cre events since the loxP sites flank exon 4), demonstrated a greater than 95% reduction in *bmp-4* expression in odontoblasts during late dentin and enamel formation in 18-day-old Bmp-4cKO mice (Fig. 1G, H). These Bmp-4cKO animals are smaller and also have osteopenia that develops rapidly within the first 2–3 weeks of life. This is noticeable in the mandible (Fig. 2). Gross teeth phenotype appears normal but there are histological differences. Bmp-4cKO mice have less collagen in the periodontium (PDL) (Fig. 1F) than wild-type (Wt) mice (Fig. 1E). Morphology of the dentin, predentin dentin tubules and odontoblasts are significantly changed in 18-day-old Bmp-4cKO mice (Fig. 1J, L) in comparison to the 18-day-old Wt mice (Fig. 1I, K) (unpublished observations). Two-day-old Bmp-4cKO mice lost practically all BMP-4 signaling through PhosphoSmad1/5/8, since quantification of protein expression in odontoblasts showed over 90% reduction of PhosphoSmad1/5/8 protein immunoreactivity (Fig. 3A, B) [31]. One of the well-known genes downstream of BMP signaling is the transcription factor, osterix

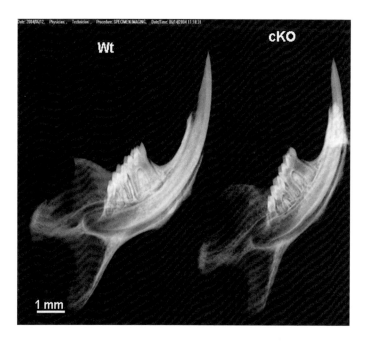

Figure 2
One-month-old Bmp-4cKO mice have reduced bone and display osteopenia in the alveolar bone compared to the Wt mice determined by x-ray.

and in the Bmp-4cKO animals, at 18 days, osterix was reduced 84%. DMP1 is expressed in odontoblasts at a fairly early stage and is involved in dentinal tubule formation [39, 40]. DMP1 expression was reduced greater than 90% in the Bmp-4cKO animals. The DSPP gene is a late marker for odontogenesis and is associated with the mineralization phase. DSPP expression was also reduced 67% in the Bmp-4cKO animals. These data are shown in (Fig. 3C–H).

X-ray analysis of Bmp-4cKO mice and BMP-4 involvement in other components of the tooth and surrounding structures

X-ray images of Bmp-4cKO mice molars show 40% reduced thickness of dentin (Fig. 4) [31]. Other experimental evidence supports BMP-4 involvement in dentin formation such as the use of BMP-4 protein for stimulation of dentin formation *in vitro* [41]. An additional observation was a complete down-regulation of DMP1 mRNA in cementoblasts following ablation of BMP-4 in Bmp-4cKO mice (Fig. 3F) compared with Wt animals (Fig. 3E). These results suggest a role of BMP-4 from odontoblasts having a paracrine function in the formation of cementum.

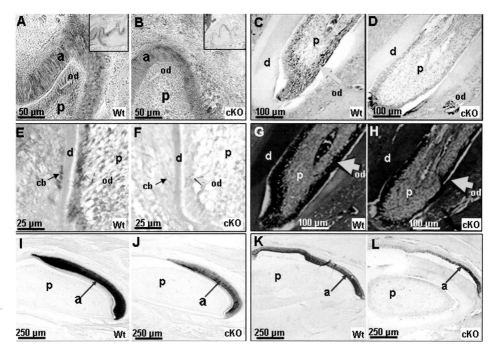

Figure 3
Expression of PhosphoSmad1/5/8 (red color) in odontoblasts of 12-day postnatal Bmp-4cKO mice (B) shows immunostaining signal of Smad1/5/8 reduced by >90% compared to Wt mice (A). In situ hybridization (blue color) in 1-month-old Bmp-4cKO mice shows an 84% reduction in osterix mRNA in odontoblasts (D) compared to Wt animals (C). In situ hybridization in 12-day-old Bmp-4cKO mice shows that expression of DMP1 mRNA (blue color) (F) is reduced by 90% in odontoblasts and cementoblasts compared with expression in Wt mice (E). One-month-old Wt mice (G) are shown, using in situ hybridization (blue color), to express high DSPP mRNA in odontoblasts, but in Bmp-4cKO mice this expression is decreased by 67% (H). High mRNA in situ hybridization expression (black) of ameloblastin and amelogenin, in 1-month-old Wt mice (I, K), was reduced in Bmp-4cKO by 59% and 5–95%, respectively (J, L). a, ameloblasts; od, odontoblasts (yellow arrows); p, dental pulp; d, dentin; cb, cementoblasts (black arrows).

As stated earlier, *BMP-4* is expressed in secretory ameloblasts but not in preameloblasts [1] (Fig. 1D). This expression suggests a role of BMP-4 in enamel development. There are virtually no data on the effects of BMP-4 protein on ameloblasts *in vivo* and *in vitro*, and on enamel formation, either BMP-4 from ameloblast directly or from odontoblasts in a paracrine type of role. Using Bmp-4cKO mice, BMP-4 was removed only in early odontoblasts, not ameloblasts. However, in Bmp-4cKO mice, a delay in amelogenesis was observed. The results from gene expression studies

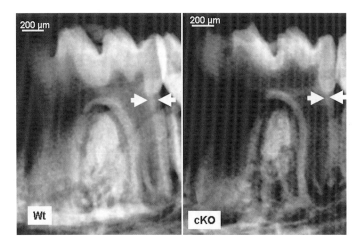

Figure 4
Dentin thickness in Bmp-4cKO mice. X-rays showed that in 1-month-old Bmp-4cKO mice dentin thickness is reduced by 40% compared to Wt mice (arrows).

of Bmp-4cKO mice also suggest that deletion of Bmp-4 in odontoblasts leads to a delayed enamel formation process. For example, the ameloblast marker, ameloblastin, is decreased over 59% in mice at day 18. Amelogenin expression is decreased anywhere from 5% to 95% as assayed in 35-day-old mice [31]. Amelogenin in 35-day-old mice was expressed evenly in amelobalsts along the tooth. In the Bmp-4cKO mice, a decrease of ameloblastin in ameloblasts showed great differences along the tooth in proximal ameloblasts and close to the tooth root. This decrease was as much as 95% in the proximal, but in the distal ameloblasts only 5%. This difference in expression pattern support the idea that odontogenesis is tightly linked to the process of amelogenesis (Fig. 3L). This involvement of Bmp-4 in ameloblast development is additionally supported by experiments using demineralization of 5-day-old teeth, a time before maturation at which protein-enriched enamel has mineralized. However, in the Bmp-4cKO animals, this mineralization process is delayed compared to the control-wild type (Wt) animals (Fig. 5). One model would be that BMP-4 secreted by odontoblasts, maybe moving through the dentinal tubules, directly stimulated the ameloblasts to differentiate. However, only a slight decrease in Phospho-Smad 1/5/8 signaling was observed in the ameloblast layer (Fig. 3A, B). It is possible that other factors made by odontoblasts stimulate the ameloblasts or that the Bmp-4 is working through other non-Smad pathways, such as the p38 pathway. As a result of the potential defects in the dentin, communication between the dentin and enamel may be compromised in the absence of BMP-4 [42]. Further studies of the effect of BMP-4 on enamel formation are awaited with interest. Additional mechanisms of BMP-4 during tooth cytodifferentiation remain to be explored.

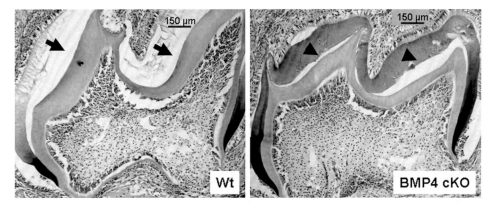

Figure 5
Delayed enamel maturation in Bmp-4cKO mice in maxillary second molar of 5-day-old mice. Hematoxylin and eosin staining, after demineralization of Wt mice showed that tooth enamel is partially lost (arrows), but is well preserved in Bmp-4cKO mice (arrowheads), since enamel of Bmp-4cKO has less mineral with delayed enamel maturation compared to Wt mice.

The following section describes the future possibilities of BMP-4 use in dental applications.

Future application of BMP-4 in teeth involving a deeper understanding of role and mechanism

The experimental evidence described above as well as studies using BMP-4 stimulation of dental pulp cells in odontoblasts [8, 41, 43] support inductive roles of BMP-4 in odontoblast differentiation, dentin formation, ameloblast differentiation, PDL formation and cementum differentiation [1, 31].

Comparing expression of *bmp-4* during tooth development in mice and humans showed basically similar expression patterns [44]. Even in zebrafish expression of BMP-4 is highly conserved in the oral region. A lack of BMP-4 expression in zebrafish was found only in the toothless oral region, supporting importance of BMP-4 in tooth development [45].

New studies showed that BMP-4 supports development of tooth germ cells into mature dental structures *in vitro*, suggesting the feasibility of using BMP-4 in tooth bioengineering [46, 47]. These possibilities are striking, and together with experimental evidence lay a strong foundation of integrating BMP-4 action into the formation of teeth from the necessary stem cell precursors. Further knowledge of BMP-4 in amelogenesis may lead to new paradigms for producing new enamel from the appropriate precursor cells.

Acknowledgements

This study was supported by National Institutes of Health research grants NIDCR R03 DE16949, AR46798 and AR054616.

References

1 Aberg T, Wozney J, Thesleff I (1997) Expression patterns of bone morphogenetic proteins (Bmps) in the developing mouse tooth suggest roles in morphogenesis and cell differentiation. *Dev Dyn* 210: 383–396
2 Chen D, Zhao M, Harris SE, Mi Z (2004) Signal transduction and biological functions of bone morphogenetic proteins. *Front Biosci* 9: 349–358
3 Gritli-Linde A, Bei M, Maas R, Zhang XM, Linde A, McMahon AP (2002) Shh signaling within the dental epithelium is necessary for cell proliferation, growth and polarization. *Development* 129: 5323–5337
4 Jernvall J, Aberg T, Kettunen P, Keranen S, Thesleff I (1998) The life history of an embryonic signaling center: BMP-4 induces p21 and is associated with apoptosis in the mouse tooth enamel knot. *Development* 125: 161–169
5 Kratochwil K, Dull M, Farinas I, Galceran J, Grosschedl R (1996) Lef1 expression is activated by BMP-4 and regulates inductive tissue interactions in tooth and hair development. *Genes Dev* 10: 1382–1394
6 Maas R, Bei M (1997) The genetic control of early tooth development. *Crit Rev Oral Biol Med* 8: 4–39
7 Neubuser A, Peters H, Balling R, Martin GR (1997) Antagonistic interactions between FGF and BMP signaling pathways: a mechanism for positioning the sites of tooth formation. *Cell* 90: 247–255
8 Ohazama A, Tucker A, Sharpe PT (2005) Organized tooth-specific cellular differentiation stimulated by BMP4. *J Dent Res* 84: 603–606
9 Tabata MJ, Fujii T, Liu JG, Ohmori T, Abe M, Wakisaka S, Iwamoto M, Kurisu K (2002) Bone morphogenetic protein 4 is involved in cusp formation in molar tooth germ of mice. *Eur J Oral Sci* 110: 114–120
10 Tucker AS, Matthews KL, Sharpe PT (1998) Transformation of tooth type induced by inhibition of BMP signaling. *Science* 282: 1136–1138
11 Yamashiro T, Tummers M, Thesleff I (2003) Expression of bone morphogenetic proteins and Msx genes during root formation. *J Dent Res* 82: 172–176
12 Zhang Z, Song Y, Zhang X, Tang J, Chen J, Chen Y (2003) Msx1/Bmp4 genetic pathway regulates mammalian alveolar bone formation *via* induction of Dlx5 and Cbfa1. *Mech Dev* 120: 1469–1479
13 Zhao X, Zhang Z, Song Y, Zhang X, Zhang Y, Hu Y, Fromm SH, Chen Y (2000) Transgenically ectopic expression of Bmp4 to the Msx1 mutant dental mesenchyme restores

downstream gene expression but represses Shh and Bmp2 in the enamel knot of wild type tooth germ. *Mech Dev* 99: 29–38

14   Bei M, Maas R (1998) FGFs and BMP4 induce both Msx1-independent and Msx1-dependent signaling pathways in early tooth development. *Development* 125: 4325–4333

15   Bei M, Kratochwil K, Maas RL (2000) BMP4 rescues a non-cell-autonomous function of Msx1 in tooth development. *Development* 127: 4711–4718

16   Vainio S, Karavanova I, Jowett A, Thesleff I (1993) Identification of BMP-4 as a signal mediating secondary induction between epithelial and mesenchymal tissues during early tooth development. *Cell* 75: 45–58

17   Ogawa T, Kapadia H, Feng JQ, Raghow R, Peters H, D'Souza RN (2006) Functional consequences of interactions between Pax9 and Msx1 genes in normal and abnormal tooth development. *J Biol Chem* 281: 18363–18369

18   Oosterwegel M, van de Wetering M, Timmerman J, Kruisbeek A, Destree O, Meijlink F, Clevers H (1993) Differential expression of the HMG box factors TCF-1 and LEF-1 during murine embryogenesis. *Development* 118: 439–448

19   Hart PS, Wright JT, Savage M, Kang G, Bensen JT, Gorry MC, Hart TC (2003) Exclusion of candidate genes in two families with autosomal dominant hypocalcified amelogenesis imperfecta. *Eur J Oral Sci* 111: 326–331

20   Holleville N, Quilhac A, Bontoux M, Monsoro-Burq AH (2003) BMP signals regulate Dlx5 during early avian skull development. *Dev Biol* 257: 177–189

21   Kratochwil K, Galceran J, Tontsch S, Roth W, Grosschedl R (2002) FGF4, a direct target of LEF1 and Wnt signaling, can rescue the arrest of tooth organogenesis in Lef1(–/–) mice. *Genes Dev* 16: 3173–3185

22   James MJ, Jarvinen E, Wang XP, Thesleff I (2006) Different roles of Runx2 during early neural crest-derived bone and tooth development. *J Bone Miner Res* 21: 1034–1044

23   Nadiri A, Kuchler-Bopp S, Perrin-Schmitt F, Lesot H (2006) Expression patterns of BMPRs in the developing mouse molar. *Cell Tissue Res* 324: 33–40

24   Nadiri A, Kuchler-Bopp S, Haikel Y, Lesot H (2004) Immunolocalization of BMP-2/-4, FGF-4, and WNT10b in the developing mouse first lower molar. *J Histochem Cytochem* 52: 103–112

25   Vaahtokari A, Aberg T, Jernvall J, Keranen S, Thesleff I (1996) The enamel knot as a signaling center in the developing mouse tooth. *Mech Dev* 54: 39–43

26   Dassule HR, McMahon AP (1998) Analysis of epithelial-mesenchymal interactions in the initial morphogenesis of the mammalian tooth. *Dev Biol* 202: 215–227

27   Bei M, Stowell S, Maas R (2004) Msx2 controls ameloblast terminal differentiation. *Dev Dyn* 231: 758–765

28   Cho SW, Lee HA, Cai J, Lee MJ, Kim JY, Ohshima H, Jung HS (2007) The primary enamel knot determines the position of the first buccal cusp in developing mice molars. *Differentiation* 75: 441–451

29   Feng JQ, Zhang J, Tan X, Lu Y, Guo D, Harris SE (2002) Identification of cis-DNA

regions controlling Bmp4 expression during tooth morphogenesis *in vivo. J Dent Res* 81: 6–10

30 Feng JQ, Chen D, Cooney AJ, Tsai MJ, Harris MA, Tsai SY, Feng M, Mundy GR, Harris SE (1995) The mouse bone morphogenetic protein-4 gene. Analysis of promoter utilization in fetal rat calvarial osteoblasts and regulation by COUP-TFI orphan receptor. *J Biol Chem* 270: 28364–28373

31 Gluhak-Heinrich J, Harris M, Martinez L, Zhang J, Guo D, Yang W, Edwards J, Kream B, Lichtler AC, Hogan B et al (2006) Role of BMP4 in cytodifferentiation and bone morphogenesis. *Abstracts of the 6th International Conference on Bone Morphogenetic Proteins, Dubrovnik* Session VII: Maxillofacial & Cranial Applications, 68–OC20

32 Winnier G, Blessing M, Labosky PA, Hogan BL (1995) Bone morphogenetic protein-4 is required for mesoderm formation and patterning in the mouse. *Genes Dev* 9: 2105–2116

33 Kulessa H, Hogan BL (2002) Generation of a loxP flanked bmp4loxP-lacZ allele marked by conditional lacZ expression. *Genesis* 32: 66–68

34 Liu F, Woitge HW, Braut A, Kronenberg MS, Lichtler AC, Mina M, Kream BE (2004) Expression and activity of osteoblast-targeted Cre recombinase transgenes in murine skeletal tissues. *Int J Dev Biol* 48: 645–653

35 Nagy A (2000) Cre recombinase: The universal reagent for genome tailoring. *Genesis* 26: 99–109

36 Rossant J, McMahon A (1999) "Cre"-ating mouse mutants-a meeting review on conditional mouse genetics. *Genes Dev* 13: 142–145

37 Kwan KM (2002) Conditional alleles in mice: Practical considerations for tissue-specific knockouts. *Genesis* 32: 49–62

38 Harris SE, Guo D, Yang W, Harris M, Gluhak-Heinrich J, Edwards J, Zhao M, Mundy GR, Anderson HC, Kream B et al (2006) BMP2 and BMP4 action and signaling in bone: Conditional deletion studies in mice. *Abstracts of the 6th International Conference on BMPs 2006*, 28–116

39 Lu Y, Xie Y, Zhang S, Dusevich V, Bonewald LF, Feng JQ (2007) DMP1–targeted Cre expression in odontoblasts and osteocytes. *J Dent Res* 86: 320–325

40 Lu Y, Ye L, Yu S, Zhang S, Xie Y, McKee MD, Li YC, Kong J, Eick JD, Dallas SL et al (2007) Rescue of odontogenesis in Dmp1-deficient mice by targeted re-expression of DMP1 reveals roles for DMP1 in early odontogenesis and dentin apposition *in vivo*. *Dev Biol* 303: 191–201

41 Nakashima M (1994) Induction of dentin formation on canine amputated pulp by recombinant human bone morphogenetic proteins (BMP)-2 and -4. *J Dent Res* 73: 1515–1522

42 Kozawa Y, Iwasa Y, Mishima H (1998) The function and structure of the marsupial enamel. *Connect Tissue Res* 39: 215–217; discussion 221–215

43 Nakashima M, Nagasawa H, Yamada Y, Reddi AH (1994) Regulatory role of transforming growth factor-beta, bone morphogenetic protein-2, and protein-4 on gene

expression of extracellular matrix proteins and differentiation of dental pulp cells. *Dev Biol* 162: 18–28
44 Lin D, Huang Y, He F, Gu S, Zhang G, Chen Y, Zhang Y (2007) Expression survey of genes critical for tooth development in the human embryonic tooth germ. *Dev Dyn* 236: 1307–1312
45 Wise SB, Stock DW (2006) Conservation and divergence of Bmp2a, Bmp2b, and Bmp4 expression patterns within and between dentitions of teleost fishes. *Evol Dev* 8: 511–523
46 Chung IH, Choung PH, Ryu HJ, Kang YH, Choung HW, Chung JH, Choung YH (2007) Regulating the role of bone morphogenetic protein 4 in tooth bioengineering. *J Oral Maxillofac Surg* 65: 501–507
47 Nakashima M, Reddi AH (2003) The application of bone morphogenetic proteins to dental tissue engineering. *Nat Biotechnol* 21: 1025–1032

Bone morphogenetic protein antagonists and kidney

Motoko Yanagita

Kyoto University Graduate School of Medicine, Kyoto 606-8501, Japan

Introduction

Bone morphogenetic proteins (BMPs) are phylogenetically conserved signaling molecules that belong to the transforming growth factor (TGF)-β superfamily [1–4]. Although these proteins were first identified by their capacity to promote endochondral bone formation [5–7], they are involved in the cascades of body patterning including nephrogenesis. Furthermore, BMPs play important roles after birth in pathophysiology of several diseases including osteoporosis [8], arthritis [5], pulmonary hypertension [9, 10], and kidney diseases [11–13]. Several BMPs are expressed in the kidney, and the expression level and pattern of each BMP varies dynamically during embryogenesis and kidney disease progression. BMP-7 is the most abundant BMP during kidney development [14], whereas the level of BMP-4, BMP-6 and BMP-7 are comparable in adult healthy kidneys (S. Yamada, unpublished data). BMP-2 is hardly detectable in developing and adult kidneys.

BMP-7 in kidney disease and development

BMP-7, also known as osteogenic protein-1 (OP-1), is a 35-kDa homodimeric protein, and kidney is the major site of BMP-7 synthesis during embryogenesis and in postnatal development [14–17]. Its genetic deletion in mice leads to severe impairment of kidney development resulting in perinatal death [18, 19]. Kidney development is essentially normal until embryonic day (E) 14.5; however, the metanephric mesenchymal cells fail to differentiate subsequently, resulting in a low number of nephrons in newborn kidneys. The mutant kidneys also suffered massive apoptosis in the uninduced mesenchymal cells, demonstrating that BMP-7 is essential for their continued survival, proliferation and differentiation. Borovecki et al. [20] demonstrated that iodinated BMP-7 (^{125}I-BMP-7) injected through the tail vein of pregnant mice passed across the placenta and localized in developing fetal organ including kidneys until E14, indicating the possibility that maternal circulating BMP-7 might rescue the lack of embryonic BMP-7 in early development.

Bmp-7 null mice also have an eye defect and minor skeletal patterning defects [18, 19]. Although Bmp-7 is expressed at diverse sites in the developing mouse embryos, the tissue defects in *bmp-7* null embryos are confined to certain organs. Dudley et al. [21] demonstrated the overlapping expression domain of BMPs and the possibility that BMP family members can functionally substitute for BMP-7 at sites where they colocalized. Oxburgh et al. [22] further supported the idea by demonstrating that the *bmp-4* knock-in allele in the *bmp-7* locus rescued the kidney development, and suggesting that BMP family members can function interchangeably.

Expression of BMP-7 in adult kidney is confined to distal convoluted tubules, collecting ducts and podocytes of glomeruli (Fig. 1) [23], and the expression decreases in several kidney disease models [24–28]. Recently, several reports indicate that the administration of pharmacological doses of BMP-7 inhibits and repairs chronic renal injury in animal models [25–27, 29–31]. The administration of BMP-7 is reported to reverse TGF-β1-induced epithelial-to-mesenchymal transition (EMT) and induce mesenchymal-to-epithelial transition (MET) *in vitro* [32, 33], inhibit the induction of inflammatory cytokine expression in the kidney [23], attenuate inflammatory cell infiltration [30], and reduce apoptosis of tubular epithelial cells in renal disease models [34]. Collectively, BMP-7 plays critical roles in repairing processes of the renal tubular damage in kidney diseases.

However, the physiological role and precise regulatory mechanism of endogenous BMP-7 remain elusive. Although many groups reported the possible actions of BMP-7 on proximal tubule epithelial cells (PTEC) in adult kidney injury [23], it is poorly understood which cells are the main source of BMP-7 in the circumstances, and how endogenous BMP-7 can be delivered to PTEC. BMP-7 might be delivered from adjacent distal nephron segments *via* the intervening interstitium, or alternatively, might be delivered from the glomerulus or *via* the circulation. Bosukonda et al. [35] reported that injected ^{125}I-BMP-7 in rats is found within glomeruli, proximal convoluted tubules, and medullary collecting tubules. *In situ* hybridization using a BMPR-II riboprobe demonstrated similar localization with ^{125}I-BMP-7 (Fig. 1A), and immunostaining of BMPR-II localized the receptor to glomeruli and proximal tubules. Further study is needed to determine whether BMP receptors are expressed in the apical or basolateral membrane of PTEC, which implies the route of endogenous BMP-7 delivery.

BMP-4 in urinary tract development

In mouse embryos, Bmp-4 is expressed in mesenchymal cells surrounding the Wolffian duct and ureter stalk [21, 36, 37]. *Bmp-4* null embryos die between E6.5 and

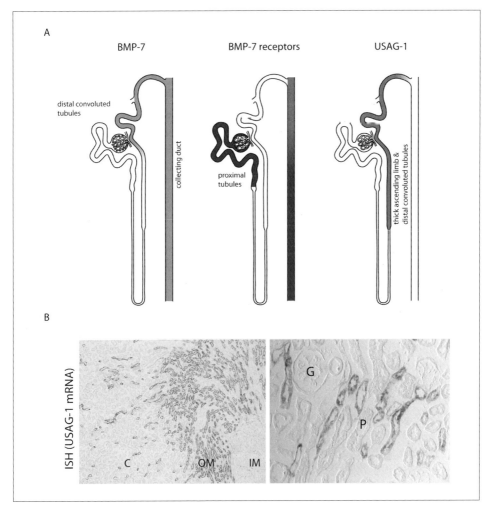

Figure 1
Expression of bone morphogenetic protein (BMP)-7, its receptor, and uterine sensitization-associated gene (USAG)-1. (A) Schematic illustration demonstrating the nephron segments in which BMP-7, its receptors and USAG-1 are expressed. (B) Localization of USAG-1 mRNA in the adult kidney.

10.0, indicating the essential role of BMP-4 in early embryonic development [38]. *Bmp-4* heterozygous null mutant mice displays abnormalities that mimic human congenital anomalies of the kidney and urinary tract (CAKUT), including hypo/dysplastic kidneys, hydroureter, ectopic ureterovesical junction, and double collecting system [36]. Further *in vivo* and *in vitro* studies clarified that BMP-4 inhibits ectopic budding from Wolffian duct and stimulates the elongation of the branching ureter within the metanephros [37].

As mentioned above, the knockin allele of *bmp-4* in the *bmp-7* locus efficiently rescues kidney development in *bmp-7* null embryos [22], indicating that BMP-4 and BMP-7, sharing only minimal sequence similarity, can function interchangeably to activate essential pathways in kidney development. The results indicate the distinct phenotypes in *bmp-4* null embryos and *bmp-7* null embryos simply reflect differences in expression domains of these two molecules.

Extracellular modification of BMP activity

The local activity of endogenous BMP is precisely regulated at multiple steps: intracellulary, at the membrane site, and extracellulary (Fig. 2). In this review, we focus on the extracellular modification of BMP signaling.

At the membrane, the transmembrane protein BAMBI (BMP and activin membrane-bound inhibitor) functions as a pseudoreceptor to interfere with BMP, activn, and TGF-β signaling in *Xenopus* [39, 40]. BAMBI and its mammalian homologue Nma are structurally related to type I serine/threonine kinase receptors in the extracellular domain, but lack the intracellular serine/threonine kinase domain. BAMBI/Nma stably associate with type II receptors, thus preventing the formation of active receptor complex.

Recently, repulsive guidance molecule (RGMA) [41], DRAGON (RGMB) [42, 43], and hemojuvelin [44] are reported to act as BMP-activating co-receptors. These are glycosyl phosphatidyl inositol (GPI)-anchored proteins, which form a complex with BMP type I receptors and enhance receptor binding to BMP-2 and BMP-4, potentiating their biological effects.

In the extracellular space, BMP signaling is precisely regulated by certain classes of molecules termed as BMP antagonists [45, 46]. BMP antagonists function through direct association with BMPs, thus prohibiting BMPs from binding their cognate receptors. The interplay between BMP and their antagonists fine-tunes the level of available BMPs, and governs developmental and cellular processes as diverse as establishment of the embryonic dorsal–ventral axis [47], induction of neural tissue [48], formation of joints in the skeletal system [5] and neurogenesis in the adult brain [49].

In addition to the modulation by BMP antagonists, high affinity binding of BMP to extracellular matrix modifies the local activity of BMP. Vukicevic et al. [50] previously showed that BMP-7 binds to basement membrane components including type IV collagen. In addition, Gregory et al. [51] recently demonstrated that the prodomain of BMP-7 targets BMP-7 complex to the extracellular matrix. In most tissues, BMP mRNA expression and BMP protein are found colocalized. Restricted diffusion of BMP proteins is considered to increase its local concentration.

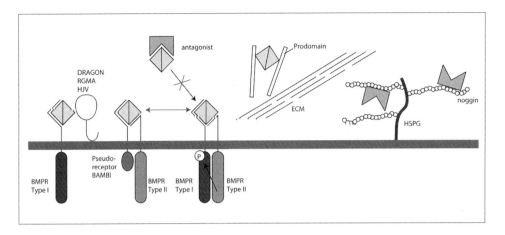

Figure 2
Extracellular modulation of BMP signaling. Modified from [45]. ECM; extracellular matrix.

Heparin sulfate proteoglycans (HSPGs) are also reported to shape the BMP gradient at the cell surface. Jiao et al. [52] recently reported that HSPGs mediate BMP-2 internalization and modulate BMP-2 osteogenic activity, while other groups reported that BMP antagonists such as chordin and noggin are retained at cell surface and regulated diffusion by binding to HSPGs [53].

BMP antagonists in the kidney

Classification and expression of BMP antagonist in the kidney

BMP antagonists have a secretory signal peptide and cysteine arrangement consistent with the formation of the cystine knot structure and represent a subfamily of cystine knot superfamily, which comprises of TGF-β, growth differentiation factors (GDFs), gonadotropins, and platelet-derived growth factors, and BMPs [54]. Recently, Avsian-Kretchmer et al. [55] classified BMP antagonists into three subfamilies based on the size of the cystine knot: the DAN family (eight-membered ring), twisted gastrulation (Tsg) (nine-membered ring) and chordin and noggin (10-membered ring). They further divided the DAN family into four subgroups based on a conserved arrangement of additional cysteine residues outside of the cystine knots: (1) PRDC and gremlin, (2) coco and Cer1 homologue of *Xenopus* Cerberus, (3) Dan, and (4) USAG-1/wise/ectodin and sclerostin. This subdivision is almost consistent with the phylogenic tree based on the overall amino acid sequence similarity shown in Figure 3.

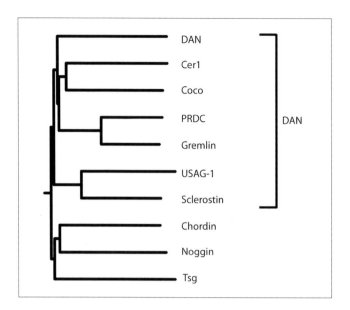

Figure 3
Phylogenetic tree of human BMP antagonists based on the overall amino acid sequence similarity of representative members from each subfamily. The GenomeNet server at http://www.genome.jp/ was used for phylogenetic tree construction. Modified from [66].

To compare the expression level of BMP antagonists in the kidney, our group utilized modified real-time PCR and demonstrated that USAG-1 is by far the most abundant BMP antagonist in adult kidney (Fig. 4A), as well as in embryogenesis (Fig. 4B). In the following section, we review the papers describing the possible role of BMP antagonists in the kidney.

USAG-1: the most abundant BMP antagonist in the kidney

Discovery and characterization of USAG-1 as a BMP antagonist

Through a genome-wide search for kidney-specific transcripts, our group found a novel gene, which encodes a secretory protein with a signal peptide and cysteine-rich domain [56]. The rat orthologue of the gene was previously reported as a gene of unknown function that was preferentially expressed in sensitized endometrium of rat uterus, termed uterine sensitization-associated gene-1 (*USAG-1*) [57]. Amino acid sequences encoded in rat and mouse cDNAs are 97% and 98% identical to the human sequence respectively, indicating high degrees of sequence conservation.

Domain search predicted this protein to be a member of the cystine-knot superfamily, and homology search revealed that USAG-1 has significant amino acid

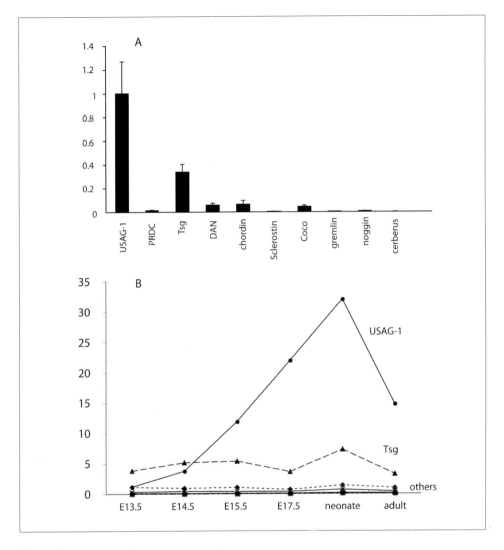

Figure 4
Expression of BMP antagonists in adult healthy kidney (A) and developing kidney (B). Tsg; twisted gastrulation.

identities (38%) to sclerostin, the product of the *SOST* gene (Fig. 4). Mutations of *SOST* are found in patients with sclerosteosis, a syndrome of sclerosing skeletal dysplasia. Sclerostin was expressed in bones and cartilages, and subsequently shown to be a new member of BMP antagonist [58–61], as well as a modulator of Wnt signaling [62–64].

USAG-1 protein is a 28–30-kDa secretory protein [56] and is heavily glycosylated (A. Yoshioka, unpublished data). USAG-1 behaves as a monomer, although a number of BMP antagonists form disulfide-bridged dimers. This is consistent with the fact that USAG-1 protein does not have the extra cysteine residues present in noggin and DAN, which are necessary to make inter-molecular disulfide bridges. Recombinant USAG-1 protein physically interacts with BMP-2, -4, -6, and -7, leading to the inhibition of alkaline phosphatase activities (ALP) induced by each BMP in C2C12 cells and MC3T3-E1 cells dose-dependently [56, 65], while sclerostin only inhibits BMP-6 and BMP-7 activities [58]. The activity of USAG-1 as a BMP modulator was also confirmed *in vivo* using *Xenopus* embryogenesis [56]. Injection of synthetic RNA encoding BMP antagonists to the ventral portion of *Xenopus* embryos inhibits the ventralizing signal of endogenous BMP, and induces dorsalizing phenotypes of the embryos including secondary axis formation and hyperdorsalization. The injection of as little as 100 pg USAG-1 mRNA was sufficient to cause secondary axis formation, and injection of increasing doses of mRNA up to 1000 pg led to a corresponding increase in the frequency of dorsalization phenotypes, while embryos developed normally when irrelevant mRNA was injected.

USAG-1 as a central regulator of renoprotective action of BMP-7

In adult tissues, the expression of USAG-1 was by far the most abundant in the kidney and is restricted to the thick ascending limb, and distal convoluted tubules (Fig. 1) [56, 66–68]. Thus, the cellular distribution of USAG-1 overlaps with that of BMP-7 in the distal convoluted tubules. Together with the fact that PTECs are the site of injury in many types of kidney diseases, and that the PTECs express the receptors for BMP-7, we hypothesized a working model of the regulation of the reno-protective action of BMP-7: in renal injury, PTECs are mainly damaged and undergo apoptosis or EMT to fibroblast-like mesenchymal cells. BMP-7 secreted from distal tubules binds to the receptors on the cell surface of PTECs, and inhibits apoptosis and EMT. USAG-1 is also secreted from distal tubules, binds to BMP-7, and inhibits the reno-protective actions of BMP-7 by reducing the amount of available BMP-7.

To evaluate this working model, our group generated *usag-1* null mice, and induced acute and chronic renal disease models in which the renal tubules were mainly damaged [68]. *usag-1* null mice exhibited prolonged survival and preserved renal function in acute and chronic renal injuries. Renal BMP signaling, assessed by phosphorylation of Smad proteins, is significantly enhanced in *usag-1* null mice during renal injury, indicating that the preservation of renal function is attributed to enhancement of endogenous BMP signaling. Furthermore, the administration of neutralizing antibody against BMP-7 abolished reno-protection in *usag-1* null mice, indicating that USAG-1 plays a critical role in the modulation of reno-protective action of BMP, and that inhibition of USAG-1 will be a promising means of development of novel treatment for kidney diseases. In addition, we demonstrated that the expression of USAG-1 in the kidney biopsy could be a diagnostic tool to predict renal prognosis [69].

USAG-1 as a context-dependent activator and inhibitor of Wnt signaling
Itasaki et al. [70] reported that *wise*, a *Xenopus* orthologue of USAG-1, functions as a context-dependent activator and inhibitor of Wnt signaling in *Xenopus* embryogenesis. They also demonstrated the physical interaction between wise/USAG-1 and Wnt co-receptor LRP6, and that Wise/USAG-1 can compete with Wnt8 for binding to LRP6. Recently, they demonstrated that the cellular localization of Wise/USAG-1 has distinct effects on the Wnt pathway readout [71]. While secreted Wise/USAG-1 either synergizes or inhibits the Wnt signals depending on the partner ligand, enndoplasmic reticulum (ER)-retained Wise/USAG-1 consistently blocks the Wnt pathway. ER-retained Wise/USAG-1 reduces LRP6 on the cell surface, making cells less susceptible to the Wnt signal. Further studies are needed to clarify the biological function of USAG-1 *in vivo*; however, it might be possible that these two proteins possess dual activities, and play as a molecular link between Wnt and BMP signaling pathway.

Gremlin: Essential for kidney development and a possible role in fibrosis

Gremlin was identified from a *Xenopus* ovarian library for activities inducing secondary axis [72]. Gremlin is a 28-kDa protein, and binds to BMP-2/4 and inhibits their binding to the receptors. *Gremlin* null mice are neonatally lethal because of the lack of kidneys and septation defects in lung [73]. In early limb buds, mesenchymal gremlin is required to establish a functional apical ectodermal ridge and the epithelial-mesenchymal feedback signaling that propagates the sonic hedgehog morphogen [74]. In the *gremlin* null embryos, metanephric development is disrupted at the stage of intiating ureteric bud outgrowth and genetic lowering of BMP-4 levels in *gremlin* null embryos completely restores ureteric bud outgrowth and branching morphogenesis, indicating that initiation of metanephric kidney development requires the reduction of BMP-4 activity by the antagonist gremlin in the mesenchyme, which in turn enables ureteric bud outgrowth and establishment of autoregulatory GDNF/WNT11 feedback signaling [75].

Gremlin is also known as DRM (down-regulated by v-*mos*) because it was identified as a gene that down-regulated in *mos*-transformed cells [76, 77]. Another name for gremlin is IHG-2 (induced in high glucose 2) because its expression in cultured kidney mesangial cells is induced by high ambient glucose, mechanical strain, and TGF-β [78]. The expression of gremlin is not detected in adult healthy kidney, but is increased in streptozotocin-induced diabetic nephropathy model [24], as well as in human diabetic nephropathy. Although some expression of gremlin is observed in occasional glomeruli, gremlin expression was prominent in areas of tubulointerstitial fibrosis, where it colocalized with TGF-β expression [79]. The authors of the study also demonstrated that gremlin expression correlated well with the tubulointerstitial fibrosis, and the result is consistent with previous reports demonstrating the up-regulation of gremlin in fibrosis of other organs [80, 81].

Recently, Sun et al. [82] reported a novel intracellular regulatory mechanism by which Gremlin interacts with BMP-4 precursor, prevents secretion of mature BMP-4, and therefore inhibits BMP-4 activity more efficiently. Furthermore, they defined a 30-amino acid peptide sequence within the Gremlin DAN domain that is essential for BMP-4 interaction. This result implies that the level of BMP-4 mRNA expression does not truly reflect BMP-4 activity when Gremlin and BMP-4 are co-expressed within the same cell. Similar regulatory mechanisms may be utilized by other DAN family proteins.

Noggin: Effective tool to inhibit BMP signaling

Noggin is a 32-kDa glycoprotein secreted by Spemann organizer of *Xenopus* embryos, and is found to rescue dorsal development in the ultraviolet-induced ventralized embryos [83]. Noggin antagonizes the action of BMPs, induces neural tissues and dorsalizes ventral mesoderm [84]. Noggin binds to BMP-2 and BMP-4 with high affinity and to BMP-7 with low affinity, and prevent BMPs from binding to its receptors. Groppe et al. [85] reported the crystal structure of Noggin bound to BMP-7, which shows that Noggin inhibits BMP signaling by blocking the molecular interfaces of the binding epitopes for both type I and type II receptors. The BMP-7-binding affinity of site-specific variants of Noggin is correlated with alterations in bone formation and apoptosis in chick limb development, showing that Noggin functions by sequestering its ligand in an inactive complex. The scaffold of Noggin contains a cystine knot topology similar to that of BMPs; thus, ligand and antagonist seem to have evolved from a common ancestral gene.

In mice, Noggin is expressed in the node, notochord, dorsal somite, condensing cartilage, and immature chondrocytes, and null mutation of Noggin results in serious developmental abnormalities including failure of neural tube formation, and dismorphogenesis of the axial skeleton and joint lesions [86–88].

In the healthy kidney, Noggin is not expressed, but its high binding affinity to BMP is utilized as a tool to inhibit BMP signaling in certain cell type. Recently, it was reported that overexpression of noggin in podocytes leads to the development of mesangial expansion, indicating the importance of endogenous BMP signaling in the maintenance of glomerular structure [89]. Because the expression of noggin is almost undetectable in healthy and diseased kidney, other negative regulator of endogenous BMP might play a role in glomerular mesangial expansion.

Crim1: A membrane-bound antagonist and a role in glomerular development

Crim1 is a transmembrane protein possessing cysteine-rich repeat (CRR), and plays a role in the tethering of growth factors at the cell surface [90]. Crim1 binds to BMP-4 and -7 *via* the CRR-containing portion, and functions as a BMP antagonist in three different ways: Crim1 binding with BMP-4 and -7 occurs when these proteins are co-expressed within the Golgi compartment of the cell and leads to (i) a reduction in the production and processing of pre-protein to mature BMP, (ii) tethering of pre-BMP to the cell surface, and (iii) an effective reduction in the secretion of mature BMP. Hence, Crim1 modulates BMP activity by affecting its processing and delivery to the cell surface.

Crim1 is expressed in a spatially and temporally restricted manner during organogenesis of the limbs, kidney, lens, pinna, erupting teeth, and testis [91–93]. During metanephric development, Crim1 is expressed in the ureteric tree, the early condensing mesenchyme and distal comma-shaped bodies. As the nephron elongates, Crim1 becomes expressed in the proximal end of the S-shaped bodies [91]. In later stages of development, Crim1 is also detected in podocyte and mesangial cells in glomeruli [94].

A gene-trap mouse line with an insertion of β-Geo cassette into intron 1 of the *Crim1* gene (*Crim1$^{KST264/KST264}$*) is a Crim1 hypomorph, and displayed perinatal lethality with defects in multiple organ systems [95]. In the kidney, *Crim1$^{KST264/KST264}$* mice displayed abnormal glomerular development, including enlarged capillary loops, podocyte effacement, and mesangiolysis [94]. When outbred, homozygotes that reached birth displayed marked albuminuria. The podocytic co-expression of Crim1 with vascular endothelial growth factor-A (VEGF-A) suggested a role for Crim1 in the regulation of VEGF-A action. Crim1 and VEGF-A were shown to interact directly, providing evidence that CRR-containing proteins can bind to non-TGF-β superfamily ligands.

In addition, a homologue of Crim1 in *Caenorhabditis elegans*, crm-1, is reported to facilitate BMP signaling to control body size in *C.s elegans* [96].

Kielin/chordin-like protein: BMP agonist with a role in kidney injury

Lin et al. [97] recently identified a cDNA clone from an embryonic kidney library that contained multiple CRRs. The entire coding lesion was similar to the *Xenopus* kielin protein, and thus was named kielin/chordin-like protein (KCP). KCP is a secretory protein with 18 CRRs, and increases the binding of BMP-7 to its receptor and enhances downstream signaling pathways. The expression of KCP was detected in developing nephrons, but not in adult healthy kidneys. *kcp* null mice developed normally. When introduced in a kidney injury model, *kcp* null mice showed reduced levels of phosphorylated Smad1, and were susceptible to developing renal interstitial fibrosis, and more sensitive to tubular injury.

In contrast to the enhancing effect on BMPs, KCP inhibits both activin A- and TGF-β1-mediated signaling through the Smad2/3 pathway. KCP binds directly to TGF-β1 and blocks the interactions with its receptors. Consistent with this inhibitory effect, primary renal epithelial cells from *KCP* null cells are hypersensitive to TGF-β1 [98].

Crossveinless 2: Another BMP agonist related to kielin

Crossveinless 2 (Cv2) is also closely related to kielin, and was first identified in a fly mutant study as a gene required for the formation of cross-veins in the fly wings [99]. Genetic studies in flies showed that the formation of these veins required high Bmp signaling activity, and that Cv2 was essential for enhancing the local Bmp signal near the receiving cells. By contrast, the *in vivo* role of the vertebrate counterpart of Cv2 remains to be elucidated, as some reports indicate that Cv2 is an anti-BMP factor [100], while others describe its pro-BMP activity [101]. Analysis of *cv2* null mice terminated the argument, and demonstrated that Cv2 is a pro-BMP factor in mouse embryogenesis [102].

In *cv2* null mouse, gastrulation occurs normally, but a number of defects are found in Cv2-expressing tissues such as the skeleton. The defects of the vertebral column and eyes in the *cv2* null mouse are substantially enhanced by deleting one copy of the *bmp-4* gene, suggesting a pro-Bmp role of Cv2 in the development of these organs. In addition, *cv2* null mice exhibit kidney hypoplasia, and the phenotype is synergistically enhanced by the additional deletion of *kcp*, that encodes a pro-Bmp protein structurally related to Cv2 (see previous section).

Conclusions

In conclusion, BMPs and their modulators play important roles in kidney injury as well as in kidney development. Because negative and positive modulators of BMP signaling regulate and define the boundaries of BMP activity, further understanding of these modulators would give valuable information about their pathophysiological functions and provide a rationale for a therapeutic approach against these proteins.

Acknowledgement

I thank Dr. Sampath and Dr. Vukicevic for giving me an opportunity to write this review. This study was supported by Grants-in Aid from the Ministry of Education, Culture, Science, Sports, and Technology of Japan (177090551), a Center of Excellence grant from the Ministry of Education, Culture, Science, Sports, and Technology of Japan, a research grant for health sciences from the Japanese Ministry of Health, Labor and Welfare, and by a grant from the Astellas Foundation for Research on Metabolic Disorders, a grant from the Novartis Foundation for the promotion of science, a grant from Kato Memorial Trust for Nambyo Research, a grant from Hayashi Memorial Foundation for Female Natural Scientists, and a grant from Japan Foundation for Applied Enzymology.

References

1 Massague J, Chen YG (2000) Controlling TGF-beta signaling. *Genes Dev* 14: 627–644
2 Canalis E, Economides AN, Gazzerro E (2003) Bone morphogenetic proteins, their antagonists, and the skeleton. *Endocr Rev* 24: 218–235
3 Reddi AH (2001) Interplay between bone morphogenetic proteins and cognate binding proteins in bone and cartilage development: noggin, chordin and DAN. *Arthritis Res* 3: 1–5
4 Attisano L, Wrana JL (1996) Signal transduction by members of the transforming growth factor-beta superfamily. *Cytokine Growth Factor Rev* 7: 327–339
5 Reddi AH (2000) Bone morphogenetic proteins and skeletal development: The kidney-bone connection. *Pediatr Nephrol* 14: 598–601
6 Urist MR (1965) Bone: formation by autoinduction. *Science* 150: 893–899
7 Wozney JM, Rosen V, Celeste AJ, Mitsock LM, Whitters MJ, Kriz RW, Hewick RM, Wang EA (1988) Novel regulators of bone formation: Molecular clones and activities. *Science* 242: 1528–1534
8 Wang EA (1993) Bone morphogenetic proteins (BMPs): Therapeutic potential in healing bony defects. *Trends Biotechnol* 11: 379–383
9 Miyazono K, Kusanagi K, Inoue H (2001) Divergence and convergence of TGF-beta/BMP signaling. *J Cell Physiol* 187: 265–276
10 Morse JH, Deng Z, Knowles JA (2001) Genetic aspects of pulmonary arterial hypertension. *Ann Med* 33: 596–603
11 Klahr S (2003) The bone morphogenetic proteins (BMPs). Their role in renal fibrosis and renal function. *J Nephrol* 16: 179–185
12 Hruska KA, Saab G, Chaudhary LR, Quinn CO, Lund RJ, Surendran K (2004) Kidney-bone, bone-kidney, and cell-cell communications in renal osteodystrophy. *Semin Nephrol* 24: 25–38
13 Zeisberg M, Muller GA, Kalluri R (2004) Are there endogenous molecules that protect kidneys from injury? The case for bone morphogenic protein-7 (BMP-7). *Nephrol Dial Transplant* 19: 759–761
14 Helder MN, Ozkaynak E, Sampath KT, Luyten FP, Latin V, Oppermann H, Vukicevic

S (1995) Expression pattern of osteogenic protein-1 (bone morphogenetic protein-7) in human and mouse development. *J Histochem Cytochem* 43: 1035–1044

15 Vukicevic S, Stavljenic A, Pecina M (1995) Discovery and clinical applications of bone morphogenetic proteins. *Eur J Clin Chem Clin Biochem* 33: 661–671

16 Vukicevic S, Kopp JB, Luyten FP, Sampath TK (1996) Induction of nephrogenic mesenchyme by osteogenic protein 1 (bone morphogenetic protein 7). *Proc Natl Acad Sci USA* 93: 9021–9026

17 Ozkaynak E, Schnegelsberg PN, Oppermann H (1991) Murine osteogenic protein (OP-1): High levels of mRNA in kidney. *Biochem Biophys Res Commun* 179: 116–123

18 Dudley AT, Lyons KM, Robertson EJ (1995) A requirement for bone morphogenetic protein-7 during development of the mammalian kidney and eye. *Genes Dev* 9: 2795–2807

19 Luo G, Hofmann C, Bronckers AL, Sohocki M, Bradley A, Karsenty G (1995) BMP-7 is an inducer of nephrogenesis, and is also required for eye development and skeletal patterning. *Genes Dev* 9: 2808–2820

20 Borovecki F, Jelic M, Grgurevic L, Sampath KT, Bosukonda D, Vukicevic S (2004) Bone morphogenetic protein-7 from serum of pregnant mice is available to the fetus through placental transfer during early stages of development. *Nephron* 97: e26–32

21 Dudley AT, Robertson EJ (1997) Overlapping expression domains of bone morphogenetic protein family members potentially account for limited tissue defects in BMP7 deficient embryos. *Dev Dyn* 208: 349–362

22 Oxburgh L, Dudley AT, Godin RE, Koonce CH, Islam A, Anderson DC, Bikoff EK, Robertson EJ (2005) BMP4 substitutes for loss of BMP7 during kidney development. *Dev Biol* 286: 637–646

23 Gould SE, Day M, Jones SS, Dorai H (2002) BMP-7 regulates chemokine, cytokine, and hemodynamic gene expression in proximal tubule cells. *Kidney Int* 61: 51–60

24 Wang SN, Lapage J, Hirschberg R (2001) Loss of tubular bone morphogenetic protein-7 in diabetic nephropathy. *J Am Soc Nephrol* 12: 2392–2399

25 Hruska KA (2002) Treatment of chronic tubulointerstitial disease: A new concept. *Kidney Int* 61: 1911–1922

26 Morrissey J, Hruska K, Guo G, Wang S, Chen Q, Klahr S (2002) Bone morphogenetic protein-7 improves renal fibrosis and accelerates the return of renal function. *J Am Soc Nephrol* 13 Suppl 1: S14–21

27 Dube PH, Almanzar MM, Frazier KS, Jones WK, Charette MF, Paredes A (2004) Osteogenic Protein-1: Gene expression and treatment in rat remnant kidney model. *Toxicol Pathol* 32: 384–392

28 Almanzar MM, Frazier KS, Dube PH, Piqueras AI, Jones WK, Charette MF, Paredes AL (1998) Osteogenic protein-1 mRNA expression is selectively modulated after acute ischemic renal injury. *J Am Soc Nephrol* 9: 1456–1463

29 Vukicevic S, Basic V, Rogic D, Basic N, Shih MS, Shepard A, Jin D, Dattatreyamurty B, Jones W, Dorai H et al (1998) Osteogenic protein-1 (bone morphogenetic protein-

7) reduces severity of injury after ischemic acute renal failure in rat. *J Clin Invest* 102: 202–214

30 Hruska KA, Guo G, Wozniak M, Martin D, Miller S, Liapis H, Loveday K, Klahr S, Sampath TK, Morrissey J (2000) Osteogenic protein-1 prevents renal fibrogenesis associated with ureteral obstruction. *Am J Physiol Renal Physiol* 279: F130–143

31 Wang S, Chen Q, Simon TC, Strebeck F, Chaudhary L, Morrissey J, Liapis H, Klahr S, Hruska KA (2003) Bone morphogenic protein-7 (BMP-7), a novel therapy for diabetic nephropathy. *Kidney Int* 63: 2037–2049

32 Zeisberg M, Hanai J, Sugimoto H, Mammoto T, Charytan D, Strutz F, Kalluri R (2003) BMP-7 counteracts TGF-beta1-induced epithelial-to-mesenchymal transition and reverses chronic renal injury. *Nat Med* 9: 964–968

33 Zeisberg M, Bottiglio C, Kumar N, Maeshima Y, Strutz F, Müller GA, Kalluri R (2003) Bone morphogenic protein-7 inhibits progression of chronic renal fibrosis associated with two genetic mouse models. *Am J Physiol Renal Physiol* 285: F1060–1067

34 Li T, Surendran K, Zawaideh MA, Mathew S, Hruska KA (2004) Bone morphogenetic protein 7: A novel treatment for chronic renal and bone disease. *Curr Opin Nephrol Hypertens* 13: 417–422

35 Bosukonda D, Shih MS, Sampath KT, Vukicevic S (2000) Characterization of receptors for osteogenic protein-1/bone morphogenetic protein-7 (OP-1/BMP-7) in rat kidneys. *Kidney Int* 58: 1902–1911

36 Miyazaki Y, Oshima K, Fogo A, Hogan BL, Ichikawa I (2000) Bone morphogenetic protein 4 regulates the budding site and elongation of the mouse ureter. *J Clin Invest* 105: 863–873

37 Miyazaki Y, Oshima K, Fogo A, Ichikawa I (2003) Evidence that bone morphogenetic protein 4 has multiple biological functions during kidney and urinary tract development. *Kidney Int* 63: 835–844

38 Dunn NR, Winnier GE, Hargett LK, Schrick JJ, Fogo AB, Hogan BL (1997) Haploinsufficient phenotypes in Bmp4 heterozygous null mice and modification by mutations in Gli3 and Alx4. *Dev Biol* 188: 235–247

39 Onichtchouk D, Chen YG, Dosch R, Gawantka V, Delius H, Massagué J, Niehrs C (1999) Silencing of TGF-beta signalling by the pseudoreceptor BAMBI. *Nature* 401: 480–485

40 Grotewold L, Plum M, Dildrop R, Peters T, Ruther U (2001) Bambi is coexpressed with Bmp-4 during mouse embryogenesis. *Mech Dev* 100: 327–330

41 Babitt JL, Zhang Y, Samad TA, Xia Y, Tang J, Campagna JA, Schneyer AL, Woolf CJ, Lin HY (2005) Repulsive guidance molecule (RGMa), a DRAGON homologue, is a bone morphogenetic protein co-receptor. *J Biol Chem* 280: 29820–29827

42 Samad TA, Rebbapragada A, Bell E, Zhang Y, Sidis Y, Jeong SJ, Campagna JA, Perusini S, Fabrizio DA, Schneyer AL et al (2005) DRAGON, a bone morphogenetic protein coreceptor. *J Biol Chem* 280: 14122–14129

43 Samad TA, Srinivasan A, Karchewski LA, Jeong SJ, Campagna JA, Ji RR, Ji RR, Fabrizio DA, Zhang Y, Lin HY, Bell E, Woolf CJ (2004) DRAGON: A member of the repul-

sive guidance molecule-related family of neuronal- and muscle-expressed membrane proteins is regulated by DRG11 and has neuronal adhesive properties. *J Neurosci* 24: 2027–2036

44 Babitt JL, Huang FW, Wrighting DM, Xia Y, Sidis Y, Samad TA, Campagna JA, Chung RT, Schneyer AL, Woolf CJ et al (2006) Bone morphogenetic protein signaling by hemojuvelin regulates hepcidin expression. *Nat Genet* 38: 531–539

45 Balemans W, Van Hul W (2002) Extracellular regulation of BMP signaling in vertebrates: A cocktail of modulators. *Dev Biol* 250: 231–250

46 Gazzerro E, Canalis E (2006) Bone morphogenetic proteins and their antagonists. *Rev Endocr Metab Dis* 7: 51–65

47 Wagner DS, Mullins MC (2002) Modulation of BMP activity in dorsal-ventral pattern formation by the chordin and ogon antagonists. *Dev Biol* 245:109–123

48 Wessely O, Agius E, Oelgeschlager M, Pera EM, De Robertis EM (2001) Neural induction in the absence of mesoderm: Beta-catenin-dependent expression of secreted BMP antagonists at the blastula stage in Xenopus. *Dev Biol* 234: 161–173

49 Lim DA, Tramontin AD, Trevejo JM, Herrera DG, Garcia-Verdugo JM, Alvarez-Buylla A (2000) Noggin antagonizes BMP signaling to create a niche for adult neurogenesis. *Neuron* 28: 713–726

50 Vukicevic S, Latin V, Chen P, Batorsky R, Reddi AH, Sampath TK (1994) Localization of osteogenic protein-1 (bone morphogenetic protein-7) during human embryonic development: High affinity binding to basement membranes. *Biochem Biophys Res Commun* 198: 693–700

51 Gregory KE, Ono RN, Charbonneau NL, Kuo CL, Keene DR, Bächinger HP, Sakai LY (2005) The prodomain of BMP-7 targets the BMP-7 complex to the extracellular matrix. *J Biol Chem* 280: 27970–2780

52 Jiao X, Billings PC, O'Connell MP, Kaplan FS, Shore EM, Glaser DL (2007) Heparan sulfate proteoglycans (HSPGs) modulate BMP2 osteogenic bioactivity in C2C12 cells. *J Biol Chem* 282: 1080–1086

53 Paine-Saunders S, Viviano BL, Economides AN, Saunders S (2002) Heparan sulfate proteoglycans retain Noggin at the cell surface: A potential mechanism for shaping bone morphogenetic protein gradients. *J Biol Chem* 277:2089–2096

54 Vitt UA, Hsu SY, Hsueh AJ (2001) Evolution and classification of cystine knot-containing hormones and related extracellular signaling molecules. *Mol Endocrinol* 15: 681–694

55 Avsian-Kretchmer O, Hsueh AJ (2004) Comparative genomic analysis of the eight-membered ring cystine knot-containing bone morphogenetic protein antagonists. *Mol Endocrinol* 18: 1–12

56 Yanagita M, Oka M, Watabe T, Iguchi H, Niida A, Takahashi S, Akiyama T, Miyazono K, Yanagisawa M, Sakurai T (2004) USAG-1: A bone morphogenetic protein antagonist abundantly expressed in the kidney. *Biochem Biophys Res Commun* 316: 490–500

57 Simmons DG, Kennedy TG (2002) Uterine sensitization-associated gene-1: A novel gene

induced within the rat endometrium at the time of uterine receptivity/sensitization for the decidual cell reaction. *Biol Reprod* 67: 1638–1645

58 Kusu N, Laurikkala J, Imanishi M, Usui H, Konishi M, Miyake A, Thesleff I, Itoh N (2003) Sclerostin is a novel secreted osteoclast-derived bone morphogenetic protein antagonist with unique ligand specificity. *J Biol Chem* 278: 24113–24117

59 Winkler DG, Sutherland MS, Ojala E, Turcott E, Geoghegan JC, Shpektor D, Skonier JE, Yu C, Latham JA (2005) Sclerostin inhibition of Wnt-3a-induced C3H10T1/2 cell differentiation is indirect and mediated by BMP proteins. *J Biol Chem* 280: 2498–2502

60 Winkler DG, Sutherland MK, Geoghegan JC, Yu C, Hayes T, Skonier JE, Shpektor D, Jonas M, Kovacevich BR, Staehling-Hampton K et al (2003) Osteocyte control of bone formation *via* sclerostin, a novel BMP antagonist. *EMBO J* 22: 6267–6276

61 Winkler DG, Yu C, Geoghegan JC, Ojala EW, Skonier JE, Shpektor D, Sutherland MK, Latham JA (2004) Noggin and sclerostin bone morphogenetic protein antagonists form a mutually inhibitory complex. *J Biol Chem* 279: 36293–36298

62 Ellies DL, Viviano B, McCarthy J, Rey JP, Itasaki N, Saunders S, Krumlauf R (2006) Bone density ligand, Sclerostin, directly interacts with LRP5 but not LRP5G171V to modulate Wnt activity. *J Bone Miner Res* 21: 1738–1749

63 Semenov M, Tamai K, He X (2005) SOST is a ligand for LRP5/LRP6 and a Wnt signaling inhibitor. *J Biol Chem* 280: 26770–26775

64 Li X, Zhang Y, Kang H, Liu W, Liu P, Zhang J, Harris SE, Wu D (2005) Sclerostin binds to LRP5/6 and antagonizes canonical Wnt signaling. *J Biol Chem* 280:19883–19887

65 Laurikkala J, Kassai Y, Pakkasjarvi L, Thesleff I, Itoh N (2003) Identification of a secreted BMP antagonist, ectodin, integrating BMP, FGF, and SHH signals from the tooth enamel knot. *Dev Biol* 264: 91–105

66 Yanagita M (2005) BMP antagonists: Their roles in development and involvement in pathophysiology. *Cytokine Growth Factor Rev* 16: 309–317

67 Yanagita M (2006) Modulator of bone morphogenetic protein activity in the progression of kidney diseases. *Kidney Int* 70: 989–993

68 Yanagita M, Okuda T, Endo S, Tanaka M, Takahashi K, Sugiyama F, Kunita S, Takahashi S, Fukatsu A, Yanagisawa M et al (2006) Uterine sensitization-associated gene-1 (USAG-1), a novel BMP antagonist expressed in the kidney, accelerates tubular injury. *J Clin Invest* 116:70–79

69 Tanaka M, Endo S, Okuda T, Economides AN, Valenzuela DM, Murphy AJ, Robertson E, Sakurai T, Fukatsu A, Yancopoulos GD, Kita T, Yanagita M (2008) Expression of BMP-7 and USAG-1 (a BMP antagonist) in kidney development and injury. *Kidney Int* 73: 181–191

70 Itasaki N, Jones CM, Mercurio S, Rowe A, Domingos PM, Smith JC, Krumlauf R (2003) Wise, a context-dependent activator and inhibitor of Wnt signalling. *Development* 130: 4295–4305

71 Guidato S, Itasaki N (2007) Wise retained in the endoplasmic reticulum inhibits Wnt signaling by reducing cell surface LRP6. *Dev Biol* 310: 250–263

72 Hsu DR, Economides AN, Wang X, Eimon PM, Harland RM (1998) The Xenopus dorsalizing factor Gremlin identifies a novel family of secreted proteins that antagonize BMP activities. *Mol Cell* 1: 673–683

73 Michos O, Panman L, Vintersten K, Beier K, Zeller R, Zuniga A (2004) Gremlin-mediated BMP antagonism induces the epithelial-mesenchymal feedback signaling controlling metanephric kidney and limb organogenesis. *Development* 131: 3401–3410

74 Khokha MK, Hsu D, Brunet LJ, Dionne MS, Harland RM (2003) Gremlin is the BMP antagonist required for maintenance of Shh and Fgf signals during limb patterning. *Nat Genet* 34: 303–307

75 Michos O, Goncalves A, Lopez-Rios J, Tiecke E, Naillat F, Beier K, Galli A, Vainio S, Zeller R (2007) Reduction of BMP4 activity by gremlin 1 enables ureteric bud outgrowth and GDNF/WNT11 feedback signalling during kidney branching morphogenesis. *Development* 134: 2397–2405

76 Topol LZ, Modi WS, Koochekpour S, Blair DG (2000) DRM/GREMLIN (CKTSF1B1) maps to human chromosome 15 and is highly expressed in adult and fetal brain. *Cytogenet Cell Genet* 89: 79–84

77 Topol LZ, Bardot B, Zhang Q, Resau J, Huillard E, Marx M, Calothy G, Blair DG (2000) Biosynthesis, post-translation modification, and functional characterization of Drm/Gremlin. *J Biol Chem* 275: 8785–8793

78 McMahon R, Murphy M, Clarkson M, Taal M, Mackenzie HS, Godson C, Martin F, Brady HR (2000) IHG-2, a mesangial cell gene induced by high glucose, is human gremlin. Regulation by extracellular glucose concentration, cyclic mechanical strain, and transforming growth factor-beta1. *J Biol Chem* 275: 9901–9904

79 Dolan V, Murphy M, Sadlier D, Lappin D, Doran P, Godson C, Martin F, O'Meara Y, Schmid H, Henger A et al (2005) Expression of gremlin, a bone morphogenetic protein antagonist, in human diabetic nephropathy. *Am J Kidney Dis* 45: 1034–1039

80 Koli K, Myllärniemi M, Vuorinen K, Salmenkivi K, Ryynänen MJ, Kinnula VL, Keski-Oja J (2006) Bone morphogenetic protein-4 inhibitor gremlin is overexpressed in idiopathic pulmonary fibrosis. *Am J Pathol* 169: 61–71

81 Boers W, Aarrass S, Linthorst C, Pinzani M, Elferink RO, Bosma P (2006) Transcriptional profiling reveals novel markers of liver fibrogenesis: gremlin and insulin-like growth factor-binding proteins. *J Biol Chem* 281: 16289–16295

82 Sun J, Zhuang FF, Mullersman JE, Chen H, Robertson EJ, Warburton D, Liu YH, Shi W (2006) BMP4 activation and secretion are negatively regulated by an intracellular gremlin-BMP4 interaction. *J Biol Chem* 281:29349–29356

83 Smith WC, Harland RM (1992) Expression cloning of noggin, a new dorsalizing factor localized to the Spemann organizer in Xenopus embryos. *Cell* 70: 829–840

84 Lamb TM, Knecht AK, Smith WC, Stachel SE, Economides AN, Stahl N, Yancopolous GD, Harland RM (1993) Neural induction by the secreted polypeptide noggin. *Science* 262: 713–718

85 Groppe J, Greenwald J, Wiater E, Rodriguez-Leon J, Economides AN, Kwiatkowski W,

Affolter M, Vale WW, Belmonte JC, Choe S (2002) Structural basis of BMP signalling inhibition by the cystine knot protein Noggin. *Nature* 420: 636–642

86 Brunet LJ, McMahon JA, McMahon AP, Harland RM (1998) Noggin, cartilage morphogenesis, and joint formation in the mammalian skeleton. *Science* 280:1455–1457

87 McMahon JA, Takada S, Zimmerman LB, Fan CM, Harland RM, McMahon AP (1998) Noggin-mediated antagonism of BMP signaling is required for growth and patterning of the neural tube and somite. *Genes Dev* 12: 1438–1452

88 Wijgerde M, Karp S, McMahon J, McMahon AP (2005) Noggin antagonism of BMP4 signaling controls development of the axial skeleton in the mouse. *Dev Biol* 286: 149–157

89 Miyazaki Y, Ueda H, Yokoo T, Utsunomiya Y, Kawamura T, Matsusaka T, Ichikawa I, Hosoya T (2006) Inhibition of endogenous BMP in the glomerulus leads to mesangial matrix expansion. *Biochem Biophys Res Commun* 340: 681–688

90 Wilkinson L, Kolle G, Wen D, Piper M, Scott J, Little M (2003) CRIM1 regulates the rate of processing and delivery of bone morphogenetic proteins to the cell surface. *J Biol Chem* 278: 34181–34188

91 Georgas K, Bowles J, Yamada T, Koopman P, Little MH (2000) Characterisation of Crim1 expression in the developing mouse urogenital tract reveals a sexually dimorphic gonadal expression pattern. *Dev Dyn* 219: 582–587

92 Kolle G, Georgas K, Holmes GP, Little MH, Yamada T (2000) CRIM1, a novel gene encoding a cysteine-rich repeat protein, is developmentally regulated and implicated in vertebrate CNS development and organogenesis. *Mech Dev* 90: 181–193

93 Lovicu FJ, Kolle G, Yamada T, Little MH, McAvoy JW (2000) Expression of Crim1 during murine ocular development. *Mech Dev* 94(1–2): 261–265

94 Wilkinson L, Gilbert T, Kinna G, Ruta LA, Pennisi D, Kett M, Little MH (2007) Crim1KST264/KST264 mice implicate Crim1 in the regulation of vascular endothelial growth factor-A activity during glomerular vascular development. *J Am Soc Nephrol* 18: 1697–1708

95 Pennisi DJ, Wilkinson L, Kolle G, Sohaskey ML, Gillinder K, Piper MJ, McAvoy JW, Lovicu FJ, Little MH (2007) Crim1KST264/KST264 mice display a disruption of the Crim1 gene resulting in perinatal lethality with defects in multiple organ systems. *Dev Dyn* 236: 502–511

96 Fung WY, Fat KF, Eng CK, Lau C (2007) crm-1 facilitates BMP signaling to control body size in *Caenorhabditis elegans*. *Dev Biol* 311: 95–105

97 Lin J, Patel SR, Cheng X, Cho EA, Levitan I, Ullenbruch M, Phan SH, Park JM, Dressler GR (2005) Kielin/chordin-like protein, a novel enhancer of BMP signaling, attenuates renal fibrotic disease. *Nat Med* 11: 387–393

98 Lin J, Patel SR, Wang M, Dressler GR (2006) The cysteine-rich domain protein KCP is a suppressor of transforming growth factor beta/activin signaling in renal epithelia. *Mol Cell Biol* 26: 4577–4585

99 Conley CA, Silburn R, Singer MA, Ralston A, Rohwer-Nutter D, Olson DJ, Gelbart W, Blair SS (2000) Crossveinless 2 contains cysteine-rich domains and is required for high

levels of BMP-like activity during the formation of the cross veins in Drosophila. *Development* 127: 3947–3959

100 Binnerts ME, Wen X, Cante-Barrett K, Bright J, Chen HT, Asundi V, Sattari P, Tang T, Boyle B, Funk W, Rupp F (2004) Human Crossveinless-2 is a novel inhibitor of bone morphogenetic proteins. *Biochem Biophys Res Commun* 315: 272–280

101 Rentzsch F, Zhang J, Kramer C, Sebald W, Hammerschmidt M (2006) Crossveinless 2 is an essential positive feedback regulator of Bmp signaling during zebrafish gastrulation. *Development* 133: 801–811

102 Ikeya M, Kawada M, Kiyonari H, Sasai N, Nakao K, Furuta Y, Sasai Y (2006) Essential pro-Bmp roles of crossveinless 2 in mouse organogenesis. *Development* 133: 4463–4473

Induction of cementogenesis and periodontal ligament regeneration by the bone morphogenetic proteins

Ugo Ripamonti, Jean-Claude Petit and June Teare

Bone Research Unit, MRC/University of the Witwatersrand, 7 York Road Medical School, 2193 Parktown Johannesburg, South Africa

Bone: Formation by autoinduction

The complex tissue morphologies of the periodontal tissues, the locking of the teeth into the alveolar bone with the overlying gingival tissues, and the temporo-mandibular joint with the associated powerful masticatory muscles are a superb example of design architecture and engineering.

The challenging theme of the emergence of the complex tissue morphologies of the periodontal tissues is the molecular basis of morphogenesis and the induction of cementogenesis, which provides the pattern formation of periodontal tissue regeneration with faithful insertion of periodontal ligament fibers into the newly formed cementum along the exposed root dentin after acute/chronic episodes of destructive periodontitis in man and other animals [1–15].

The final goal of periodontal therapy is the generation of functionally oriented periodontal ligament fibers inserted into the newly formed cementum and alveolar bone [2, 4, 6, 7, 15]. Regenerative phenomena invoked by tissue engineering and regenerative medicine, are still learning the secrets of its principles from the phenomenon of bone: formation by autoinduction [16]; indeed, the rules that govern the induction of bone formation are being applied to the regeneration of the periodontal tissues, that is the induction of cementogenesis with insertion of Sharpey's fibers into the newly formed cementum, the essential ingredients to engineer periodontal tissue regeneration [2, 15].

These developments have arisen from a desire to understand fundamental developmental processes, the control of cell differentiation and the generation of form or morphogenesis. One expectation of this research is the discovery of regulatory morphogens with novel biological activities and therapeutic potential in clinical context [1]. One of the most exciting advances in morphogenesis and tissue engineering has been the recognition of the extracellular matrix of bone as a multifactorial repository of locally active pleiotropic morphogenetic proteins that initiate and modulate the cascade of endochondral bone formation by induction [1, 8, 17].

Since antiquity, during speciation and phylogeny of the *Australopithecinae* and *Homo* species at the Pleio-pleistocene boundary in Southern Africa, the hard evidence of periodontal bone loss has been dramatically shown to paleoanthropologists and paleopathologists alike (Fig. 1) [15]. After millions of years of evolution and further speciation, extant *Homo sapiens* has had the possibility to discover and deploy the osteogenic proteins (OPs) of the transforming growth factor-β (TGF-β) superfamily, engineering thus not only the induction of bone formation but also the induction of periodontal tissue regeneration and cementogenesis [15].

The aim of this chapter is to present a concise account of the effects of bone morphogenetic proteins (BMPs) and OPs on periodontal tissue regeneration highlighting the molecular mechanisms for such regeneration and reporting the pleiotropism of the BMPs that singly, or in combination, induce cementogenesis, periodontal ligament regeneration and the assembly of a regenerated periodontal ligament system with Sharpey's fibers attached to both newly formed cementum and alveolar bone [15, 17].

The initiation of *de novo* endochondral bone formation by induction in heterotopic extraskeletal sites upon implantation of alcohol-extracted and/or demineralized bone matrices has prompted a concerted major research effort in the isolation and identification of a class of soluble molecular signals responsible for the induction of bone formation [8, 13, 17]. During last century research, a stumbling block was the realization that the bone matrix is in the solid state [8, 18]. A major advancement for the recognition of the critical role of the soluble signals initiating the induction of bone formation has been the recognition that the extracellular matrix of bone could be dissociatively extracted by chaotropic agents yielding an insoluble component, mainly insoluble collagenous bone matrix and soluble signals, the putative BMPs tightly bound to the extracellular matrix of bone [19, 20].

We now know that the molecular and morphogenetic cascade of events initiated by the heterotopic extraskeletal implantation of demineralized bone matrix or by purified OPs of the TGF-β superfamily are highly reminiscent of embryonic development [8, 10, 11, 13, 17, 21]. Embryonic development is thus a template for periodontal tissue regeneration [15] since soluble molecular signals exploited in embryogenesis are re-exploited and re-deployed postnatally to engineer tissue morphogenesis and the induction of periodontal tissue regeneration [11, 13, 15, 17].

We have asked previously [2] which are the cell populations originating from both the periodontal ligament space including the cementum and the residual bone housing that respond to the soluble osteogenic molecular signals of the TGF-β superfamily? Only recently, studies have started to identify and/or to discuss progenitor stem cells capable of differentiation towards periodontal ligament and cementoblast-like cells [22–24].

The fundamental work of Senn [25], Sacerdotti and Frattin [26], Huggins [27], Levander [28, 29], Levander and Willestaedt [30], Ray and Holloway [31], Bridges and Pritchard [32], Sharrard and Collins [33], Burger et al. [34], Trueta [35], Urist

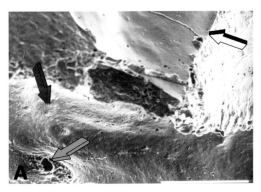

Figure 1
Hard evidence of alveolar bone loss in fossilized gnathic remains of adult (A) and juvenile (B) Australopithecus africanus *specimens unearthed at Sterkfontein Transvaal South Africa, 2–3 million years before the present. (A) Lipping of the remaining buccal alveolar bone (red arrow) exposing the furcation defect of a mandibular molar. Below the lipping of the alveolar bone, there is the presence of a perforating vascular canal (blue arrow) penetrating the remaining bony housing. (B) Prepubertal periodontitis in a juvenile A. africanus specimen with horizontal and cuneiform bone loss (magenta arrow) affecting the supporting alveolar bone of a deciduous maxillary molar [15]. White arrows in (A) and (B) indicate the cementoenamel junction.*

[16], Urist et al. [36], Urist and Strates [37], and Reddi and Huggins [38] has shown that several mineralized and non-mineralized extracellular matrices of mammalian tissues including the extracellular matrices of uroepithelium, bone and dentine contain an "osteogenic activity/potency", which is endowed with the striking prerogative of initiating bone formation by induction in heterotopic extraskeletal sites of recipient animal models [13, 15, 17]. This classic work has indicated that the "osteogenic activity" resides within the extracellular matrices of different tissues and, when implanted in heterotopic extraskeletal sites of animal models, this osteogenic activity diffuses out of the implanted matrices interacting with transforming resident mesenchymal cells capable of differentiation into chondroblastic and osteoblastic cell lines initiating bone differentiation by induction [8, 13, 17].

In 1889, Senn used well-decalcified antiseptic bone to treat skull and excavated long bone defects in canine models [25]. Senn demonstrated the value of implantation of a disk or plate of decalcified bone in canine calvarial defects [25]. Astutely, he thus noticed that the implanted decalcified matrix was incorporated and removed by a large mass of embryonic tissue, a condition he stated "... *is favorable to the formation of new bone at the site of the operation*" [25]. His classic studies concluded "*antiseptic decalcified bone is the best substitute for living bone grafts in restoration of a loss of substance in bone*" [25].

Although the above-mentioned experiments hypothesized the presence of morphogens, firstly described by Turing as *"forms generating substances"* [39] and later as a BMP complex within the tested extracellular matrices [16, 36, 37], its identification has been hindered by the fact that the extracellular matrix of bone exists in the solid state and that the organic components of the matrix, which includes small quantities of putative OPs, are tightly bound to the inorganic mineral phase of the bone matrix [8, 19].

The paper of Ray and Holloway [31] recognized the heterogeneity of the two components of the bone matrix, i.e., a *"heterogeneous system of organic matrix and inorganic salts"* and thus the study was set to determine which of the various components of the bone would be *"most readily replaced"* [31]. In a series of experiments in calvarial defects in the parietal bone of adult Long-Evans rats, Ray and Holloway prepared and implanted frozen intact homogenous bone, deproteinized bone and decalcified homogenous bone matrix [31]. Histological examination showed that decalcified bone matrix, i.e., the organic matrix of bone devoid of its organic salts, was the best substitute for fresh autogenous bone, also showing that several endothelial elements invaded the particulate fragments of the implanted demineralized matrix [31]. Importantly, it was also reported that the presence of inorganic salts in the implant would appear to impede rather than accelerate the process of bone replacement. Ray and Holloway [31] concluded, like Senn [25], *"antiseptic decalcified bone is the best substitute for living bone grafts and the restoration of loss of substance in bone"* [25]. The experiments of Ray and Holloway [31] recognized the importance of the various components of the bone matrix for the process of bone replacement after implantation of differently treated bone matrices. Such experiments were thus critical for the later understanding of the chaotropic dissociative extraction and reconstitution of the extracellular matrix components restoring the biological activity of the extracted proteins [8, 19].

Further to the studies of Ray and Holloway [31], Sharrard and Collins [33] presented controlled observations on the implantation of decalcified autogenous bone alongside the vertebral spine of three scoliotic children together with biopsy material harvested 6 weeks after implantation [33]. The results showed that the decalcified autogenous bone *"was a good scaffold for the appositional growth of new reparative bony callus"* [33]. Burger et al. [34], in a number of studies using demineralized bone matrices transplanted with or without chondroitin sulfate in rodents calvarial defects, also confirmed the findings of Ray and Holloway [31] and Senn [25] as described above. Burger et al. [34] reported osteoblastic activity in newly formed bone as early as 7 days after implantation of demineralized bone matrices in rodent calvarial defects, discussing the hypothesis that bone contains an extractable substance, or substances, capable of inducing new bone formation [34].

Building thus on significant published experimental work by Levander [28, 29], Levander and Willestaedt [30], Ray and Holloway [31], Bridges and Pritchard [32], Burger et al. [34], Urist, borrowing the term *'induction'* from Spemann and

Levander [29], produced reproducible evidence that implantation of demineralized bone matrices in heterotopic intramuscular sites in rodents results in the induction of bone differentiation [16]. Which are the molecular signals or morphogens that initiate the cascade of bone formation by induction? Solubilization of the putative OPs from demineralized bone matrix yielded soluble molecular signals and insoluble signals or substrata, mainly insoluble collagenous bone matrix that, when reconstituted with the extracted soluble osteogenic molecular signals, restored the biological activity of the intact demineralized bone matrix [19, 20]. The combination of an insoluble signal or substratum, with solubilized osteogenic soluble signals was a key experiment that propelled the bone induction principle into the pre-clinical and clinical arena culminating in the isolation and purification of an entirely new family of protein initiators, the BMP/OP members of the TGF-β supergene family [40]. Amino acid sequences of tryptic peptides resulted in the molecular cloning and expression of several recombinant human BMPs/OPs [8, 17, 40, 41].

Highly purified naturally derived and recombinant human BMPs/OPs induce heterotopic endochondral bone differentiation in a variety of animal models including non-human and human primates (Fig. 2) [8–11, 17]. The purification and molecular cloning of the BMPs/OPs have set the stage for novel therapeutic approaches to correct congenital and acquired orthopedic, craniofacial and periodontal defects [8–11, 15, 17]. The premises of BMPs/OPs as molecular therapeutics for regenerative medicine and tissue engineering rest on the evidence that induction and regeneration of endochondral bone differentiation in postnatal models recapitulate events that occur in the normal course of embryonic development [8–11, 15, 17]. Thus, therapeutic osteogenesis induced by local administration of naturally derived and/or recombinant human BMPs/OPs exploits a functionally conserved process originally deployed in embryonic development and recapitulated in postnatal tissue induction and morphogenesis [42].

Structure/activity profile and apparent redundancy of osteogenic soluble molecular signals

The distinct patterns of expression of BMP/OP transcripts in several developing organs and tissues not limited to skeletal elements, strongly indicates multiple and specialized roles in vertebrate organogenesis [8, 17, 43, 44]. The fact that several recombinant hBMPs/OPs initiate endochondral bone formation in the heterotopic bioassay in rodents raises important questions on the biological significance of this apparent redundancy of soluble molecular signals that singly are endowed with the striking prerogative of initiating the cascade of bone differentiation by induction [8, 9, 17, 21]. Limited promiscuity amongst receptors in binding different BMPs/OPs also suggests that different morphogenetic proteins may have different functions *in vivo* [11, 17]. An important question for tissue engineering of bone and regenera-

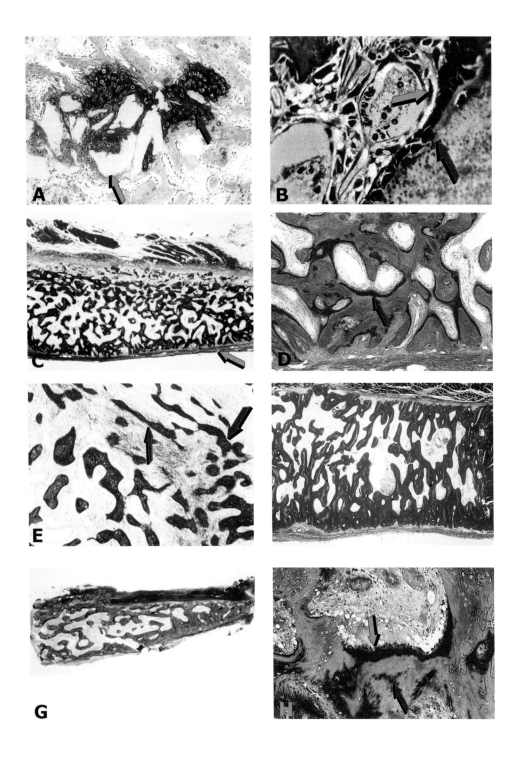

tive medicine at large is whether the presence of molecularly different isoforms of BMPs/OPs has a therapeutic significance [9–11, 15, 17]. The most obvious proposal for the role of BMPs/OPs in mammalian systems is that of molecular initiators of bone formation. This can be exploited for the induction of bone regeneration of craniofacial and axial skeletal defects in clinical contexts. Ample evidence has been accrued on the efficacy and safety of naturally derived BMPs/OPs and recombinant hOP-1 and hBMP-2 for the regeneration of craniofacial and appendicular bone defects in animal models including non-human primates which have thus initiated clinical trials and commercialization of the recombinant proteins [45, 46].

The expression pattern of different BMPs/OPs during embryonic development suggests novel strategies of therapeutic intervention outside the mere osseous domain [17, 43, 44]. Thus, the critical role of BMP-2, BMP-3, BMP-4 and OP-1 as soluble mediators of epithelial mesenchymal interactions during tooth morphogenesis [43, 44] has indicated that specific BMPs/OPs can also be used for tissue engineering of the periodontal tissues [2, 4, 15].

Indeed, periodontal regenerative studies in several animal models including canines and non-human primates have shown that naturally derived and recom-

Figure 2
Bone: Formation by autoinduction [16] in rodents (A, B), the non-human primate Papio ursinus *(C–F) and the human primate* Homo sapiens *(G, H) 11 (A), 7 (B), 30 (C, D, E), and 90 (F, G, H) days after implantation of soluble osteogenic molecular signals. (A) Endochondral bone induction as a recapitulation of embryonic development 11 days after heterotopic subcutaneous implantation of baboon bone-derived bone morphogenetic proteins (BMPs)/osteogenic proteins (OPs) purified greater than 50 000-fold after heparin-Sepharose and S-200 gel filtration chromatography. The blue arrow in (A) points to the insoluble collagenous bone matrix as carrier for the osteogenic activity of the highly purified OPs. The magenta arrow indicates islands of induced endochondral anlage recapitulating development. (B) Very intimate relationship between invading capillaries and differentiating cells (magenta arrow) attached to the matrix compartment of the implanted insoluble collagenous carrier reconstituted with naturally derived osteogenin purified to homogeneity from baboon bone matrices showing the invading capillary with endothelial cells (blue arrow) migrating from the vascular basement membrane compartment. (C, D) Low and high power views of calvarial regeneration 30 days after implantation of highly purified baboon BMPs/OPs after gel filtration Sephacryl S-200 chromatography of extracts of baboon bone matrices. The blue arrow in (C) indicates the dural layer, and the magenta arrow in (D) points to osteoid seams populated by contiguous osteoblasts. (E) Induction of prominent osteoid seams (magenta arrows) after implantation of 100 μg hOP-1/g gamma irradiated bovine insoluble collagenous bone matrix in the rectus abdominis of* Papio ursinus *on day 30. (F) Complete regeneration of a calvarial defect of an adult baboon after implantation of 0.5 mg hOP-1/g carrier matrix on day 90. (G, H) Bioptic material of undecalcified sections of newly formed bone in* Homo sapiens *after implantation of highly purified bovine osteogenic fractions implanted in a mandibular defect, showing in (G) mineralized bone in blue surfaced by osteoid seams in red/orange (blue arrow in H) with remnants of collagenous matrix as carrier (magenta arrow in H).*

binant hBMPs/OPs induce periodontal tissue regeneration [15]. It is important to point out that different morphological results can be ascribed to the specific isoform used as well in the animal model [15]. As predicted more than 10 years ago [2], the choice of a suitable morphogen is still a formidable challenge for the practicing periodontologist [2]. The future direction will depend on optimal combinations of OPs and, more importantly, on developing a structure/activity profile of the now available recombinant hBMPs/OPs [15, 17].

To date, several molecularly different proteins with BMP/OP-like sequences and activities have been sequenced and cloned [8, 15, 17, 40, 41]. Little is known about their interactions during bone formation by induction or about the biological and therapeutic significance of this apparent redundancy. The biological significance of redundancy in controlling multiple specialized functions beyond bone, rests on the homologies and amino acid sequence variations of the C-terminal domain of each morphogenetic protein [13–15]. The therapeutic implications of the biological significance of redundancy will be ascribed by developing a structure activity profile amongst homologous members of the BMP/OP family [13–15]. It will be important to evaluate *in vivo* and in primate models the periodontal morphological impulse of each single and structurally molecularly different recombinant hBMP/OP; *in vivo* studies in primate models are mandatory since we have shown that primate tissues react differently from other animal species, such as rodents and canines, to the implanted soluble molecular signals resulting in the induction of different tissue morphologies by structurally related but molecularly different morphogens when applied to primate tissues [9–11, 17].

Amino acid variations in the C-terminal domain of each isoform confer specialized activities to a BMP/OP isoform, and this is the molecular basis that determines the structure/activity profile of BMP/OP proteins [14, 15, 17]. *In vivo* studies in primate models including humans are bound to discover additional pleiotropic activities of different protein isoforms based on specific amino acid sequences. Our studies reported below, the first to address in primate tissues the structure/activity profile amongst BMP/OP family members, have indicated that tissue induction and morphogenesis by either hOP-1 or hBMP-2 are qualitatively different when applied singly to exposed root surfaces in the non-human primate *Papio ursinus* [47].

Single administration of relatively low doses of hOP-1 (0.1 and 0.5 mg hOP-1/g collagenous matrix) preferentially induced cementogenesis as evaluated 60 days after implantation in surgically induced mandibular furcation defects in *Papio ursinus* (Fig. 3) [3, 48]. This seemingly specific cementogenic function of hOP-1 has indicated that a structure/activity profile exists amongst BMP/OP family members to control tissue morphogenesis of disparate tissues and organs [3, 4, 9, 15, 21]. The inductive specificity is additionally regulated at least in part by the extracellular matrix micro-environment, i.e., bone and/or dentine [3, 15, 17].

Based on gene transcript expressions and on immunolocalization patterns of BMP-2, BMP-3 and OP-1 during tooth root morphogenesis, the memory of developmental events during cementogenesis and periodontal ligament formation could be re-deployed therapeutically by combining OP-1 and BMP-3 or OP-1 and BMP-2 [44, 47]. Co-localization of OP-1 and BMP-2 mRNA transcripts during tooth morphogenesis indicates that the two expressed gene products of the BMP/OP family cooperatively induce tissue morphogenesis during development [49]. Mandibular furcation defects prepared in *Papio ursinus* were used to assess whether qualitative morphological patterns of periodontal tissue regeneration could be enhanced and tissue morphogenesis modified by binary or single applications of recombinant hOP-1 and hBMP-2 [47]. Undecalcified sections prepared 60 days after implantation of 100 µg hOP-1 in mandibular furcation defects showed substantial cementogenesis as reported in previous studies using identical and higher doses of the recombinant protein (Fig. 3) [3, 48]. On the other hand, recombinant hBMP-2 applied singly induced greater amounts of mineralized bone and osteoid when compared to hOP-1 (Fig. 3). Cementogenesis was limited and histomorphometry showed a temporal enhancement of alveolar bone regeneration and remodeling [47]. Synchronous but spatially different *op-1* and *bmp-2* expression during murine root development and tooth morphogenesis points to specific functions of the isoforms in periodontal tissue morphogenesis and thus regeneration in postnatal life [11, 15, 17, 49]. It is noteworthy that hBMP-2 induced also substantial alveolar bone regeneration in canine models [50–53], whereas cementogenesis was limited when compared to the extent of bone regeneration [50–53]. Clear-cut morphological differences in periodontal tissue induction point to the structure/activity profile amongst members of the BMP/OP family, highlighting morphogenetic pleiotropic functions in relation to minor but specific amino acid sequence variation in the active C-terminal domains [13–15, 17, 47].

Cementogenesis and periodontal ligament regeneration by the osteogenic proteins of the TGF-β superfamily

Periodontal tissue regeneration entails the rapid colonization of the treated root surface by cementoblasts and the induction of cementogenesis along the denuded root surface with the insertion of Sharpey's fibers into the newly regenerated cementum [2, 7, 15, 54]. Cementogenesis with the faithful insertion of Sharpey's fibers is the true key of periodontal tissue engineering even before the induction of alveolar bone.

Compositional and structural modifications of extracellular matrices and of basement membrane components play major roles in the differentiation and maintenance of specific phenotypes [17]. It is highly possible that, as a recapitulation of embryonic development, exposed dentin may release chemoattractants or act as chemoattractant to direct cell migration of pre-cementoblasts/cementoblasts toward the dentinal surface [55]. Dentin matrix retains BMPs/OPs as a memory of develop-

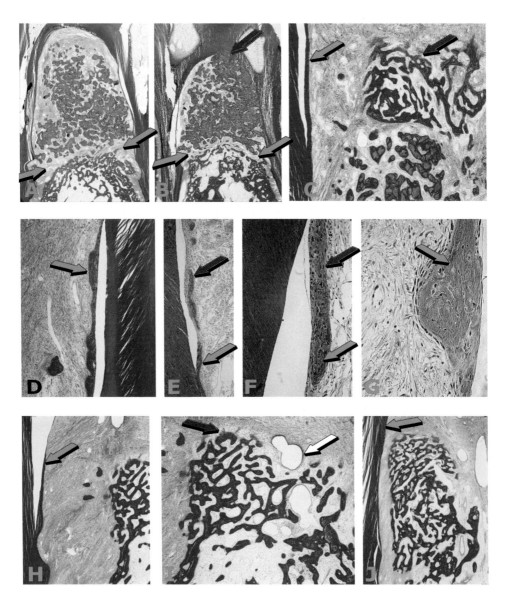

Figure 3
Tissue induction, morphogenesis and regenerative phenomena as a recapitulation of embryonic development in furcation defects of mandibular molars of the non-human primate Papio ursinus *after implantation of recombinant hBMPs/OPs. (A, B) Extensive cementogenesis induced by 0.1 and 0.5 mg hOP-1/g insoluble collagenous bone matrix 60 days after implantation in class II furcation defects of* Papio ursinus. *Blue arrows point to the newly formed pseudo-ligament space between the newly formed mineralized matrix implanted in the furcation defect and the remaining alveolar bone housing after preparation of the notches*

mental events in embryogenesis during tooth morphogenesis as highlighted by the osteogenic activity of demineralized dentin matrix in the *rectus abdominis* muscle of *Papio ursinus* [56]. Exposed dentin surfaces additionally root planed and citric acid conditioned may release BMPs/OPs as chemoattractants and inducers that direct pre-cementoblasts and cementoblasts to attach and migrate coronally along the root surface.

Which are the cell populations originating from both the periodontal ligament space and the residual bony housing that respond to the soluble molecular signals of the TGF-β superfamily? [2] Classic studies have suggested that osteoblasts, cementoblasts and their progenitors that are found in the periodontal ligament space may have their origin from the endosteal spaces of the alveolar process [57]. Expression of different phenotypes and generation of cementum or alveolar bone may be dependent on whether a common lineage of progenitor cells attach on residual cementum or exposed dentin or stay in the alveolar bony side of the periodontal ligament space [57].

Recent work by Seo et al. [24] has suggested that the periodontal ligament space contains stem cells that have the potential to generate cementum and periodontal ligament *in vivo* [24]. The isolated periodontal ligament cells are clonogenic and highly proliferative, indicating those are multipotent stem cells capable of developing into cementoblast-like cells *in vivo* [24]. Later studies identified putative stem cells in the periodontal ligament space in paravascular and extra-vascular regions, with the latter in close proximity to the cementum [22]. Localization of putative periodontal ligament stem cells by selected antigenic markers in paravascular regions around capillaries within the periodontal ligament space highly suggests that the paravascular cells recently reported by Chen et al. [22] are the responding cells to the BMPs as reported by Urist in his classic studies on the "*bone induction principle*" [36].

The magnitude and quality of new connective tissue attachment formation observed after application of naturally derived highly purified BMPs/OPs predomi-

---

on the exposed root surfaces. (B) Extensive attachment of the newly formed and mineralized cementogenic matrix (magenta arrow) to the root dentin of the exposed furcation defect. (C) High-power view showing newly formed cementum (blue arrow) and osteoid seams (magenta arrow) surfacing newly formed mineralized bone in blue. (D–G) Details of tissue induction and morphogenesis of furcation defects implanted with 100 and 500 μg hOP-1/g matrix and harvested on day 60. Blue arrows point to mineralized cementum in blue with areas of cementoid matrix (magenta arrows) still to be mineralized. (H, I) Details of tissue induction by 100 μg hBMP-2/g collagenous matrix as carrier showing limited cementogenesis (blue arrow in H) but prominent osteogenesis and angiogenesis (magenta and white arrows in I). (J) Binary application of 100 μg each of hOP-1 and hBMP-2 showing prominent osteogenesis with osteoid synthesis in the newly formed alveolar bone with restoration of cementogenesis (blue arrow) along the exposed root surface.

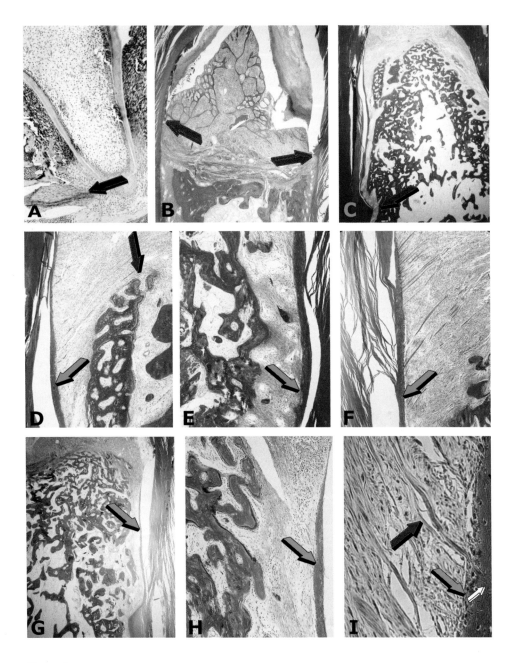

Figure 4
Recapitulation of embryonic development after implantation of highly purified naturally derived BMPs/OPs purified greater than 50 000-fold with respect to crude extracts of bovine bone matrices predominantly containing BMP-3/osteogenin and OP-1/BMP-7 in furcation

nantly containing BMP-3 (osteogenin) and OP-1 has shown that highly purified OPs also initiate cementogenesis and the assembly of a functionally oriented periodontal ligament space (Fig. 4) [58]. Thus, the presence of multiple molecularly different BMPs/OPs reflects a biological significance locally regulating the regeneration of other tissues including the cementum and the assembly of the periodontal ligament space [58].

BMPs/OPs have been used extensively by researchers to induce periodontal tissue regeneration in a variety of animal models, including rodents, canines, and non-human primates with varying degrees of success [14, 15, 47, 50–53, 58]. Naturally derived BMPs/OPs, purified more than 50 000-fold with respect to crude extract of bovine bone matrices, applied to mandibular furcation defects of the non-human primate *Papio ursinus* induce cementogenesis with the insertion of periodontal ligament fibers into newly formed cementum (Fig. 4) [58].

Several studies have proposed that BMPs/OPs possess a structure/activity profile with BMP-2 exhibiting mainly osteogenic properties, whereas OP-1, also known as BMP-7, was mainly cementogenic when applied to periodontal defects and in contact with dentin extracellular matrices (Tab. 1) [3, 14, 15, 47, 48]. Periodontal regenerative studies using hBMP-2 [47, 50, 53] have demonstrated enhanced alveolar bone formation and limited cementum formation [15, 47]. On the other hand, regenerative studies using hOP-1 have shown limited osteogenesis but a superior cementogenic induction, clearly indicating that the induction of different tissue morphologies is due to the structure/activity profile of the tested recombinant proteins [14, 15, 17, 47, 48]. Long-term studies using hOP-1 in canine and non-human

---

defects of Papio ursinus *and harvested on day 60. (A) BMP-3 immunolocalization in the developing murine periodontal tissues including the cementum, alveolar bone and periodontal ligament system. The blue arrow points to bmp-3 expression within the inferior alveolar nerve. (B) Low-power view of an untreated class II periodontal defect in* Papio ursinus *showing the extent of the defect with granulation tissue and apical migration of the junctional epithelium (magenta arrows). (C) Extensive regeneration of newly formed cementum, periodontal ligament and alveolar bone coronally to the notch (magenta arrow), indicating the remaining of the alveolar bony housing after surgical creation of a class II furcation defect. (D–F) Details of periodontal tissue regeneration in furcation defects treated with 250 μg highly purified bovine BMPs/OPs fractions [58]. (D) Induction of cementogenesis along the exposed root surface (blue arrow) with the induction of mineralized bone in blue covered by osteoid seams in red/orange (magenta arrow). (E, F) Assembly of a regenerated periodontal ligament structure with Sharpey's fibers uniting the newly formed cementum (blue arrows) with the newly formed alveolar bone. (G–I) Regeneration of the three components of the periodontal tissues, i.e., cementum (blue arrows), periodontal ligament fibers and alveolar bone in blue surfaced by osteoid seams populated by contiguous osteoblasts. The blue arrow in (I) indicates cementoblasts with a surface layer of newly deposited still-to-be-mineralized cementoid with attached generated Sharpey's fibers (magenta arrow) deeply embedded (white arrow) within the newly formed mineralized cementum.*

Table 1 - Overview of BMP performance in periodontal defects in animal models, with emphasis on osteogenesis and cementogenesis. With the exception of the canine study by Kinoshita et al., (1997), implantation with rhBMP-2 demonstrates mainly osteogenic activities while rhBMP-7 exhibits mainly cementogenic capabilities.

Animal model	Time period	Carrier	Naturally-derived BMPs	BMP-2	BMP-6	BMP-7 (OP-1)	BMP-12	Defect type & size	Alveolar bone	Cementum	Refs.
Rat	28 days	Collagen sponge			0.1 µg/µl 0.3 µg/µl 1.0 µg/µl			Fenestration 1.5 x 3 mm	28.1±4.5 46.7±5.7 29.2±4.5 (Area 10^{-2} mm^2)	4.88±0.63 8.43±0.76 4.82±0.60 (Area µm^2)	Huang et al., 2005
Dog	56 days	PLGA + autologous blood	20 µl/ 100 µl					Class III 3.7±0.3 (Height mm)	3.5±0.6 (Height mm)	1.6±0.6 (Height mm)	Sigurdsson et al., 1995
Dog	56 days	Collagen matrix				0.75 mg/g 2.5 mg/g 7.5 mg/g		Class III 4.82±0.65 5.19±0.90 5.18±0.92 (Height mm)	2.26±1.66 2.35±1.97 3.91±1.70 (Height mm)	2.38±1.65 2.07±1.34 3.94±1.51 (Height mm)	Giannobile et al., 1998
Dog				0.2 mg/ml				Class III 5.8±0.4 (Height mm)	4.1±1.6 (Height mm)	2.5±1.4 (Height mm)	Wikesjö et al., 2004
Dog	56 days	Collagen sponge					0.04 mg/ml 0.2 mg/ml 1.0 mg/ml	Class III 6.0±0.7 5.9±0.8 5.9±0.5 (Height mm)	3.1±1.9 3.3±2.2 3.4±1.3 (Height mm)	2.2±1.0 2.4±1.3 1.4±0.8 (Height mm)	

Species	Time	Carrier	Dose	Defect	Measurement (Height mm)	Measurement (Height mm)	Measurement (Height mm)	Reference
Dog	84 days	PLGA + gelatin	40 μl/ 100 μl	Class III	3.61±0.48	0.68±0.63	1.80±0.51	Kinoshita et al., 1997
Dog	56 days	Collagen sponge	0.2 mg/ml	3-wall intrabony	5.5±0.3	3.6±0.6	3.0±1.4	Choi et al., 2002
	168 days			3-wall intrabony	5.0±0.4	3.4±0.6	2.4±1.2	
Baboon	60 days	ICBM	250 μl/ 150 mg	Class II *Distal* *Mesial*	7.79±0.17 7.46±0.33	*Distal* 4.91±0.57 *Mesial* 5.05±0.45	*Distal* 5.79±0.52 *Mesial* 5.16±0.43	Ripamonti et al., 1994
			100 μg/g	Class II *Distal* *Mesial*	7.8±0.4 8.4±0.2	*Distal* 4.2±0.2 *Mesial* 3.6±0.1	*Distal* 3.2±0.3 *Mesial* 3.7±0.4	Ripamonti et al., 2001
Baboon	60 days	ICBM	100 μg/g	Class II *Distal* *Mesial*	8.6±0.1 7.6±0.6	*Distal* 3.7±0.4 *Mesial* 2.6±0.4	*Distal* 5.7±0.3 *Mesial* 5.1±0.9	

Table 1 (continued)

Animal model	Time period	Carrier	Naturally-derived BMPs	BMP-2	BMP-6	BMP-7 (OP-1)	BMP-12	Defect type & size	Alveolar bone	Cementum	Refs.
Baboons with P. gingivalis	6 months	ICBM				0.5 mg/g 2.5 mg/g		Class II 1st Molar 7.0 6.8 (Height mm)	1st Molar 5.5 4.9 (Height mm)	1st Molar 6.4 4.9 (Height mm)	Ripamonti et al., 2002
						0.5 mg/g 2.5 mg/g		Class II 2nd Molar 7.6 7.2 (Height mm)	2nd Molar 7.1 4.6 (Height mm)	2nd Molar 6.9 4.8 (Height mm)	

primate models have also demonstrated osteogenesis with a superior cementogenic component in canine [59] and non-human primate models with inflammatory-induced chronic attachment loss [60].

Periodontal tissue regeneration, induced by hBMP-2 is often accompanied by ankylosis [52, 53, 61]. Further studies have revealed that acid conditioning of root surfaces prior to hBMP-2 application enhanced ankylosis when compared with non-acid conditioning [62]. Saito [63] maintained that applying hBMP-2 directly to a root-planed surface may cause ankylosis and found that the incorporation of a spacer eliminated the incidence of ankylosis but also limited the amount of bone regeneration [63]. Takahashi et al. [64] applied rhBMP-2 to class III periodontal defects that were allowed to heal for 3, 6 and 12 weeks. Ankylosis was observed at 3 and 6 weeks, but not at 12 weeks, which led the investigators to conclude that resolution of ankylosis by osteoclastic activity had taken place [64].

Recently, in a rat fenestration defect study, BMP-6 was applied *via* an absorbable collagen sponges and resulted in new bone and cementum formation in a dose-dependent manner [65]. In a canine study, Wikesjö et al. [52] evaluated hBMP-2 and hBMP12 in periodontal regenerative studies. Recombinant hBMP12 and hBMP-2 were implanted *via* absorbable collagen sponges in periodontal defects and results compared after 8 weeks of healing [52]. More regenerated bone was observed in implants that had received hBMP-2 but ankylosis was noted [52]. Defects implanted with hBMP12 showed less bone regeneration but exhibited a functionally oriented periodontal ligament system inserted into newly formed cementum and alveolar bone [52]. However, in a tooth replantation study using hBMP12, Sorensen et al. [66] noted that a topical application of rhBMP12 to roots, which had been previously denuded of periodontal ligament, failed to re-establish a new periodontal attachment apparatus.

To summarize, the cementum with the insertion of newly generated Sharpey's fibers is at the crux of periodontal tissue engineering [15, 54]; the multiple effects of cemental homeostasis within the periodontal ligament space including bone and the periodontal ligament itself will need to be further studied [52] to ascertain whether cemental matrices of different animal models including primates retain BMP/OP activities as a memory of developmental events [11–15]. The presence of cementum specific proteins still remain questionable [54] though several studies have reported that partially purified cemental extracts contain mitogenic and differentiating proteins which may effect various cell populations within the periodontal ligament space [67, 68]. The chaotropic extraction of cemental matrices and the reconstitution of the solubilized protein components with collagenous matrices followed by implantation in the rat subcutaneous bioassay are still lacking to date; whether cementum contains BMP/OP activities that can initiate the induction of bone formation [8, 15] is as yet unknown. This information will cast additional regulatory roles to the assemblage of the new attachment apparatus with generation of periodontal ligament fibers inserting into the newly formed cementum.

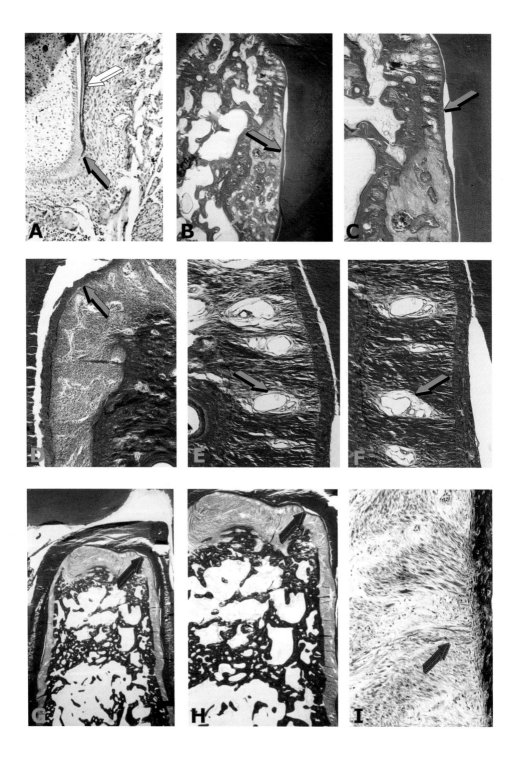

Long-term experiments in the non-human primate *Papio ursinus* have shown that relatively high doses of recombinant hOP-1 induce the regeneration of the three essential components of the periodontium, the cementum, the periodontal ligament, and alveolar bone (Fig. 5) [60]. The study also demonstrated that a single recombinant protein can induce a cascade of molecular and morphological events leading to the regeneration of the alveolar bone, the cementum, and the assembly of a functionally oriented periodontal ligament system [15, 60]. It will be, however, necessary to gain insights into the potential distinct spatial and temporal patterns of expression of TGF-β superfamily members during morphogenesis and regeneration elicited by single application of hOP-1 and/or hBMP-2 to periodontal defects so as to design therapeutic approaches based on information of gene regulation by recombinant hBMPs/OPs as a recapitulation of embryonic development [15, 17, 69].

Acknowledgements

This work is supported by the South African Medical Research Council, the University of the Witwatersrand, Johannesburg, the National Research Foundation, and by *ad hoc* grants to the Bone Research Unit. We thank Jamie Kemler, David Rueger and Stryker Biotech for the supply of recombinant human osteogenic protein-1. Special thanks to Marshall Urist and Hari A. Reddi for having initiated the Bone Research Laboratory to the fascinating scenario of bone: formation by autoinduction.

Figure 5
Embryonic cementogenesis, induction of chronic periodontitis in the non-human primate Papio ursinus *and induction of cementogenesis and periodontal tissue regeneration after long-term application of high doses of hOP-1 delivered by bovine collagenous bone matrices 6 months after implantation in inflammatory-induced furcation defects [60]. (A) Immunolocalization of OP-1 in a developing root of a mandibular molar of a 13-day-old mouse pup with the expression of op-1 in mantle dentin (blue arrow) and during coronal cementogenesis in cementoblasts and cementoid matrix (white arrow) from which there is generation of Sharpey's fibers of the forming periodontal ligament space. (B, C) Low-power views of cementogenesis (blue arrows) and alveolar bone regeneration in periodontal defects induced by inflammatory-infective periodontitis in* Papio ursinus *treated with 0.5 mg hOP-1 (B–F) and 2.5 mg hOP-1 (G–I) delivered by 1 g gamma-irradiated bovine collagenous bone matrix as carrier. (E, F) Regeneration of the periodontal ligament system with prominent vascular canals (arrows) between Sharpey's fibers inserting into newly formed alveolar bone and cementum. (G, H) Restitutio ad integrum of the periodontal tissues after implantation of 2.5 mg hOP-1/g of carrier matrix in chronically induced periodontal defects of* Papio ursinus *showing complete cementogenesis across the furca of the treated mandibular molar. (I) Generation of Sharpey's fibers (arrow) inserting into newly formed highly cellular cementoid induced along the exposed root surface.*

References

1. Ripamonti U, Reddi AH (1992) Growth and morphogenetic factors in bone induction: Role of osteogenin and related bone morphogenetic proteins in craniofacial and periodontal bone repair. *Crit Rev Oral Biol Med* 3: 1–14
2. Ripamonti U, Reddi AH (1994) Periodontal regeneration: potential role of bone morphogenetic proteins. *J Periodontal Res* 29: 225–235
3. Ripamonti U (1996) Induction of cementogenesis and periodontal ligament regeneration by bone morphogenetic proteins. In: TS Lindholm (ed): *Bone morphogenetic proteins: Biology, biochemistry and reconstructive surgery*. Academic Press, Austin, 189–198
4. Ripamonti U, Reddi AH (1997) Tissue engineering, morphogenesis, and regeneration of the periodontal tissues by bone morphogenetic proteins. *Crit Rev Oral Biol Med* 8: 154–163
5. Reddi AH (1988) Role of morphogenetic proteins in skeletal Tissue Engineering and regeneration. *Nat Biotechnol* 16: 247–252
6. Cochran DL, Wozney JM (1999) Biological mediators for periodontal regeneration. *Periodontol 2000* 19: 40–58
7. Bartold PM, McCulloch AG, Narayanan AS, Pitaru, S (2000) Tissue engineering: A new paradigm for periodontal regeneration based on molecular and cell biology. *Periodontol 2000* 24: 253–269
8. Reddi AH (2000) Morphogenesis and tissue engineering of bone and cartilage: Inductive signals, stem cells, and biomimetic biomaterials. *Tissue Eng* 6: 351–359
9. Ripamonti U, Ramoshebi LN, Matsaba T, Tasker J, Crooks J, Teare J (2001) Bone Induction by BMPs/OPs and related family members. The critical role of delivery systems. *J Bone Joint Surg Am* 83-A (Suppl 1): 116–127
10. Ripamonti U (2003) Osteogenic proteins of the TGF-β superfamily. In: HL Henry, AW Norman (eds): *Encyclopedia of hormones*. Academic Press, San Diego, 80–86
11. Ripamonti U, Herbst N-N, Ramoshebi LN (2005) Bone morphogenetic proteins in craniofacial and periodontal tissue engineering: Experimental studies in the non-human primate *Papio ursinus*. *Cytokine Growth Factor Rev* 16: 357–368
12. Ripamonti U, Renton L (2006) Bone morphogenetic proteins and the induction of periodontal tissue regeneration. *Periodontol 2000* 41: 73–87
13. Ripamonti U, Ferretti C, Heliotis M (2006) Soluble and insoluble signals and the induction of bone formation: molecular therapeutics recapitulating development. *J Anat* 209: 447–468
14. Ripamonti U, Teare J, Petit J-C (2006) Pleiotropism of bone morphogenetic proteins: From bone induction to cementogenesis and periodontal ligament regeneration. *J Int Acad Periodontol* 8: 23–32
15. Ripamonti U (2007) Recapitulating development: A template for periodontal Tissue Engineering. *Tissue Eng* 13: 51–71
16. Urist MR (1965) Bone: Formation by autoinduction. *Science* 220: 680–686

17 Ripamonti U (2006) Soluble osteogenic molecular signals and the induction of bone formation. *Biomaterials* 27: 807–822
18 Reddi AH (1994) Bone and cartilage differentiation. *Curr Opin Genet Dev* 4: 737–744
19 Sampath TK, Reddi AH (1981) Dissociative extraction and reconstitution of extracellular matrix components involved in local bone differentiation. *Proc Natl Acad Sci USA* 78: 7599–7603
20 Sampath TK, Reddi AH (1983) Homology of bone-inductive proteins from human, monkey, bovine and rat extracellular matrix. *Proc Natl Acad Sci USA* 80: 6591–6595
21 Ripamonti U (2004) Soluble, insoluble and geometric signals sculpt the architecture of mineralized tissues. *J Cell and Mol Med* 8: 169–180
22 Chen SC, Marino V, Gronthos S, Bartold PM (2006) Location of putative stem cells in human periodontal ligament. *J Periodontal Res* 41: 547–553
23 Hou L-T, Li T-I, Liu C-M, Liu B-Y, Liu C-L, Mi H-W (2007) Modulation of osteogenic potential by recombinant human bone morphogenetic protein-2 in human periodontal ligament cells: effect of serum, culture medium, and osteoinductive medium. *J Periodontal Res* 42: 244–252
24 Seo B-M, Miura M, Gronthos S, Bartold PM, Batouli S, Brahim J, Young M, Robey PG, Wang C-Y, Shi S (2004) Investigation of multipotent postnatal stem cells from human periodontal ligament. *Lancet* 364: 149–155
25 Senn N (1889) On the healing of aseptic bone cavities by implantation of antiseptic decalcified bone. *Am J Med Sci* 98: 219–243
26 Sacerdotti C, Frattini G (1901) Sulla produzione eteroplastica dell'osso. *Riv Accad Med Torino* 825–836
27 Huggins CB (1931) The formation of bone under the influence of epithelium of the urinary tract. *Arch Surg* 22: 377–408
28 Levander G (1938) A study of bone regeneration. *Surg Gyn Obst* 67: 705–714
29 Levander G (1945) Tissue induction. *Nature* 155: 148–149
30 Levander G, Willestaedt H (1946) Alcohol-soluble osteogenetic substance from bone marrow. *Nature* 3992: 587
31 Ray RD, Holloway JA (1957) Bone implants: Preliminary report of an experimental study. *J Bone Joint Surg* 39A: 1119–1128
32 Bridges JB, Pritchard JJ (1958) Bone and cartilage induction in the rabbit. *J Anat* 92: 28–38
33 Sharrard WJW, Collins DH (1961) The fate of human decalcified bone grafts. *Proc R Soc Med* 54: 1101–1102
34 Burger M, Sherman BS, Sobel AE (1962) Observations of the influence of chondroitin sulphate on the rate of bone repair. *J Bone Joint Surg* 44B: 675–687
35 Trueta J (1963) The role of the vessels in osteogenesis. *J Bone Joint Surg* 45B: 402–418
36 Urist MR, Silverman BF, Büring K, Dubuc FL, Rosenberg JM (1967) The bone induction principle. *Clin Orthop Relat Res* 53: 243–283

37 Urist MR, Strates BS (1971) Bone morphogenetic protein. *J Dent Res* 50: 1392–1406
38 Reddi AH, Huggins CB (1972) Biochemical sequences in the transformation of normal fibroblasts in adolescent rats. *Proc Natl Acad Sci USA* 69: 1601–1605
39 Turing AM (1952) The chemical basis of morphogenesis. *Philos Trans R Soc Lond* 237: 37–41
40 Wozney JM, Rosen V, Celeste AJ, Mitsock LM, Whitters MJ, Kriz RW, Hewick RM, Wang EA (1988) Novel regulators of bone formation: Molecular clones and activities. *Science* 242: 1528–1534
41 Celeste AJ, Ianazzi JA, Taylor RC, Hewick RM, Rosen V, Wang EA, Wozney JM (1990) Identification of transforming growth factor beta family members present in bone inductive proteins purified from bovine bone. *Proc Natl Acad Sci USA* 87: 9843–9847
42 Ripamonti U, Ma S, Cunningham N, Yeates L, Reddi AH (1992) Initiation of bone regeneration in adult baboons by osteogenin, a bone morphogenetic protein. *Matrix* 12: 369–380
43 Hogan BLM (1996) Bone morphogenetic proteins: Multifunctional regulators of vertebrate development. *Genes Dev* 10: 1580
44 Thomadakis G, Ramoshebi LN, Crooks J, Rueger CD, Ripamonti U (1999) Immunolocalization of bone morphogenetic protein-2 and -3 and osteogenic protein-1 during murine tooth root morphogenesis and in other craniofacial structures. *Eur J Oral Sci* 107: 368–377
45 Friedlander GE, Perry CR, Cole JD, Cook SD, Cierny G, Muschler GF, Zych CA, Calhoun JH, LaForte AJ, Yin S (2001) Osteogenic protein-1 (bone morphogenetic protein-7) in the treatment of tibial nonunions. *J Bone Joint Surg* 83A: S151–S158
46 Govender S, Csimma C, Genant HK, Valentin-Opran A, Amit Y, Arbel R, Aro H, Atar D, Bishay M, Börner MG et al (2002) Recombinant human bone morphogenetic protein-2 for treatment of open tibial fractures: a prospective, controlled, randomized study of four hundred and fifty patients. *J Bone Joint Surg* 84A: 2123–2134
47 Ripamonti U, Crooks J, Petit J-C, Rueger D (2001) Periodontal tissue regeneration by combined applications of recombinant human osteogenic protein-1 and bone morphogenetic protein-2. A pilot study in Chacma baboons (*Papio ursinus*). *Eur J Oral Sci* 109: 241–248
48 Ripamonti U, Heliotis M, Rueger DC, Sampath TK (1996) Induction of cementogenesis by recombinant human osteogenic protein-1 (hOP-1/BMP-7) in the baboon. *Arch Oral Biol* 41: 121–126
49 Åberg T, Wozney J, Thesleff I (1997) Expression patterns of bone morphogenetic proteins (*Bmps*) in the developing mouse tooth suggest roles in morphogenesis and cell differentiation. *Dev Dyn* 210: 383–396
50 Sigurdsson TJ, Lee MB, Kubota K, Turek TJ, Wozney JM, Wikesjö UME (1995) Periodontal repair in dogs: recombinant human bone morphogenetic protein-2 significantly enhances periodontal regeneration. *J Periodontol* 66: 131–138
51 Wikesjö UME, Guglielmoni P, Promsudthi A, Cho KS, Trombelli L, Selvig KA, Jin L, Wozney JM (1999) Periodontal repair in dogs: effect of rhBMP-2 concentration

on regeneration of alveolar bone and periodontal attachment. *J Clin Periodontol* 6: 392–400

52 Wikesjö UME, Sorenson RG, Kinoshita A, Li J, Wozney JM (2004) Effect of recombinant human bone morphogenetic protein-12 (rhBMP-12) on regeneration of alveolar bone and periodontal attachment. *J Clin Periodontol* 31: 662–670

53 Choi S-H, Kim C-K, Cho K-S, Huh JS, Sorenson RG, Wozney JM, Wikesjö UM (2002) Effect of recombinant human bone morphogenetic protein-2/absorbable collagen sponge (rhBMP-2/ACS) on healing in 3-wall intrabony defects in dogs. *J Periodontol* 73: 63–72

54 Zeichner-David M (2006) Regeneration of periodontal tissues: Cementogenesis revisited. *J Periodontol* 41: 196–217

55 Cho M-I, Garant PR (1988) Ultrastructural evidence of directed cell migration during initial cementoblast differentiation in root formation. *J Periodont Res* 23: 268–276

56 Moehl T, Ripamonti U (1992) Primate dentine extracellular matrix induces bone differentiation in heterotopic sites of the baboon (*Papio ursinus*). *J Periodontal Res* 27: 92–96

57 Melcher AH, McCulloch CAG, Cheong T, Nemeth E, Shiga A (1987) Cells from bone synthesize cementum-like and bone-like tissue *in vitro* and may migrate into periodontal ligament *in vivo*. *J Periodontal Res* 22: 246–247

58 Ripamonti U, Heliotis M, van den Heever B, Reddi AH (1994) Bone morphogenetic proteins induce periodontal regeneration in the baboon (*Papio ursinus*). *J Periodontal Res* 29: 439–445

59 Giannobile WV, Ryan S, Shih MS, Su DL, Kaplan PL, Chan TC (1998) Recombinant human osteogenic protein-1 (OP-1) stimulates periodontal wound healing in class III furcation defects. *J Periodontol* 69: 129–137

60 Ripamonti U, Crooks J, Teare J, Petit J-C, Rueger DC (2002) Periodontal tissue regeneration by recombinant human osteogenic protein-1 in periodontally-induced furcation defects of the primate *Papio ursinus*. *S Afr J Sci* 98: 361–368

61 Kinoshita A, Oda S, Takahashi K, Yokota S, Ishikawa I (1997) Periodontal regeneration by application of recombinant human bone morphogenetic protein-2 to horizontal circumferential defects created by experimental periodontitis in beagle dogs. *J Periodontol* 68: 103–109

62 King GN, King N, Cruchley AT, Woaney JM, Hughes FJ (1997) Recombinant human bone morphogenetic protein-2 promotes wound healing in rat periodontal fenestration defects. *J Dent Res* 76: 1460–1470

63 Saito E (2003) Favourable healing following space creation in rhBMP-2–induced periodontal regeneration of horizontal circumferential defect in dogs with experimental periodontitis *J Periodontol* 74: 1808–1815

64 Takahashi D, Odajima T, Morita M, Kawanami M, Kato H (2005) Formation and resolution of ankylosis under application of recombinant human bone morphogenetic protein-2 (rhBMP-2) to Class III furcation defects in cats. *J Periodontal Res* 40: 299–305

65 Huang K-K, Shen C, Chiang C-Y, Hsieh Y-D, Fu E (2005) Effects of bone morphogenet-

ic protein-6 on periodontal wound healing in a fenestration defect of rats. *J Periodontal Res* 40: 1–10
66 Sorensen RG, Polimeni G, Kinoshita A, Wozney JM, Wikesjö UME (2004) Effect of recombinant human bone morphogenetic protein-12 (rhBMP-12) on regeneration of periodontal attachment following tooth implantation in dogs. *J Clin Periodontol* 31: 654–661
67 Nakae H, Narayanan AS, Raines E, Page RC (1991) Isolation and partial characterization of mitogenic factors from cementum. *Biochemistry* 30: 7047–7052
68 Narayanan SA, Yonemura K (1993) Purification and characterization of a novel growth factor from cementum. *J Periodontal Res* 28: 563–565
69 Ripamonti U (2005) Bone induction by recombinant human osteogenic protein-1 (hOP-1, BMP-7) in the primate *Papio ursinus* with expression of mRNA of gene products of the TGF-β superfamily. *J Cell Mol Med* 9: 911–928

Control of bone mass by sclerostin: Inhibiting BMP- and WNT-induced bone formation

David J. J. de Gorter[1], Carola Krause[1], Clemens W. G. M. Löwik[2], Rutger L. van Bezooijen[2] and Peter ten Dijke[1]

[1]Department of Molecular Cell Biology, Leiden University Medical Center, Einthovenweg 20, 2300 RC Leiden, The Netherlands; [2]Department of Endocrinology and Metabolic Diseases, Leiden University Medical Center, Albinusdreef 2, 2300 RC Leiden, The Netherlands

Bone remodeling and osteoporosis

Bone is continuously replacing itself by the actions of bone-resorbing osteoclasts and bone-forming osteoblasts, a process called bone remodeling. Because both cell types control each other's activity, there is a tight balance between bone resorption and bone formation. However, when this delicate balance is disturbed by increased osteoclast or decreased osteoblast activity, it can lead to diseases characterized by low bone mass such as osteoporosis. Osteoporosis is a common disorder characterized by decreasing bone-mineral density (BMD), degenerative microarchitectural changes in bone tissue, and an increased fracture risk, and has become an important public health problem. In recent years, sclerostin was postulated to be an attractive target for treatment of osteoporotic patients to restore lost bone. In this chapter, our current knowledge on the function of sclerostin in bone formation and its potential as a target for anabolic treatment is discussed.

Sclerostin's function in development and disease

SOST/sclerostin characteristics

The sclerostin protein is encoded by the *SOST* gene, and belongs to the DAN family of secreted proteoglycans, a specific subfamily of cystine knot-containing proteins. This family also includes DAN (differential screening-selected gene aberrant in neuroblastoma), Cerberus, Gremlin, PRDC (protein related to DAN and Cerberus), Coco, Caronte, Dante, and USAG-1, which function as antagonists of bone morphogenetic protein (BMP) activity [1, 2]. Within this family, sclerostin is most closely related to USAG-1, which is also known as Ectodin or Wise [3–6]. Sclerostin and USAG-1 share 38% amino acid homology and define a subfamily

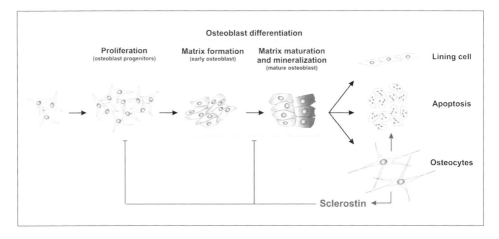

Figure 1
Schematic model illustrating the possible multiple inhibitory effects of sclerostin on osteoblast proliferation and differentiation and its stimulatory effect on osteoblast apoptosis. Direct regulatory effects of sclerostin on osteocytes cannot be excluded. (Adapted from ten Dijke et al. [87], with permission from the authors.)

among the DAN family members. It is assumed that the genes encoding these proteins are derived of the same ancestral gene [7]. Both sclerostin and USAG-1 lack a cysteine residue that is present in the other DAN family members that is thought to be involved in dimerization.

Sclerostin was found to inhibit osteoblast development *in vitro* by decreasing proliferation and inhibiting early and late stages of osteoblast differentiation of mouse and human osteogenic cells (Fig. 1) [8, 9]. In addition, human mesenchymal stem cells treated with sclerostin displayed increased caspase activity and fragmented, histone-associated DNA levels, indicating that sclerostin also stimulates apoptosis [10]. Moreover, conditional expression of sclerostin in bone under transcriptional control of the Osteocalcin promoter induced osteopenia; bone mineral content, thickness of cortex, amount of trabecular bone, and bone strength were all reduced [8]. This was the consequence of impaired bone formation, since calcein-labeling, osteoblast surface, and bone formation rate were decreased, while bone resorption markers remained unaffected.

Using *in situ* hybridization and immunohistochemistry, *SOST* mRNA and sclerostin protein was detected in adult mouse and human bones, specifically in osteocytes [8, 9], which are terminally differentiated osteoblasts entrapped in the matrix during bone formation. Moreover, newly embedded osteocytes do not yet express sclerostin, but become positive for sclerostin when the matrix mineralizes [11]. This suggests that osteocytes limit further bone formation by secreting sclerostin that is transported to osteoblasts on the bone surface. Osteocytes have mechanosensory properties, and mechanical loading triggers them to modulate bone homeostasis. *In vivo* mechanical loading was found to reduce sclerostin expression, thereby providing a possible mechanism by which bone formation upon mechanical loading is increased [12].

During mouse embryogenesis *SOST* mRNA expression was found in all mineralized bones [8, 9]. At this early stage of bone development, however, it is difficult to determine whether expression is restricted to osteocytes or is expressed by osteoblasts as well. *SOST*/sclerostin is not expressed by osteoclasts in mouse and human bones [9]. Due to the lack of methods for isolating osteocytes from bones, it is technically difficult to study *SOST*/sclerostin expression in osteogenic cells *in vitro*. Currently, osteogenic cell cultures that form mineralized bone nodules are the only available method to generate osteocyte-like cells *in vitro* [13]. In mouse osteogenic cultures, *SOST* expression is induced at low levels after onset of bone nodule mineralization, suggesting *SOST* expression by only a few cells, possibly osteocytes within the bone nodules [9]. In human primary osteogenic cultures that do not form bone nodules, *SOST* mRNA expression can be detected at the undifferentiated stage and is found at very low levels at the differentiated mineralized stage [8, 9, 14, 15].

SOST transcription is controlled by Runx2 (also known as Cbfa1), a transcription factor essential for osteoblast differentiation [16, 17], which binds directly to a Runx2 binding element within the proximal promoter of the *SOST* gene [18]. Overexpression of Runx2 in SAOS-2 and MG-63 osteosarcoma cells, of which the latter normally do not express *SOST*, increased transcription controlled by this *SOST* promoter segment. Also MyoD overexpression stimulated transcriptional activity of the proximal *SOST* promoter, presumably by binding an E-box motif within this region [18]. Furthermore, Osterix, another crucial transcription factor in osteoblast differentiation [19], was found to regulate *SOST* expression [20]. *SOST* and *Osx* (Osterix) mRNA expression co-localized in embryonic bone, and RNAi-mediated knock-down of *Osx* down-regulated *SOST* expression [20]. However, it remains to be established whether Osterix binds regulatory elements of the *SOST* gene directly or controls *SOST* expression in an indirect manner.

Besides sclerostin expression in bone, *SOST* mRNA expression was detected in several other organs, including cartilage, bone marrow, kidney, heart, aorta, lung, pancreas, skeletal muscle, liver, skin, and placenta [14, 21]. However, these data require confirmation by immunohistochemistry to determine whether sclerostin protein is actually expressed in these tissues. For example, although *SOST* mRNA was found in human kidney [14, 21], no sclerostin protein expression was detected in any of the human kidney samples analyzed thus far, including a human kidney sample that was previously described to express *SOST* mRNA expression in glomeruli [22, 23]. During embryonic development of the mouse heart, *SOST* mRNA expression was detected in the medial vessel wall of the great arteries containing smooth muscle cells [24]. However, thus far the function of sclerostin in cardiovascular development remains to be established.

In humans, hypertrophic chondrocytes and cementocytes are, in addition to osteocytes, the only cells known in which sclerostin protein is expressed [8, 25]. Within the growth plate, only the mineralized hypertrophic chondrocytes are positive for sclerostin expression, while all other stages of chondrocytes differentiation are negative [23]. Also chondrocytes within articular cartilage do not express sclerostin [9]. Cementocytes, which are found in the cementum at the roots of teeth, are also positive for sclerostin protein expression [25]. Thus, in humans, sclerostin expression appears to be restricted to terminally differentiated cell types embedded in a mineralized matrix.

Sclerosteosis

In 2001, two groups independently identified mutations in a gene they named *SOST* as the cause of sclerosteosis (OMIM 269500) [14, 21]. Mutations in the *SOST* gene introducing a premature stop-codon or causing expression of an improperly spliced message have been reported in patients with sclerosteosis. Consequently, osteocytes of patients with sclerosteosis do not express the sclerostin protein [9]. These patients display overgrowth and sclerosis of the skeleton affecting particularly the skull, resulting in enlargement of the jaw and entrapment of cranial nerves with consequent facial palsy, hearing loss, loss of smell, optic atrophy and raised intracranial pressure. Sudden death as a result of impaction of the brainstem in the foramen magnum has also been described. These features can present in early childhood, but at birth only syndactyly of mostly the second (index) and third (middle) finger are indicative for the disease [26]. The increase in bone mass in sclerosteosis is thought to be due to increased bone formation. In bone biopsies of affected individuals there is a predominance of cuboibal, active appearing osteoblasts, increased double tetracycline label spacing, and increased osteoid levels that mineralize normally [9, 27, 28].

Sclerosteosis occurs mainly in Afrikaners, the white farmers who separated from the Dutch East India Company in the 1600s and moved into the backcountry of South Africa. Sixty-three patients from 38 families, 8 of which known to be consanguineous, have been described so far in South Africa but affected individuals and families from Spain, Brazil, USA, Germany, Japan, Switzerland, and Senegal have also been reported [26]. The estimated carrier rate for a single founder-derived mutation in the Afrikaner population is 1 in 100 [23], with 10 000 clinically normal heterozygotes in this population [29] and one or two affected births per year [26]. Sclerosteosis is reported to be transmitted as an autosomal recessive trait [29], but radiological features of heterozygotes suggested that heterozygotes may have increased bone mass as well [30].

Table 1 - *Common and distinctive features between sclerosteosis, Van Buchem disease and HBM caused by LRP5 mutations.*

Sclerosteosis	Van Buchem disease	LRP5 **HBM**

Progressive bone overgrowth
- Jaw enlargement
- Entrapment of nerves
  Facial palsy
  Facial distortion
  Hearing loss
  Loss of sense of smell
Impaction of brainstem leading to sudden death
	Progressive bone overgrowth
	- Jaw enlargement
	- Entrapment of nerves
	  Facial palsy
	  Facial distortion
	  Hearing loss
	  Loss of sense of smell
		High bone density
		- Jaw enlargement

Hyperostosis of tubular bones	Hyperostosis of tubular bones	Hyperostosis of tubular bones
Tall stature		
Syndactyly		
Nail hypoplasia		
Dental malocclusion		
Torus palatinus		Torus palatinus
Autosomal recessive pattern of inheritance	Autosomal recessive pattern of inheritance	Autosomal dominant pattern of inheritance
Mutations in *SOST* gene	52-kb deletion 32 kb downstream of *SOST* gene	Mutations in *LRP5* gene

Van Buchem disease

The clinical features of Van Buchem disease (OMIM 239100) are similar to those of sclerosteosis but the disease runs a more benign course and syndactyly is not present (Tab. 1). Similar to sclerosteosis, Van Buchem disease is radiologically characterized by progressive bone thickening that is most pronounced in calvaria, mandible, and the base of the skull, but the whole skeleton is dense with typical thickening of the cortexes of the long bones [23, 31]. Bone mineral content of metacarpals of patients with Van Buchem disease is increased and the calculated bone volume and polar moment of inertia are markedly elevated suggesting increased bone strength [31]. Consistent with these changes, there are no reports of fractures in patients with either Van Buchem disease or sclerosteosis.

A few Van Buchem disease patients, mainly from a small fishing village in the Netherlands, have been described. Because of the clinical similarities between the two disorders, it was speculated that sclerosteosis and Van Buchem disease were allelic conditions caused by the same abnormal gene. This hypothesis was supported by the co-localization of the sclerosteosis gene and the Van Buchem disease gene to the same linkage interval on chromosome 17q12-21 [32]. However, no mutations within the coding region of the *SOST* gene were found in patients with Van Buchem disease, but further analysis of the region surrounding the *SOST* gene revealed that these patients had a 52-kb deletion 35 kb downstream of the *SOST* gene [14, 33, 34]. Since the deleted region did not appear to encode for any gene, this deletion was suggested to suppress *SOST* expression. Indeed, analysis of transgenic mice expressing either normal human *SOST* alleles or genetically modified Van Buchem disease alleles showed that an enhancer element within the Van Buchem disease deletion drives *SOST* expression in the mouse skeleton [35]. In addition, no sclerostin expression was found in bone biopsies of patients with Van Buchem disease [23], which is consistent with the idea that Van Buchem disease is the result of down-regulated sclerostin expression.

Molecular mechanisms underlying sclerostin-mediated inhibition of bone formation

Sclerostin inhibits BMP-induced bone formation

BMPs belong to the TGF-β superfamily and were originally identified as the active components in bone extracts capable of inducing bone formation at ectopic sites [36]. BMPs are required for skeletal development and maintenance of adult bone homeostasis, and play an important role in fracture healing [37, 38]. Furthermore, BMP signaling is essential during embryogenesis, by controlling neural development, hematopoiesis and induction of somite formation. Mice deficient in BMP ligands, BMP receptors or Smad proteins (the intracellular mediators of BMP signaling) are frequently not vital and often display skeletal defects [37, 38]. In addition, several naturally occurring mutations in BMPs or their receptors cause inherited disorders, including fibrodysplasia ossificans progressiva (FOP) [37], in which bone is formed at ectopic sites, such as muscles, tendons, ligaments and other connective tissues. Together, this demonstrates that the BMP signaling pathway fulfills a key role in skeletal development and bone remodeling. Indeed, at present BMP-2 and BMP-7 have demonstrated clinical utility for bone regeneration, and are commercially available through the use of recombinant DNA technology [39]. BMPs promote bone formation by blocking mesenchymal stem cell differentiation into myoblast and stimulating their differentiation towards the osteoblast lineage [40]. The osteogenic potential of the different BMPs varies, and BMP-2, -6, -7, and

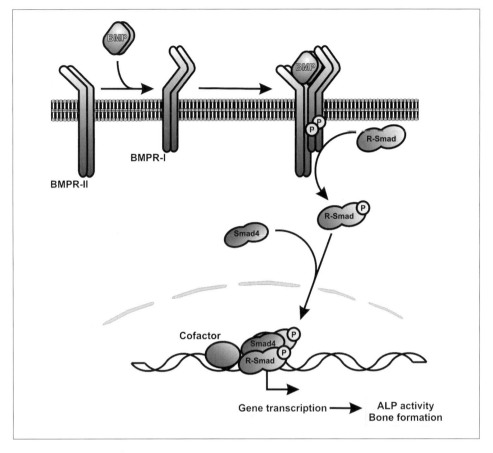

Figure 2
Schematic model illustrating the BMP signaling pathway in osteoblasts. BMP signals via heteromeric complexes of BMP type I and type II transmembrane serine/threonine kinase receptors and specific intracellular Smad effector proteins. Activated heteromeric complexes between R-Smad (Smad1, Smad5 and Smad8) and Smad4 can act as transcription factors and regulate gene transcriptional responses, ultimately stimulating alkaline phosphatase activity (early marker for osteoblast differentiation) and bone formation.

-9 were found to be the most potent inducers of osteogenesis both *in vitro* and *in vivo* [41, 42].

BMPs bind as dimers to type I and type II serine/threonine receptor kinases, forming an oligomeric complex (Fig. 2). The type II receptors are constitutively active and phosphorylate and consequently activate the type I receptors upon oligomerization. Subsequently, the activated type I receptors phosphorylate receptor-regulated Smads, Smad1, -5 and -8, at their extreme C termini. Upon activation, the receptor-regulated Smads associate with the Co-Smad, Smad4, and translocate into the nucleus, where they together with other transcription factors bind promoters of target genes and control their expression (Fig. 2) [43, 44]. For comprehensive overviews on BMP signaling in bone, the reader is referred to the chapters written by Harris et al. and by Seemann et al. of this book, and on Smad signaling to Massagué et al. [43], and Feng and Derynck [44].

The activity of BMPs is tightly regulated by BMP antagonists, including members of the DAN family [22, 45]. They control the level of BMP signaling by sequestering BMP ligands, which prevents receptor activation. Sclerostin was also found to bind BMPs. It bound BMP-2, -4, -5, -6 and -7, all with relative low affinity and comparable kinetics, whereas it did not bind TGF-β isoforms [8]. Furthermore, type I and type II BMP receptor-Fc fusion proteins disturbed the binding of BMPs to sclerostin, suggesting that sclerostin and BMP receptors compete for BMP binding [8]. Moreover, sclerostin suppressed BMP-induced differentiation and mineralization of osteoblastic cells [8, 9]. BMP-2, -4, and -6 induce *SOST* expression in mouse and human osteogenic cells, suggesting a negative feedback mechanism to prevent excessive exposure of skeletal cells to BMP [15, 20]. In addition, Retinoic acid and 1,25-dihydroxyvitamin D3 potentiate BMP-induced *SOST* expression, while dexamethasone mitigates this response [15]. Dexamethasone is required to induce human osteogenic cells to differentiate into mature mineralizing cells, and may be the cause of the low *SOST* expression levels in these cultures.

Sclerostin is an antagonist of the WNT signaling pathway

WNT signaling has a crucial function in embryonic development and homeostatic self renewal, and deregulation of this pathway underlies a wide variety of diseases, ranging from cancer to osteoporosis [46]. WNT proteins bind to the Frizzled (Fzd) 7-transmembrane spanning receptors and the co-receptor low-density lipoprotein (LDL) receptor-related protein-5 or -6 (LRP5/6), and surface expression of both receptors is required to initiate signaling [46]. Canonical WNT signaling is mediated by cytoplasmic β-catenin, which in absence of WNT is recruited to the destruction complex consisting of the tumor suppressor proteins APC and axin, and the kinases glycogen synthase kinase 3 (GSK3) and casein kinase I (CK1) [46]. Subsequently, β-catenin becomes phosphorylated on several highly conserved serine/threonine residues, ubiquitinated and targeted for proteosomal degradation (Fig. 3). In the presence of WNT, the destruction complex dissociates, and, as a consequence, β-catenin is not targeted for degradation, accumulates, and translocates into the nucleus [46]. Here β-catenin binds members of the lymphoid enhancing factor (LEF)/T cell factor (TCF) family of transcription factors, displaces the transcriptional repressor Groucho, and drives transcription of WNT target genes (Fig. 3). An additional level of control is provided by Dickkopf (DKK) proteins, which can bind LRP5/6 and

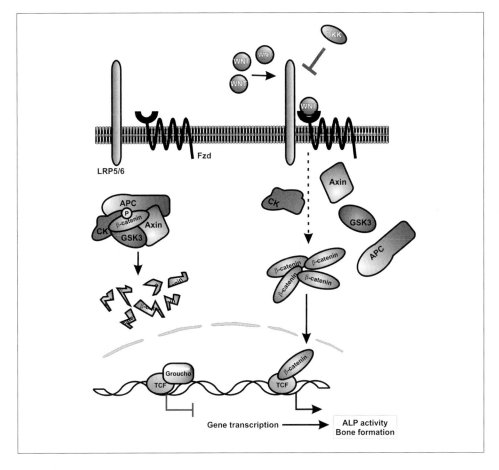

Figure 3
Schematic model illustrating the canonical WNT signaling pathway in osteoblasts. In absence of WNT, β-catenin forms a complex with APC, Axin, GSK3 and CK1, be-comes phosphorylated, ubiquitinated and targeted for proteosomal degradation. WNT binding to Frizzled (Fzd) receptors and LRP5/6 co-receptors, prevents formation of the complex, leading to stabilization of β-catenin and regulation of gene expression through LEF/TCF transcription factors, ultimately stimulating alkaline phosphatase activity (early marker for osteoblast differentiation) and bone formation. Dickkopf (DKK) proteins antagonize WNT responses by binding to LRP5/6.

act as antagonists of canonical WNT signaling [47–49]. To efficiently inhibit WNT signaling, DKK also requires binding to Kremen proteins [50].

Recently, canonical WNT signaling was demonstrated to be indispensable for osteoblastogenesis in the mouse embryo [51–53]. Conditional deletion of β-catenin led osteochondroprogenitor cells to differentiate into chondrocytes instead of osteoblasts during both intramembranous and endochondral ossification, whereas ectopical WNT signaling enhanced osteoblast differentiation [51–53]. This indicates WNT signaling controls differentiation of osteochondroprogenitor cells towards the osteoblast lineage. Moreover, overexpression in osteoblasts of DKK1 resulted in severe osteopenia in mice [54]. Conversely, heterozygous DKK1-deficient mice exhibited an increase in all bone formation parameters, with no change in bone resorption, leading to an increase in bone mass [55]. Notably, expression of DKK1 and DKK2 are upregulated during osteoblast differentiation, and down-regulation of WNT signaling seems to be required for terminal osteogenic differentiation and mineralized matrix formation [56, 57]. Using TOPGAL mice, transgenic mice that express the lacZ reporter gene under transcriptional control of multimerized TCF-binding elements, active WNT signaling was detected in osteoblasts and osteocytes in the fetal and neonatal skeleton [58]. Remarkably, although active WNT signaling persisted in osteocytes with aging of the mice, it could no longer be detected in osteoblasts. However, WNT signaling was activated in osteoblasts by physical deformation, suggesting WNT signaling may be involved in the coupling of mechanical force to anabolic activity in the skeleton. Furthermore, both DKK1 overexpression and deficiency in β-catenin also inhibited BMP-2-induced bone formation [59]. Thus, WNT signaling plays an important role in osteoblast differentiation and bone formation. For comprehensive overviews on WNT signaling in bone, the reader is referred to Baron and Rawadi [60].

The importance of WNT signaling in bone homeostasis is also illustrated by the fact that loss-of-function mutations in the *LRP5* gene are associated with the autosomal recessive disorder osteoporosis-pseudoglioma (OPPG) syndrome [61], whereas several missense mutations in *LRP5* result in the autosomal dominant high bone mass (HBM) trait [62–64]. Since the HBM phenotype caused by these *LRP5* mutations resembles that of sclerosteosis and Van Buchem disease patients (Tab. 1), it was suggested that sclerostin might act as a WNT antagonist. Indeed, it was found that when co-injected with *Xwnt8* mRNA in *Xenopus* embryos, *SOST* mRNA blocked axis duplication induced by *Xwnt8* mRNA [65]. Moreover, sclerostin is able to inhibit WNT-induced activity of WNT-luciferase reporter constructs in mammalian cells [65–67], showing that sclerostin functions as a inhibitor of WNT signaling. Similar to DKK, sclerostin binds LRP5/6 [47, 48, 65, 66]. Furthermore, addition of anti-sclerostin antibodies or introduction of amino acid substitutions that alters its binding to LRP6 rescued from sclerostin-mediated attenuation of WNT signaling [7]. Interestingly, sclerostin binds to the first two β-propeller domains of LRP5/6 and this includes the region in which all HBM *LRP5* mutations are located, i.e., the first β-propeller domain [7, 68]. Indeed, several of the HBM-causing mutations in *LRP5* were described to reduce the ability of LRP5 to bind sclerostin. Therefore, these mutations might make LRP5 insensitive to sclerostin-mediated inhibition of WNT signaling in osteoblasts, resulting in the HBM phenotype. However, it was also found that the *LRP5-G171V* HBM mutation disrupts the interaction of LRP5 with the chaperone protein Mesd [69], which controls the transport of LRP5/6 to the cell surface [70]. Therefore, it was hypothesized that, although the LRP5-G171V mutant failed to mediate paracrine WNT signaling, autocrine WNT signaling in cells expressing this mutant still occurs [69]. Since LRP5-G171V surface expression is disturbed, this mutant protein will not encounter secreted antagonists and, as a consequence, cannot be inhibited by sclerostin. Nevertheless, cell surface expression of other LRP5 HBM mutants was shown to be unaffected, whereas binding of sclerostin to these mutants was clearly reduced, and rendered LRP5 more resistant to sclerostin-mediated inhibition [68]. The phenotype of patients with the HBM trait caused by *LRP5* mutations is, although resembling, less severe than that of sclerosteosis or Van Buchem disease patients (Tab. 1). This is most likely due to that HBM LRP5 mutants display reduced sclerostin binding, whereas in patients with sclerosteosis or Van Buchem disease sclerostin is absent. In addition, the HBM mutations are only found in *LRP5*, and not in *LRP6*, so that sclerostin is still able to affect LRP6-mediated WNT signaling.

Is sclerostin a BMP antagonist, a WNT antagonist or both?

Genome-wide transcriptional analysis of mouse osteoblastic KS483 cells has recently indicated that sclerostin specifically affects the BMP and WNT signaling pathways [67]. USAG-1, the DAN family member most closely related to sclerostin, also shows functional similarity to sclerostin because it can also inhibit WNT [4] as well as BMP signaling [3]. A similar feature, the inhibiting of more than just one signaling pathway, has also been described for the DAN family members Cerberus and Coco, which bind and inhibit Nodal, BMP and WNT proteins [71, 72]. This indicates that this property is shared by several secreted antagonists of the DAN family. An argument against sclerostin being a genuine BMP antagonist, however, is that, although sclerostin binds BMPs, it does so with relatively low affinity (Kd ~100–1 nM) compared to BMP binding to the classical BMP antagonists Noggin (Kd ~20 pM) and Chordin (Kd ~30 pM) [8, 73–75]. Since BMPs have high affinities for their receptors (Kd ~1–0.1 nM) [76, 77], sclerostin should be present in a large excess to block BMP signaling effectively. In addition, of the genes modulated by BMP4 or by sclerostin in KS483 cells, expression of less than 25% was affected by both proteins, and the combined treatment with BMP and sclerostin affected 88% of the genes in a cumulative manner [67]. Moreover, sclerostin did not inhibit direct transcriptional activation of BMP target genes or of a BMP luciferase reporter, and had no effect on BMP-induced Smad phosphorylation [9, 67]. Together, this indicates that sclerostin has a function different from that of antagonizing BMP signaling. Interestingly, BMP-induced osteogenesis is mediated by WNT signaling [59], and BMP-stimulated alkaline phosphatase activity is inhibited by DKK1 in a similar manner as sclerostin [67]. Moreover, activation of a WNT luciferase reporter by

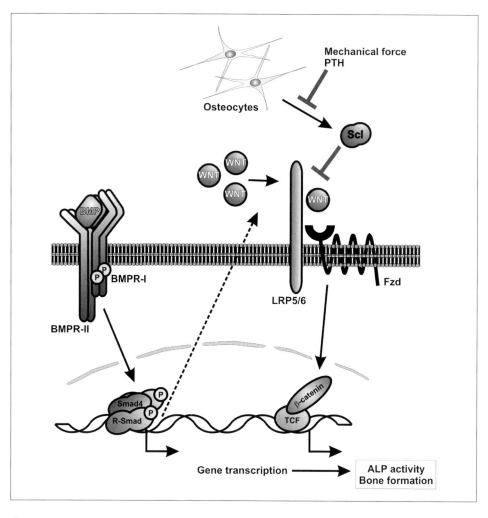

Figure 4
Schematic model illustrating the interplay between BMP, WNT and sclerostin in osteoblasts. Sclerostin (Scl), a protein that is secreted by osteocytes, antagonizes WNT responses by binding to LRP5/6. BMP and WNT pathways cooperate with each other in stimulating alkaline phosphatase activity and bone formation. Therefore, by antagonizing WNT signaling, sclerostin can inhibit both BMP- and WNT-induced bone formation. Sclerostin expression is down-regulated by mechanical force and PTH. (Adapted from ten Dijke et al. [87], with permission from the authors.)

exogenously added BMPs or expression of constitutively active BMP type I receptors is antagonized by sclerostin [67]. This suggests that sclerostin inhibits late BMP responses by antagonizing WNT signaling (Fig. 4).

Nevertheless, it cannot be excluded that sclerostin under specific circumstances can act as a direct inhibitor of BMP signaling. Possibly, similar to Twisted gastrulation (Tsg) for the BMP antagonist chordin, which, controlled by the metalloprotease Xolloid, can switch its activity from a BMP antagonist to a pro-BMP signal [78], sclerostin may require co-factors to disturb BMP signaling. Such a co-factor could either modulate the affinity of sclerostin for BMPs or increase the local concentration of sclerostin. However, although experiments have been performed to find evidence for the existence of a sclerostin co-factor, data supporting this are currently lacking [67]. Therefore, it would be interesting to determine in which manner inhibition of sclerostin affects phosphorylation of Smad1 and 5, nuclear β-catenin and reporter gene expression in BMP-Smad reporter [79] and TOPGAL mice [58].

Considerations and future perspectives

In sclerosteosis and Van Buchem disease, the inhibitory effect of sclerostin is absent leading to increased bone formation. In the absence of increased osteoclast function, this results in a positive balance between bone formation and resorption, and subsequently in an increased bone mass. It should be noted that, although *SOST* mRNA is expressed in human kidney [14, 21] and within the medial vessel wall of the great arteries containing smooth muscle cells during mouse embryogenesis [24], patients with sclerosteosis or Van Buchem disease do not have complications with the renal and cardiovascular systems [26]. The lack of extraskeletal features in patients with sclerosteosis and Van Buchem disease indicates a restricted expression of sclerostin in bone. Therefore, targeting sclerostin in osteoporotic patients is considered to be a promising therapeutic approach to stimulate good quality bone formation. Furthermore, heterozygous carriers of sclerosteosis have BMD values consistently higher than healthy subjects without any of the bone complications encountered in homozygotes [80]. This suggests that the production and/or activity of sclerostin can be titrated *in vivo*, leading to variable increases in bone mass without any unwanted skeletal effects. Consistent with these results a monoclonal antibody against sclerostin was shown to be successful in stimulating bone formation in rodents and primates [81, 82]. In addition, intermittent parathyroid hormone (PTH)/teriparatide application, which has strong anabolic effects and is an established pharmacological principle to stimulate bone formation in osteoporotic patients [83, 84], was found to reduce *SOST*/sclerostin expression both *in vitro* and *in vivo* [85, 86]. These observations suggest that the stimulatory effect of PTH on bone formation is, at least in part, due to the relief of sclerostin-mediated inhibition of osteoblast differentiation.

However, one could argue that the disturbed regulation of bone remodeling found in osteoporosis can lead to a decrease in number and/or function of osteocytes, resulting in diminished sclerostin secretion. If this is the case, treatment by blocking the actions of the low abundant sclerostin could be less effective. Thus far, no studies have explored whether *SOST*/sclerostin expression is affected in osteoporotic patients. Nonetheless, treatment with sclerostin-neutralizing antibodies increased bone mass and improved bone quality in aged ovariectomized rats with established osteopenia [81], suggesting that blocking sclerostin function can be a good anabolic treatment of post-menopausal osteoporosis.

In conclusion, the restricted expression pattern of sclerostin to bone combined with the good quality bone phenotype in patients with sclerosteosis and Van Buchem disease, and in rodents and primates treated with sclerostin-blocking antibodies, make inhibition of sclerostin an ideal target for new therapies that restore lost bone in patients with diseases characterized by low bone mass such as osteoporosis.

Acknowledgements

Research in our laboratories on the function and mechanism of action of sclerostin is supported by grants from the Dutch Organization for Scientific Research (NWO 918.66.606) and the European Commission (LSHM-CT-2003-503020).

References

1 Hsu DR, Economides AN, Wang X, Eimon PM, Harland RM (1998) The *Xenopus* dorsalizing factor Gremlin identifies a novel family of secreted proteins that antagonize BMP activities. *Mol Cell* 1: 673–683
2 Pearce, JJ, Penny G, Rossant J (1999) A mouse cerberus/Dan-related gene family. *Dev Biol* 209: 98–110
3 Laurikkala J, Kassai Y, Pakkasjarvi L, Thesleff I, Itoh N (2003) Identification of a secreted BMP antagonist, ectodin, integrating BMP, FGF, and SHH signals from the tooth enamel knot. *Dev Biol* 264: 91–105
4 Itasaki N, Jones CM, Mercurio S, Rowe A, Domingos PM, Smith JC, Krumlauf R (2003) Wise, a context-dependent activator and inhibitor of Wnt signalling. *Development* 130: 4295–4305
5 Avsian-Kretchmer O, Hsueh AJ (2004) Comparative genomic analysis of the eight-membered ring cystine knot-containing bone morphogenetic protein antagonists. *Mol Endocrinol* 18: 1–12
6 Yanagita M, Oka M, Watabe T, Iguchi H, Niida A, Takahashi S, Akiyama T, Miyazono K, Yanagisawa M, Sakurai T (2004) USAG-1: A bone morphogenetic protein antagonist abundantly expressed in the kidney. *Biochem Biophys Res Commun* 316: 490–500
7 Ellies DL, Viviano B, McCarthy J, Rey JP, Itasaki N, Saunders S, Krumlauf R (2006) Bone density ligand, Sclerostin, directly interacts with LRP5 but not LRP5G171V to modulate Wnt activity. *J Bone Miner Res* 21: 1738–1749

8   Winkler DG, Sutherland MK, Geoghegan JC, Yu C, Hayes T, Skonier JE, Shpektor D, Jonas M, Kovacevich BR, Staehling-Hampton K et al (2003) Osteocyte control of bone formation *via* sclerostin, a novel BMP antagonist. *EMBO J* 22: 6267–6276

9   van Bezooijen RL, Roelen BA, Visser A, Wee-Pals L, de Wilt E, Karperien M, Hamersman H, Papapoulos SE, ten Dijke P, Lowik CW (2004) Sclerostin is an osteocyte-expressed negative regulator of bone formation, but not a classical BMP antagonist. *J Exp Med* 199: 805–814

10 Sutherland MK, Geoghegan JC, Yu C, Turcott E, Skonier JE, Winkler DG, Latham JA (2004) Sclerostin promotes the apoptosis of human osteoblastic cells: A novel regulation of bone formation. *Bone* 35: 828–835

11 Poole KE, van Bezooijen RL, Loveridge N, Hamersma H, Papapoulos SE, Lowik C W, Reeve J (2005) Sclerostin is a delayed secreted product of osteocytes that inhibits bone formation. *FASEB J* 19: 1842–1844

12 Robling AG, Bellido T, Turner CH (2006) Mechanical stimulation *in vivo* reduces osteocyte expression of sclerostin. *J Musculoskelet Neuronal Interact* 6: 354

13 Pockwinse SM, Wilming LG, Conlon DM, Stein GS, Lian JB (1992) Expression of cell growth and bone specific genes at single cell resolution during development of bone tissue-like organization in primary osteoblast cultures. *J Cell Biochem* 49: 310–323

14 Balemans W, Ebeling M, Patel N, Van Hul E, Olson P, Dioszegi M, Lacza C, Wuyts W, Van Den EJ, Willems P et al (2001) Increased bone density in sclerosteosis is due to the deficiency of a novel secreted protein (SOST). *Hum Mol Genet* 10: 537–543

15 Sutherland MK, Geoghegan JC, Yu C, Winkler DG, Latham JA (2004) Unique regulation of SOST, the sclerosteosis gene, by BMPs and steroid hormones in human osteoblasts. *Bone* 35: 448–454

16 Ducy P, Zhang R, Geoffroy V, Ridall AL, Karsenty G (1997) Osf2/Cbfa1: A transcriptional activator of osteoblast differentiation. *Cell* 89: 747–754

17 Ducy P, Starbuck M, Priemel M, Shen J, Pinero G, Geoffroy V, Amling M, Karsenty G (1999) A Cbfa1-dependent genetic pathway controls bone formation beyond embryonic development. *Genes Dev* 13: 1025–1036

18 Sevetson B, Taylor S, Pan Y (2004) Cbfa1/RUNX2 directs specific expression of the sclerosteosis gene (SOST). *J Biol Chem* 279: 13849–13858

19 Nakashima K, Zhou X, Kunkel G, Zhang Z, Deng JM, Behringer RR, de Crombrugghe B (2002) The novel zinc finger-containing transcription factor osterix is required for osteoblast differentiation and bone formation. *Cell* 108: 17–29

20 Ohyama Y, Nifuji A, Maeda Y, Amagasa T, Noda M (2004) Spaciotemporal association and bone morphogenetic protein regulation of sclerostin and osterix expression during embryonic osteogenesis. *Endocrinology* 145: 4685–4692

21 Brunkow ME, Gardner JC, Van Ness J, Paeper BW, Kovacevich BR, Proll S, Skonier JE, Zhao L, Sabo PJ, Fu Y et al (2001) Bone dysplasia sclerosteosis results from loss of the SOST gene product, a novel cystine knot-containing protein. *Am J Hum Genet* 68: 577–589

22 Balemans W, Van Hul W (2002) Extracellular regulation of BMP signaling in vertebrates: a cocktail of modulators. *Dev Biol* 250: 231–250

23 van Bezooijen RL, ten Dijke P, Papapoulos SE, and Lowik CW (2005) SOST/sclerostin, an osteocyte-derived negative regulator of bone formation. *Cytokine Growth Factor Rev* 16: 319–327

24 van Bezooijen RL, Deruiter MC, Vilain N, Monteiro RM, Visser A, Wee-Pals L, van Munsteren CJ, Hogendoorn PC, Aguet M, Mummery CL et al (2007) SOST expression is restricted to the great arteries during embryonic and neonatal cardiovascular development. *Dev Dyn* 236: 606–612

25 van Bezooijen RL, Bronkers AL, Dikkers FG, Gortzak RA, Balemans W, van der Wee-Pals L, Visser A, Van Hul W, Hamersma H, ten Dijke P et al (2007) Sclerostin expression is absent in cementocytes in teeth and in osteocytes in bone biopsies of patients with Van Buchem disease. *17th Scientific Meeting of the International Bone and Mineral Society, June 24–29 2007 in Montréal*

26 Hamersma H, Gardner J, Beighton P (2003) The natural history of sclerosteosis. *Clin Genet* 63:192–197

27 Stein SA, Witkop C, Hill S, Fallon MD, Viernstein L, Gucer G, McKeever P, Long D, Altman J, Miller R et al (1983) Sclerosteosis: Neurogenetic and pathophysiologic analysis of an American kinship. *Neurology* 33: 267–277

28 Hill SC, Stein SA, Dwyer A, Altman J, Dorwart R, Doppman J (1986) Cranial CT findings in sclerosteosis. *AJNR Am J Neuroradiol* 7: 505–511

29 Beighton P (1988) Sclerosteosis. *J Med Genet* 25: 200–203

30 Beighton P, Davidson J, Durr L, Hamersma H (1977) Sclerosteosis – An autosomal recessive disorder. *Clin Genet* 11: 1–7

31 Wergedal JE, Veskovic K, Hellan M, Nyght C, Balemans W, Libanati C, Vanhoenacker FM, Tan J, Baylink DJ, Van Hul W (2003) Patients with Van Buchem disease, an osteosclerotic genetic disease, have elevated bone formation markers, higher bone density, and greater derived polar moment of inertia than normal. *J Clin Endocrinol Metab* 88: 5778–5783

32 Balemans W, Van Den EJ, Freire Paes-Alves A, Dikkers FG, Willems PJ, Vanhoenacker F, Almeida-Melo N, Alves CF, Stratakis CA, Hill SC et al (1999) Localization of the gene for sclerosteosis to the van Buchem disease-gene region on chromosome 17q12–q21. *Am J Hum Genet* 64: 1661–1669

33 Balemans W, Patel N, Ebeling M, Van Hul E, Wuyts W, Lacza C, Dioszegi M, Dikkers FG, Hildering P, Willems PJ et al (2002) Identification of a 52 kb deletion downstream of the SOST gene in patients with van Buchem disease. *J Med Genet* 39: 91–97

34 Staehling-Hampton K, Proll S, Paeper BW, Zhao L, Charmley P, Brown A, Gardner J C, Galas D, Schatzman RC, Beighton P et al (2002) A 52-kb deletion in the SOST-MEOX1 intergenic region on 17q12–q21 is associated with van Buchem disease in the Dutch population. *Am J Med Genet* 110: 144–152

35 Loots GG, Kneissel M, Keller H, Baptist M, Chang J, Collette NM, Ovcharenko D,

Plajzer-Frick I, Rubin EM (2005) Genomic deletion of a long-range bone enhancer misregulates sclerostin in Van Buchem disease. *Genome Res* 15: 928–935

36 Urist MR (1965) Bone: Formation by autoinduction. *Science* 150: 893–899

37 Chen D, Zhao M, Mundy GR (2004) Bone morphogenetic proteins. *Growth Factors* 22: 233–241

38 Gazzerro E, Canalis E (2006) Bone morphogenetic proteins and their antagonists. *Rev Endocr Metab Disord* 7: 51–65

39 ten Dijke P (2006) Bone morphogenetic protein signal transduction in bone. *Curr Med Res Opin* 22 Suppl 1: S7–11

40 Katagiri T, Yamaguchi A, Komaki M, Abe E, Takahashi N, Ikeda T, Rosen V, Wozney JM, Fujisawa-Sehara A, Suda T (1994) Bone morphogenetic protein-2 converts the differentiation pathway of C2C12 myoblasts into the osteoblast lineage. *J Cell Biol* 127: 1755–1766

41 Cheng H, Jiang W, Phillips FM, Haydon RC, Peng Y, Zhou L, Luu HH, An N, Breyer B, Vanichakarn P et al (2003) Osteogenic activity of the fourteen types of human bone morphogenetic proteins (BMPs). *J Bone Joint Surg Am* 85-A: 1544–1552

42 Luu HH, Song WX, Luo X, Manning D, Luo J, Deng ZL, Sharff KA, Montag AG, Haydon RC, He TC (2007) Distinct roles of bone morphogenetic proteins in osteogenic differentiation of mesenchymal stem cells. *J Orthop Res* 25: 665–677

43 Massague J, Seoane J, Wotton D (2005) Smad transcription factors. *Genes Dev* 19: 2783–2810

44 Feng XH, Derynck R (2005) Specificity and versatility in TGF-β signaling through Smads. *Annu Rev Cell Dev Biol* 21: 659–693

45 Canalis E, Economides AN, Gazzerro E (2003) Bone morphogenetic proteins, their antagonists, and the skeleton. *Endocr Rev* 24: 218–235

46 Clevers H (2006) Wnt/β-catenin signaling in development and disease. *Cell* 127: 469–480

47 Mao B, Wu W, Li Y, Hoppe D, Stannek P, Glinka A, Niehrs C (2001) LDL-receptor-related protein 6 is a receptor for Dickkopf proteins. *Nature* 411: 321–325

48 Bafico A, Liu G, Yaniv A, Gazit A, Aaronson SA (2001) Novel mechanism of Wnt signalling inhibition mediated by Dickkopf-1 interaction with LRP6/Arrow. *Nat Cell Biol* 3: 683–686

49 Glinka A, Wu W, Delius H, Monaghan AP, Blumenstock C, Niehrs C (1998) Dickkopf-1 is a member of a new family of secreted proteins and functions in head induction. *Nature* 391: 357–362

50 Mao B, Wu W, Davidson G, Marhold J, Li M, Mechler BM, Delius H, Hoppe D, Stannek P, Walter C et al (2002) Kremen proteins are Dickkopf receptors that regulate Wnt/beta-catenin signalling. *Nature* 417: 664–667

51 Hu H, Hilton M J, Tu X, Yu K, Ornitz DM, Long F (2005) Sequential roles of Hedgehog and Wnt signaling in osteoblast development. *Development* 132: 49–60

52 Day TF, Guo X, Garrett-Beal L, Yang Y (2005) Wnt/β-catenin signaling in mesenchymal

progenitors controls osteoblast and chondrocyte differentiation during vertebrate skeletogenesis. *Dev Cell* 8: 739–750

53   Hill TP, Spater D, Taketo MM, Birchmeier W, Hartmann C (2005) Canonical Wnt/β-catenin signaling prevents osteoblasts from differentiating into chondrocytes. *Dev Cell* 8: 727–738

54   Li J, Sarosi I, Cattley RC, Pretorius J, Asuncion F, Grisanti M, Morony S, Adamu S, Geng Z, Qiu W et al (2006) Dkk1–mediated inhibition of Wnt signaling in bone results in osteopenia. *Bone* 39: 754–766

55   Morvan F, Boulukos K, Clement-Lacroix P, Roman RS, Suc-Royer I, Vayssiere B, Ammann P, Martin P, Pinho S, Pognonec P et al (2006) Deletion of a single allele of the Dkk1 gene leads to an increase in bone formation and bone mass. *J Bone Miner Res* 21: 934–945

56   van der Horst G, van der Werf SM, Farih-Sips H, van Bezooijen RL, Lowik CW, Karperien M (2005) Downregulation of Wnt signaling by increased expression of Dickkopf-1 and -2 is a prerequisite for late-stage osteoblast differentiation of KS483 cells. *J Bone Miner Res* 20: 1867–1877

57   Li, X, Liu P, Liu W, Maye P, Zhang J, Zhang Y, Hurley M, Guo C, Boskey A, Sun L et al (2005) Dkk2 has a role in terminal osteoblast differentiation and mineralized matrix formation. *Nat Genet* 37: 945–952

58   Hens JR, Wilson KM, Dann P, Chen X, Horowitz MC, Wysolmerski JJ (2005) TOPGAL mice show that the canonical Wnt signaling pathway is active during bone development and growth and is activated by mechanical loading *in vitro*. *J Bone Miner Res* 20: 1103–1113

59   Chen Y, Whetstone H C, Youn A, Nadesan P, Chow EC, Lin AC, Alman BA (2007) β-catenin signaling pathway is crucial for bone morphogenetic protein 2 to induce new bone formation. *J Biol Chem* 282: 526–533

60   Baron R, Rawadi G (2007) Targeting the Wnt/β-Catenin pathway to regulate bone formation in the adult skeleton. *Endocrinology* 148: 2635–2643

61   Gong Y, Slee RB, Fukai N, Rawadi G, Roman-Roman S, Reginato AM, Wang H, Cundy T, Glorieux FH, Lev D et al (2001) LDL receptor-related protein 5 (LRP5) affects bone accrual and eye development. *Cell* 107: 513–523

62   Little RD, Carulli JP, Del Mastro RG, Dupuis J, Osborne M, Folz C, Manning SP, Swain PM, Zhao SC, Eustace B et al (2002) A mutation in the LDL receptor-related protein 5 gene results in the autosomal dominant high-bone-mass trait. *Am J Hum Genet* 70: 11–19

63   Boyden LM, Mao J, Belsky J, Mitzner L, Farhi A, Mitnick MA, Wu D, Insogna K, Lifton RP (2002) High bone density due to a mutation in LDL-receptor-related protein 5. *N Engl J Med* 346: 1513–1521

64   Van Wesenbeeck L, Cleiren E, Gram J, Beals RK, Benichou O, Scopelliti D, Key L, Renton T, Bartels C, Gong Y et al (2003) Six novel missense mutations in the LDL receptor-related protein 5 (LRP5) gene in different conditions with an increased bone density. *Am J Hum Genet* 72: 763–771

65 Semenov M, Tamai K, He X (2005) SOST is a ligand for LRP5/LRP6 and a Wnt signaling inhibitor. *J Biol Chem* 280: 26770–26775
66 Li X, Zhang Y, Kang H, Liu W, Liu P, Zhang J, Harris SE, Wu D (2005) Sclerostin binds to LRP5/6 and antagonizes canonical Wnt signaling. *J Biol Chem* 280: 19883–19887
67 van Bezooijen RL, Svensson JP, Eefting D, Visser A, van der Horst G, Karperien M, Quax PH, Vrieling H, Papapoulos SE, ten Dijke P et al (2007) Wnt but not BMP signaling is involved in the inhibitory action of sclerostin on BMP-stimulated bone formation. *J Bone Miner Res* 22: 19–28
68 Semenov MV, He X (2006) LRP5 mutations linked to high bone mass diseases cause reduced LRP5 binding and inhibition by SOST. *J Biol Chem* 281: 38276–38284
69 Zhang Y, Wang Y, Li X, Zhang J, Mao J, Li Z, Zheng J, Li L, Harris S, Wu D (2004) The LRP5 high-bone-mass G171V mutation disrupts LRP5 interaction with Mesd. *Mol Cell Biol* 24: 4677–4684
70 Hsieh JC, Lee L, Zhang L, Wefer S, Brown K, DeRossi C, Wines ME, Rosenquist T, Holdener BC (2003) Mesd encodes an LRP5/6 chaperone essential for specification of mouse embryonic polarity. *Cell* 112: 355–367
71 Piccolo S, Agius E, Leyns L, Bhattacharyya S, Grunz H, Bouwmeester T, De Robertis EM (1999) The head inducer Cerberus is a multifunctional antagonist of Nodal, BMP and Wnt signals. *Nature* 397: 707–710
72 Bell E, Munoz-Sanjuan I, Altmann CR, Vonica A, Brivanlou AH (2003) Cell fate specification and competence by Coco, a maternal BMP, TGFβ and Wnt inhibitor. *Development* 130: 1381–1389
73 Kusu N, Laurikkala J, Imanishi M, Usui H, Konishi M, Miyake A, Thesleff I, Itoh N (2003) Sclerostin is a novel secreted osteoclast-derived bone morphogenetic protein antagonist with unique ligand specificity. *J Biol Chem* 278: 24113–24117
74 Zimmerman LB, Jesus-Escobar JM, Harland RM (1996) The Spemann organizer signal noggin binds and inactivates bone morphogenetic protein 4. *Cell* 86: 599–606
75 Piccolo S, Sasai Y, Lu B, De Robertis EM (1996) Dorsoventral patterning in *Xenopus*: Inhibition of ventral signals by direct binding of chordin to BMP-4. *Cell* 86: 589–598
76 Koenig BB, Cook JS, Wolsing DH, Ting J, Tiesman JP, Correa PE, Olson CA, Pecquet AL, Ventura F, Grant RA et al (1994) Characterization and cloning of a receptor for BMP-2 and BMP-4 from NIH 3T3 cells. *Mol Cell Biol* 14: 5961–5974
77 Iwasaki S, Tsuruoka N, Hattori A, Sato M, Tsujimoto M, Kohno M (1995) Distribution and characterization of specific cellular binding proteins for bone morphogenetic protein-2. *J Biol Chem* 270: 5476–5482
78 Larrain J, Oelgeschlager M, Ketpura NI, Reversade B, Zakin L, De Robertis EM (2001) Proteolytic cleavage of Chordin as a switch for the dual activities of Twisted gastrulation in BMP signaling. *Development* 128: 4439–4447
79 Monteiro RM, Sousa Lopes SM, Korchynskyi O, ten Dijke P, Mummery CL (2004) Spatiotemporal activation of Smad1 and Smad5 *in vivo*: monitoring transcriptional activity of Smad proteins. *J Cell Sci* 117: 4653–4663
80 Gardner JC, van Bezooijen RL, Mervis B, Hamdy NA, Lowik CW, Hamersma H, Beigh-

ton P, Papapoulos SE (2005) Bone mineral density in sclerosteosis; affected individuals and gene carriers. *J Clin Endocrinol Metab* 90: 6392–6395

81 Ominsky MS, Warmington KS, Asuncion FJ, Tan HL, Grisanti MS, Geng Z, Stephens P, Henry A, Lawson A, Lightwood D et al (2006) Sclerostin monoclonal antibody treatment increases bone strength in aged osteopenic ovariectomized rats. *28th Annual Meeting of the American Society for Bone and Mineral Research*, Sept 15–19, 2006, Philadelphia, PA, USA

82 Ominsky M, Stouch B, Doellgast G, Gong J, Cao J, Gao Y, Tipton B, Haldankar R, Winters A, Chen Q et al (2006) Administration of sclerostin monoclonal antibodies to female cynomolgus monkeys result in increased bone formation, bone mineral density and bone strength. *28th Annual Meeting of the American Society for Bone and Mineral Research*, Sept 15–19, 2006, Philadelphia, PA, USA

83 Thomas T (2006) Intermittent parathyroid hormone therapy to increase bone formation. *Joint Bone Spine* 73: 262–269

84 Girotra M, Rubin MR, Bilezikian JP (2006) The use of parathyroid hormone in the treatment of osteoporosis. *Rev Endocr Metab Disord* 7: 113–121

85 Keller H, Kneissel M (2005) SOST is a target gene for PTH in bone. *Bone* 37: 148–158

86 Bellido T, Ali AA, Gubrij I, Plotkin L I, Fu Q, O'Brien CA, Manolagas SC, Jilka RL (2005) Chronic elevation of parathyroid hormone in mice reduces expression of sclerostin by osteocytes: a novel mechanism for hormonal control of osteoblastogenesis. *Endocrinology* 146: 4577–4583

87 ten Dijke P, Krause C, de Gorter DJJ, Lowik CWGM, van Bezooijen RL (2008) Osteocyte-derived Sclerostin inhibits bone formation: Its role in BMP and Wnt signaling. *J Bone Joint Surg* 90: 31–35

Bone morphogenetic proteins in cartilage biology

Susan Chubinskaya[1], Mark Hurtig[2] and David C. Rueger[3]

[1]Department of Biochemistry, Rush University Medical Center, Chicago, IL, 60612, USA; [2]Guelph University, Guelph, Ontario, Canada; [3]Stryker Biotech Division, Hopkinton, MA 01748, USA

Introduction

Cartilage repair and regeneration is a major obstacle in orthopedic medicine. The importance is enormous since osteoarthritis (OA) is a major cause of disability among the adult population in the United States and degenerative disc disease (DDD) is responsible for a significant amount of the chronic back pain. OA is considered a process of attempted, but gradually failing, repair of damaged cartilage extracellular matrix, as the balance between synthesis and breakdown of matrix components is disturbed and shifted towards catabolism. In recent times, members of the bone morphogenetic protein (BMP) family of proteins have demonstrated a great potential as anabolic factors for cartilage repair because of their ability to induce matrix synthesis and promote repair in cartilage.

BMPs are members of the transforming growth factor-β (TGF-β) superfamily [1, 2] and are found in species ranging from worms and insects to mammals. They have wide-ranging biological activities, including the regulation of cellular proliferation, apoptosis, differentiation and migration, embryonic development and the maintenance of tissue homeostasis during adult life. [3–7]. BMPs were originally purified and identified from bone as proteins capable of inducing ectopic endochondral bone formation in subcutaneous implants [8–10]. However, it is now clear that they are expressed in a variety of tissues, including cartilage [11–13]. The main focus of our studies for many years has been the seventh member of the BMP family, BMP-7, also called osteogenic protein-1 (OP-1).

Since the first *BMP* genes were identified in the late 1980s, the corresponding recombinant proteins have been produced and two of these early BMPs, BMP-7 and BMP-2, have been extensively characterized both biochemically and biologically. *In vivo* characterization has involved a variety of animal models to evaluate the therapeutic potential in bone repair applications. These studies led to the demonstration of bone repair in humans and eventually in BMP-7 and BMP-2 receiving regulatory approval as the first commercial BMPs.

The past 10 years have seen a new era opening for BMPs in the cartilage field. Although we attempt to provide an update on key BMP studies, the main focus of this review is on BMP-7 since our data have demonstrated that BMP-7 is a very important cartilage repair factor in addition to its well-known application as a bone-induction factor. This chapter reviews the information accumulated thus far from *in vitro* studies as well as from studies of repair in various animal models. The data show significant promise for BMPs in cartilage repair and suggest that both articular and disc cartilage applications could become very important for BMPs in orthopedics.

In vitro **studies**

The *in vitro* studies covering BMPs in cartilage repair are reviewed in two parts. First, the studies that address the anabolic activity of recombinant BMPs on chondrocytes are described. In this section, the activity of exogenous BMP preparations is characterized using cells either embedded in native cartilage matrix and cultured as explants or isolated from the extracellular matrix and cultured as monolayers, pellets, or embedded in different scaffolds and polymeric matrices. Secondly, studies investigating endogenous BMPs expressed by chondrocytes are described.

Anabolic activity

The ability of BMPs to induce an anabolic response in cartilage *in vitro* has been documented using different BMPs in multiple species, including human, bovine, rat, rabbit, and mouse, and a variety of culture conditions. Among the BMPs, BMP-7 has been by far the most extensively studied. In the last few years, we have gained a great amount of information about BMP-7 in articular cartilage, its role in the maintenance of normal tissue homeostasis, and the interaction BMP-7 displays with various anabolic and catabolic pathways. Anabolic activity of BMP-7 has been studied for more than a decade in different species and various culture systems. Unlike TGF-β and other BMPs, BMP-7 up-regulates chondrocyte metabolism and protein synthesis without creating uncontrolled cell proliferation and formation of osteophytes [11, 14–17]. Importantly, in all *in vitro* experimentations with chondrocytes from different species and humans, BMP-7 has been shown to stimulate only cartilage-specific extracellular proteins: collagens type II and VI, aggrecan, decorin, fibronectin, hyaluronan (HA), etc. [14, 15, 18, 19]. Furthermore, BMP-7 generated normal, functional proteoglycans (PGs), with a hydrodynamic size unaltered by the treatment [20]. It induced anabolic responses in normal and osteoarthritic (OA) chondrocytes from both young and old donors. Extracellular matrices deposited by these chondrocytes did not substantially vary in size and composition in the presence of BMP-7. Stimulating HA, its receptor CD 44 and enzyme responsible for hyaluronan synthesis HAS-2, BMP-7 promoted the formation and retention of the extracellular matrix [11, 15–21]. BMP-7 did not induce chondrocyte hypertrophy or changes in chondrocytic phenotype when chondrocytes were cultured as explants or as isolated cells in high-density monolayers or alginate beads. The effect was consistent with or without serum [14, 16, 18, 29, 22]. Noteworthy, immortalized chondrocytes [22] or autologous chondrocytes prepared for implantation after initial expansion [23] responded to BMP-7 treatment in a similar manner as primary cells. In regard to other BMPs, BMP-4 was shown to induce PG synthesis in bovine explant cultures in a dose-dependent manner under short-term conditions; longer cultures led to a decrease in PG synthesis and collagen metabolism [24–26]. BMP-2 and BMP-9 were shown to induce a similar anabolic response in the absence of serum, while in the presence of serum the bovine chondrocytes became unresponsive [27, 28]. In addition, BMP-2 has been reported to induce PG synthesis and maintain bovine chondrocytic phenotype through induction of the genes for aggrecan and type II collagen and by depressing the type X collagen genes in long-term monolayers [29, 30].

BMP-6 has been also reported to be potent for stimulation of matrix synthesis in normal and OA chondrocytes [31]. It was shown that BMP-6 is capable of increasing PG synthesis and this response to BMP-6 declined with age. BMP-6 had no effect on proliferation of human chondrocytes. These data were similar to that observed with BMP-7 treatment of human cells [32]. In our laboratory, we recently compared the response of human normal adult articular chondrocytes to recombinant BMP-2, -4, -6, -7, and cartilage-derived morphogenetic proteins (CDMP) 1 and 2 with regard to PG synthesis and content [33]. Isolated cells were cultured in alginate for 9 days in the presence of 10% fetal bovine serum. Each growth factor was added at the same concentration (100 ng/ml) every other day and the values were normalized to the DNA content. Within the first 5 days of treatment, all studied BMPs induced a similar response; however, by day 9, the highest levels of PG synthesis were identified in the BMP-7-treated group. Importantly, all BMPs were able to stimulate an anabolic response in human chondrocytes above serum levels. These findings are similar to previous data where BMP-7 was found to be more potent than TGF-β and activin [14]. The effect of tested BMPs on PG content in the chondrocyte matrix was less pronounced; no increase in PG accumulation was observed until days 5–7, when BMP-2, -4, and -7 caused a 1.5–2-fold increase in PGs accumulated in the matrix ($p<0.02$). The most pronounced effect was seen by day 9, when BMP-2, 4, and -7 increased PG content by 2–2.5-fold ($p<0.001$) and no differences between these proteins were detected. CDMPs (GDF-5, -6 and -7) were found to be less potent than selected BMPs. The strongest effect of CDMPs was observed by day 9, when they stimulated PG accumulation by about 1.7-fold ($p<0.001$). None of the BMPs caused cell death or induced chondrocyte proliferation under the described experimental conditions [33].

Anabolic effects of BMP-7 extend beyond stimulation of cartilage extracellular proteins. BMP-7 modulates the expression of receptors for certain matrix components, for instance CD44, [15] and various signaling pathways: growth factors [insulin-like growth factor-1 (IGF-1), platelet-derived growth factor (PDGF), fibroblast growth factor (FGF)], TGF-β/BMP-specific pathway and catabolic mediators [interleukin-1 (IL-1), IL-5, IL-6, IL-6, IL-13, tumor necrosis factor-α (TNF-α)] [22, 34]. BMP-7 has also been shown to regulate the synthesis of chondrocyte cytoskeleton proteins (talin, paxillin and focal adhesion kinase [35]) and enhance gene expression of the anabolic molecule tissue inhibitor of metalloproteinase (TIMP) [36], an effect which was recently confirmed by gene array analysis [34] and in a study exploring different anabolic and catabolic genes in normal and OA chondrocytes [17]. In the latter, the expression of anabolic genes in OA chondrocytes was significantly up-regulated by BMP-7 when compared with normal chondrocytes. In neither normal nor OA cells was a stimulating effect on matrix degrading enzymes (MMP-1, -3, -13 and ADAMTS-4) observed. On the contrary, the expression of these enzymes was inhibited by BMP-7 in human articular chondrocytes [34] and in rat nucleus pulposus, annulus fibrosis and end-plate cells in the animal model of disc degeneration [37].

Our recent findings highlight another important aspect of the anabolic activity of BMP-7; it promotes cell survival [16, 18, 23]. This could be due to either a direct effect on signaling mechanisms that control apoptosis (caspases for example) and cell death/cell survival pathways or indirectly *via* stimulation of the IGF-1 pathway involved in cell survival and proliferation. In addition to the reparative responses enthused in chondrocytes of different ages and different health status, BMP-7 has been implicated in the regulation of other anabolic pathways active in articular cartilage. Thus, the presence of BMP-7 in chondrocyte cultures caused an activation of the IGF-1 signaling pathway, i.e., expression of IGF-1, IGF-1 receptor, IGF-1 binding proteins, etc. We believe that such an effect on IGF-1 pathway leads (at least in part) to the restoration of the responses to IGF-1, which are lost with aging of human cartilage [16, 22]. Activation of the IGF-1 signaling pathway by BMP-7 was also seen by gene array analysis confirmed by real-time PCR [34]. Furthermore, combined application of BMP-7 and IGF-1 had a synergistic effect on cell survival and matrix synthesis. Both aged and OA chondrocytes responded to BMP-7 or combined BMP-7 and IGF-1 with enhanced anabolic activity and viability, suggesting that these growth factors may be essential in the repair of articular cartilage [16]. Intriguingly, the combination of these two growth factors induced responses that were not observed under the treatment with each individual factor: they caused at least a 2-fold increase in chondrocyte proliferation rate [16], indicating perhaps that BMP-7 restores or enhances a mitogenic activity of IGF-1. The underlying mechanism of this effect is currently unknown, but our unpublished data suggest that PI3 kinase downstream in IGF-1 signaling pathway and Id (inhibitors of differentiation) proteins, transcription factors in the BMP signaling pathway, may be involved. Unexpectedly, combination of these two growth factors with basic FGF caused an opposite effect; addition of basic FGF inhibited PG synthesis induced by BMP-7, IGF-1 or their combination [18]. We also found that BMP-7 is able to modulate TGF-β/BMP signaling pathway by regulating gene expression of related growth factors, their receptors, binding proteins and transcription factors [34]. To our surprise, by gene array, we found that treatment with BMP-7 led to more than 2.5 fold ($p < 0.001$) inhibition of endogenous *BMP-2* gene expression in human adult chondrocytes, which was confirmed by *in vitro* verification experiments [34]. Since BMP-7 and BMP-2 share the same signaling partners, increased levels of one growth factor may lead to the decrease in a second factor perhaps in a negative feedback loop to maintain a balance within signaling pathway and prevent hyperstimulation.

Anti-catabolic activity

One of the most critical properties of BMP-7 is its anti-catabolic activity that has been demonstrated in numerous studies with different experimental approaches. BMP-7 inhibits endogenous expression of various catabolic agents *in situ* and counteracts their deleterious effects on cartilage homeostasis. BMP-7 has been shown to block not only baseline levels of MMP-13 protein present in chondrocytes but also cytokine-induced expression of MMP-1 and MMP-13 [22]. A novel anti-catabolic effect of BMP-7 was recently shown in an acute cartilage trauma model in sheep, where the number of cells dying by apoptosis (positive TUNEL stain) was significantly reduced by the injection of BMP-7 [38] and correlated with the overall improvement in joint morphology and cartilage structural integrity. The role of BMP-7 in the inhibition of apoptosis was supported by gene array studies with human chondrocytes, in which the treatment with BMP-7 caused a decrease in the expression of caspases, enzymes responsible for apoptotic cell death [34, 39]. As we mentioned above, in addition to a direct inhibitory effect on endogenous expression of pro-inflammatory cytokines in chondrocytes *in situ* (especially IL-6 family of chemokines: IL-6, IL-8, IL-11, leukemia inhibitory factor), BMP-7 modulates their downstream signaling molecules: receptors, transcription factors, and mitogen-activated kinases [34].

Multiple studies documented that BMP-7 effectively counteracts chondrocyte catabolism (inhibition of PG and HA synthesis) induced by various catabolic mediators, such as proinflammatory cytokines (IL-1 [40] and IL-6) and fragments of cartilage matrix proteins (fibronectin fragments [41] or HA hexasaccharides [42]). To our knowledge, BMP-7 is the only BMP studied thus far in cartilage that exhibits broad pro-anabolic and anti-catabolic activities. Moreover, we showed that under the same conditions BMP-7 is a more potent stimulator of PG and collagen synthesis than TGF-β1 and activin [14]; only BMP-7 was able to stimulate chondrocytes above the levels of synthesis obtained in the presence of serum. As we mentioned above, BMP-7 is a stronger stimulator of PG synthesis than BMP-2, -4, -6, and CDMP-1 and -2 and demonstrated a better ability to counteract or protect IL-1β-induced inhibition of PG synthesis [33]. In a murine *in vivo* model, BMP-2 was not able to counteract IL-1-induced inhibition of PG synthesis [43], when both factors were co-injected into normal knees. IL-1 abrogated an anabolic activity of BMP-2, whereas TNF-α at the same concentration had no effect [44, 45]. However, BMP-2 was not evaluated in any *in vitro* cell culture study nor was BMP-7 used in this *in vivo* injection model. Therefore, we undertook the comparative studies described above to directly compare BMPs with a similar mode of action and found more pronounced effects induced by BMP-7 than by other BMPs. Nonetheless, this was only the initial step and more studies are warranted to understand key differences in the properties of the members of the same family.

Detailed analyses suggest that the inhibition of transcription factors NF-κB and AP-1 by BMP-7 [22] or up-regulation of inhibitors of matrix proteinases [34, 36] may be part of the underlying mechanisms responsible for the anti-catabolic activity of BMP-7. Earlier findings were confirmed by our recent studies [46] with the *in vitro* model of inflammation where chondrocytes were treated with a relatively high dose of IL-1β (10 ng/ml). For comparison, low picogram quantities are considered to be within physiological range for synovial fluid (SF) and plasma in normal human articular joints [47]. IL-1β activates NF-κB by inducing the phosphorylation and degradation of IκB-α, which allows the translocation to the nucleus of the active NF-κB dimers [48]. We found that BMP-7 inhibited IL-1β induced phosphorylation of NF-κB, and thus diminished NF-κB-mediated transcriptional responses even at high, pathophysiological dose of IL-1β. These data are in line with molecular results of Im et al. [22], who reported that the inhibition of NF-κB promoter activity by BMP-7 or a combination of BMP-7 and IGF-1 down-regulated *IL-1β* and the fibronectin fragment-induced expression and activation of MMP-1 and MMP-13. In contrast, BMP-7 did not affect IL-1β-stimulated NF-κB activity and IκB-α degradation in human mesangial cells, while inhibiting AP-1 transcription factor [49]. The inhibition of AP-1 in articular chondrocytes was also observed by Im et al. [22].

Furthermore, in our inflammation model we discovered a novel mechanism by which BMP-7 counteracts IL-1β signaling in chondrocytes (in addition to NF-κB inhibition discussed above). We found that IL-1β could induce alternative phosphorylation of receptor Smads 1/5/8 (R-Smads) within the linker region (unpublished data) through the activation of mitogen activated protein kinases (MAPK) of the ERK family, specifically JNK and P38. This alternative phosphorylation of R-Smads compromises canonical BMP-induced activation of R-Smads, leading to the reduced responses to BMP-7. Linker region phosphorylation of Smads-2/-3 has also been shown to negatively control responses to TGF-β [50, 51]. According to our newly acquired data, the mechanism responsible for the ability of BMP-7 to counteract effects of catabolic cytokines may lie in its potential to reverse MAPK signaling *via* the inhibition of IL-1β-induced P38 phosphorylation. The inhibition of p38 completely abrogated IL-1β-induced linker region phosphorylation and restored BMP-7-mediated phosphorylation of R-Smads. In fact, p38 might be a common mechanism responsible for cartilage degradation stimulated by not only IL-1β, but also by a lipopolysaccharide (LPS). In a recent study, Bobacz et al. [52] showed that the suppressive effects of LPS on cartilage anabolism are governed by an up-regulation of IL-1β *via* activation of p38 kinase and are antagonized by the application of BMP-7, thus indirectly implying a connection between BMP-7 and the inhibition of p38. This may be yet another example of cross-talk between different signaling pathways, which appears to be common in signal transduction, especially considering that (i) TGF-β and BMP-4 have already been implicated in TAK-1, ERK, and JNK of MAPK signaling [53, 54]; (ii) BMP-7 was shown to inhibit IL-1β-induced phosphorylation of JNK in mesangial cells [49] and inhibit P38 and endogenous activating transcription factor 2 (ATF2) in murine inner medullary collecting duct [55]. Understanding how anabolic and catabolic pathways are coordinated *in vivo* is an important question in chondrocyte signaling that remains to be investigated.

Endogenous expression

Although clinical application of recombinant BMPs still remains the primary focus, the understanding of the regulation and function of the BMPs, which are endogenously expressed in adult articular cartilage, offers significant supporting data. Knowledge of the mechanisms that control their synthesis, activation, induction, signaling, and interaction with other pathways in articular cartilage provides critical missing information that is necessary to develop correct strategies for the application of recombinant BMPs for cartilage restoration and repair in post-traumatic and degenerative arthritis.

Unfortunately, there are only a few studies with the focus on endogenously expressed BMPs. Fukui et al. [56] reported the expression of BMP-2, -4, -6, and CDMP-1 (GDF-5) in normal and OA cartilage. They showed that BMP-2 is expressed in both normal and OA tissue and that it is greatly induced by inflammatory cytokines, IL-1β and TNF-α. No differences were found among other BMPs. Similarly, Bobacz et al. [31] reporting the expression of BMP-6 in normal and OA cartilage also found no changes in its expression with aging or OA. The authors concluded that it is unlikely that BMP-6 is involved in the pathogenesis of OA but may be important for the maintenance of tissue homeostasis in articular cartilage. Recently, one comparative study was performed where differential expression of various BMPs (BMP-1–11, TGF-βs and CDMPs) and their receptors was reported in fetal, adult, and OA articular cartilage [57]. However, only limited amount of adult samples was entered into the analysis and those were from older individuals undergoing hip hemiarthroplasty for proximal femoral fractures. Some BMPs (1, 2, 3–6, and 11) were found to be constitutively expressed in all cartilages, whereas others were primarily expressed either in fetal tissue (BMP-3 and -8), or in adult and OA cartilages (TGF-β).

In our laboratory, about a decade ago, we identified for the first time that BMP-7 is synthesized by human knee and ankle articular chondrocytes [11], obtained from organ donors through the Gift of Hope Tissue & Organ Donor Network (Elmhurst, IL). Access to human tissues allowed the detection of endogenously expressed BMP-7 in cartilage samples of different age (fetal, newborn, young, and adult) and various morphological appearance (from normal to early and advanced degeneration including OA). We found that *BMP-7* gene and protein expression are dramatically reduced with cartilage aging and degeneration [11, 12]. Our most recent study (submitted for publication) suggests that one of the mechanisms responsible for a decrease in BMP-7 production with aging is the methylation of *BMP-7* promoter, which was primarily identified in older individuals. This might provide an explanation for the apparent conflict in results on BMP-7 expression reported by different groups [17, 57]. The inability to detect appreciable amounts of BMP-7 expression in these studies could be explained by the choice of cartilage samples from aged donors, in which *BMP-7* gene was silenced due to the structural changes of the promoter.

Through the use of antibodies to the pro- and mature domains of the BMP-7 molecule, we found by immunohistochemistry that BMP-7 is present in normal and OA cartilage in two forms: the pro-form (inactive) and the mature (active) form. These forms of BMP-7 have a distinct localization in cartilage tissue: mature BMP-7 is primarily localized in the upper cartilage zone, while the pro-form is mostly identified in the deep cartilage layer. In OA cartilage mature BMP-7 is selectively localized in cell clusters. Preferential distribution of mature BMP-7 in the upper cartilage zone was similar to that shown in the previously published data by Erlacher et al. [58] for CDMP-1 and -2. Both of these proteins were also primarily detected in the superficial layer of normal cartilage and in chondrocyte clusters of OA cartilage. Quantitatively, there is much less mature and pro-BMP-7 found in older and OA cartilage tissues than in normal young adult cartilages. The exact mechanism that controls activation of pro-BMP-7 into the mature form is not yet clearly defined, although by analogy to TGF-β and BMP-4, the role of furin convertases is implied [59].

In addition to human adult cartilage, endogenous expression of BMP-7 has been detected in adult bovine, rabbit, and goat tissues [43, 58, 60, 61]. The distribution of mature (active) BMP-7 protein in normal goat and rabbit articular cartilage is similar to that in human tissues, while in bovine cartilage mature BMP-7 is strongly detected throughout the entire cartilage thickness. In rat cartilage, Anderson et al. [61] detected BMP-7 in the middle and deep cartilage zones. Notably, the use of different antibodies and different procedures for treatment of tissues may account for the conflicting results we obtained for BMP-7, but it is also possible that in rat cartilage, the localization does indeed differ from cartilage of other species.

To quantify physiological and pathological levels of BMP-7 in different connective tissues and body fluids, we developed a quantitative sandwich ELISA method that allowed the detection of the BMP-7 protein at picogram quantities [11]. We found that in normal cartilage, the concentration of BMP-7 protein is around 50 ng/g dry tissue, which is within the physiological range of the anabolic activity of recombinant human BMP-7 (50–200 ng/ml) [62]. We also detected BMP-7 in SF from normal joints and from patients with OA and rheumatoid arthritis (RA) [63]. The origin of BMP-7 in the SF is unknown; however, recent detection of BMP-7 in cartilage, synovium, ligament, tendon and menisci suggests that it could be synthesized and released by all these tissues [62].

Notably, the levels of endogenous BMP-7 detected in SF from normal joints (about 50 ng/ml) were comparable with those extractable from normal articular cartilage (about 50 ng/g dry weight or about 50–200 ng/ml). Furthermore, in SF from RA patients, the levels of BMP-7 protein were at least twofold higher than in SF from OA patients and normal donors. Although we did not identify quantitative differences in the levels of endogenous BMP-7 in normal SF samples relative to osteoarthritic samples, there were specific qualitative differences; whereas normal SF had no detectable or barely detectable active (mature) BMP-7, the SF from OA and RA joints had both pro-BMP-7 and active (mature) BMP-7, suggesting that, at least in part, the active form of BMP-7 could be generated as the response to catabolic cytokines/inflammatory processes typical for synovitis [64, 65]. It is also possible that the cleavage and activation of BMP-7 in joint disease reflects the action of matrix proteinases induced by catabolic mediators active in RA and the late stages of OA. Numerous data indicate that BMP-7 is an inducible protein that can be up-regulated in response to various catabolic insults. Our previous animal and human studies support this statement [12, 66]. In a well-recognized animal model of OA, the intra-articular injection of chymopapain into the rabbit knee joint induced the activation of BMP-7 in cartilage [66]. Moreover, the activation and release of mature BMP-7 protein was detected in organ cultures of normal human adult articular chondrocytes in response to IL-1β, TNF-α treatment [12], or to biomechanical loading [67]. The finding that SF levels of BMP-7 were higher in RA patients than in OA patients or asymptomatic donors is also consistent with recent reports that IL-1, which is present at higher concentrations in RA joints than in OA joints, is an effective modulator and/or stimulator of *BMP-2* and *BMP-7* mRNA expression by normal and OA human articular chondrocytes [12, 46, 56]. These data are also in line with previous findings that documented an elevation of TGF-β in SF of RA patients [65]. Furthermore, we believe that elevated levels of BMP-7 protein in RA SF may be due to the release of BMP-7 residing in the extracellular matrix rather than to an increase in its synthesis. This belief is because matrix metalloproteinases activated by cytokines present in SF of RA patients induce the depletion of the extracellular matrix [68], thus promoting the release of growth factors bound to its latent domains or to the matrix components [69]. Up-regulation of the expression of endogenous BMPs in response to catabolic mediators may be qualified as an attempt to repair and appears to be a common feature of these factors. The latter was further confirmed in the goat model of osteochondral defect, in which endogenous pro-BMP-7 protein was synthesized in response to the defect [60]. The detection of endogenously expressed and synthesized BMP-7 in adult articular cartilage of different species provides support for the concept that this growth factor is not only important for skeletal development and bone healing, but it also has a functional role in the maintenance of normal cartilage homeostasis. This was confirmed by our recent studies with the *BMP-7* antisense probe [13], where transfection of human adult articular chondrocytes with *BMP-7* antisense oligonucleotides led to about 70% inhibition in *BMP-7* gene expression. Down-regulation of *BMP-7* mRNA induced significant inhibition of *aggrecan* mRNA expression and about 50% decrease in PG synthesis. Recovery (add-back) experiments with recombinant BMP-7 were able to restore, at least partially, PG synthesis indicating a direct role of autocrine BMP-7 in the regulation of PG metabolism. Histological evaluation of cartilage explants cultured in the presence of *BMP-7* antisense oligonucleotides revealed a remarkable depletion of PGs, paucity of chondrocytes, initial fibrillation of cartilage surface and the decrease in Safranin O staining in the upper and middle cartilage zones. These data together with previous results provide strong evidence for endogenous BMP-7 being a critical factor that controls cartilage matrix integrity and is involved in the maintenance of normal cartilage homeostasis. Similar results were obtained with the inhibition of *BMP-7* gene expression by siRNA approach (Fig. 1). Silencing of the *BMP-7* gene led to a significant down-regulation of the BMP-7 protein synthesis, which in turn strongly affected PG synthesis (Fig. 1B). Inhibitory effect was seen already after 24 and, even more, after 48 h of chondrocytes transfection. After 48 h of culture, PG synthesis was reduced by 56% when lipid was chosen as a carrier and by 72% in the presence of amine. For comparison BMP-7 protein synthesis was reduced by more than threefold in the presence of the lipid carrier and was essentially shut down in the presence of the amine transfection agent (Fig. 1A).

In addition, the data further suggest that the lack of endogenous BMP-7 could predispose cartilage to degenerative processes and make tissue more susceptible to the influence of catabolic agents. To confirm this hypothesis Affimetrix analysis with gene array (which contained 22 000 genes) was performed on chondrocytes treated with *BMP-7* antisense. The results clearly demonstrated that the lack of *BMP-7* gene expression negatively affected a number of matrix proteins and anabolic pathways (mostly those that were up-regulated by recombinant BMP-7) and stimulated factors associated with cartilage catabolism (proteinases with various mode of action including caspases and cathepsins, inflammatory cytokines and their signaling mediators, etc.). Moreover, gene array studies suggested interplay between BMP-7 and IL-6 and other members of the IL-6 family (LIF, IL-8, and IL-11). Our further studies are aimed at elucidating the mechanisms of the relationship between

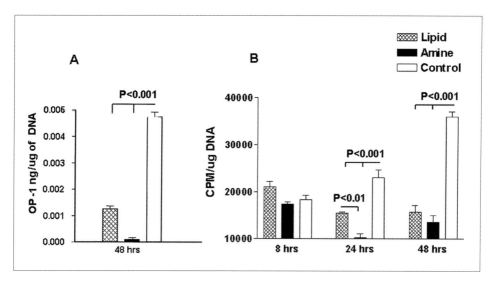

Figure 1
Effect of the inhibition of bone morphogenetic protein-7 (BMP-7/osteogenic protein-1, OP-1) gene expression by siRNA approach on BMP-7 protein (A) and proteoglycan (PG) synthesis (B). (A) BMP-7 protein is measured by ELISA after 48 h of culture in the presence of 25 nM siRNA probe with a sequence of 3'-ATGCCATCTCCGTCCTCTACCCTGTCTC-5'. The probe was delivered either with the lipid-based or amine-based carrier. (B) PG synthesis measured after 8, 24, and 48 h of culture in the presence of siRNA probe.

the BMP-7 and other anabolic and catabolic signaling pathways that are important in cartilage homeostasis in health and disease.

In summary, knowledge accumulated in our laboratory over the last decade unequivocally points to the crucial role BMPs play in the regulation of overall cartilage homeostasis. For example, in addition to the pro-anabolic properties of BMP-7 as a repair factor, it also acts as an anti-catabolic agent affecting numerous catabolic pathways (Fig. 2). New information developed over recent years has significantly advanced our understanding of the mechanisms of action of BMPs in cartilage and provided a strong scientific basis for considering members of the BMP family as therapeutic agents for cartilage repair and regeneration.

Animal studies

The pivotal role of BMPs in the development and regeneration process of the skeleton had originally suggested a role in articular cartilage repair. Furthermore, the accumulation of data from *in vitro* studies clearly demonstrated that certain BMPs

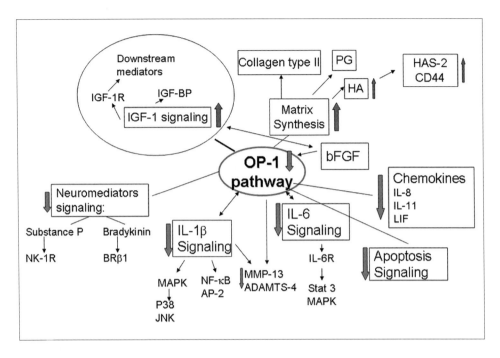

Figure 2
Schematic summary of the in vitro *effects of recombinant BMP-7 and the changes endogenous BMP-7 undergoes with aging, osteoarthritis (OA) and under the influence of various catabolic mediators.*

have an important role in chondrocyte differentiation and extracellular matrix production as well as the maintenance of adult chondrocyte phenotype. For the most part, two BMPs, BMP-7 and BMP-2, have been the subject of animal models of cartilage repair, although in recent years another BMP, GDF-5 (also called CDMP-1 or MP-52) has also begun to be investigated. Studies have been reported in a wide range of animal species and have involved both articular cartilage models, as well as non-articular cartilage tissue, particularly the intervertebral disc. This section summarizes the animal studies thus far reported, describing the use of BMPs for repairing cartilage defects, including investigations evaluating repair with the BMP proteins as well as the BMP genes.

Articular cartilage repair

Studies describing the use of BMPs for repairing articular cartilage defects are reviewed in three parts. First, studies addressing the repair of both bone and carti-

Table 1 - Articular cartilage animal studies. Focal osteochondral defects.

Citation	Year	Species	BMP	Carrier	Model
Cook and Rueger [73]	1996	Rabbit	BMP-7	Collagen particles	12-week evaluation
Grgic et al. [74]	1997	Rabbit	BMP-7	Collagen particles	8-week evaluation
Sellers et al. [70]	1997	Rabbit	BMP-2	Collagen sponge	4-, 8-, 24 week evaluation
Mattioli-Belmonte et al. [75]	1999	Rabbit	BMP-7	Chitosan	7-, 14-, 60-day evaluation
Louwerse et al. [81]	2000	Goat	BMP-7	Collagen particles	Mixed with blood clot and contained with periosteal membrane; 1-, 2-, 4-month evaluation
Sellers et al. [71]	2000	Rabbit	BMP-2	Collagen sponge	1-year evaluation
Frenkel et al. [42]	2000	Rabbit	BMP-2	Collagen sponge	In combination with allogenic chondrocytes; 6-week evaluation
Mason et al. [77]	2000	Rabbit	BMP-7	Polyglycolic acid	Gene delivery using transfected periosteal-derived allogenic mesenchymal cells: 4-, 8-, 12-week evaluation
Cook et al. [80]	2003	Dog	BMP-7	Collagen particles	Mixed with carboxymethyl-cellulose; 6-, 12-, 16-, 24-, 52-week evaluation
Hunziker et al. [82]	2003	Minipig	BMP-2	Fibrin-gelatin	Mixed with anti-angiogenic factor in upper layer and BMP-2 liposome-encapsulated; 8-week evaluation
Shimmin et al. [84]	2003	Sheep	BMP-7	Collagen particles	Mixed with carboxymethylcellulose mosaicplasty augmentation; 3-, 6-, 12-week evaluation
Simank et al. [76]	2004	Rabbit	GDF-5	Collagen sponge	4-, 8-, 24-week evaluation
Katayama et al. [78]	2004	Rabbit	GDF-5	Collagen gel	Gene delivery using transfected bone marrow-derived allogenic mesenchymal cells; 2-, 4-, 8-week evaluation
Jung et al. [83]	2006	Minipig	GDF-5	Collagen/hyaluronate	3-, 12-month evaluation
Di Cesare et al. [79]	2006	Rabbit	BMP-2	Collagen sponge	Gene delivery using naked DNA; 12-week evaluation

Table 2 - Articular cartilage animal studies - Focal chondral defects

Citation	Year	Species	BMP	Carrier	Model
Jelic et al. [85]	2001	Sheep	BMP-7	Liquid	Minipump delivery to joint; 3-, 6-month evaluation
Hunziker et al. [86]	2001	Minipig	BMP-2, GDF-6	Fibrin	BMPs were encapsulated in liposomes; 6-week evaluation
Hidaka et al. [88]	2003	Horse	BMP-7	Fibrin	Gene delivery using transfected allogenic chondrocytes; 1-, 8-month evaluation
Kuo et al. [87]	2006	Rabbit	BMP-7	Collagen sponge	Microfracture augmentation study; 24-week evaluation
Park et al. [89]	2006	Rat	BMP-2	Fibrin	Gene delivery comparing 3 types of mesenchymal cell sources; 8-week evaluation

lage tissue using focal osteochondral defect are reviewed (Tab. 1). Secondly, studies describing chondral or partial thickness repair are reviewed (Tab. 2). Lastly, the use of BMPs in animal models of OA is only in the very early stage, but what information is available is also reviewed.

Numerous studies have been done using deep osteochondral defects in articular cartilage with the BMPs delivered locally into the defect site on a collagen scaffold that was press-fitted into the defect site. However, a few studies have also been reported using the more difficult chondral (partial thickness) defect models where the defect did not penetrate the calcified cartilage layer. In these studies, the BMP was delivered by a variety of methods to the defects, including *via* a minipump to the SF. More recently, articular cartilage disease models are being pursued and animal models of OA and methods to deliver the BMP to the joint are being developed.

Focal osteochondral defect models

Repair of osteochondral defects, which involve both the cartilage tissue and the underlying bone, is known to occur to a limited extent promoted by the presence of both stem cells and growth and differentiation factors brought into the defect by the blood/marrow. In animal studies these defects undergo repair with formation of a new layer of bone and cartilage, but the macromolecular organization and the bio-chemical characteristics of the cartilage matrix are imperfect. High levels of type I instead of type II collagen and PGs that are not cartilage specific, such as dermatan sulfate containing PGs, make up the repair tissue and result in fibrillations and degenerative changes over time. In contrast to the osteochondral defects, the repair of cartilage defects that do not penetrate into the subchondral bone (chondral defects) is not believed to occur even to a limited extent.

Most investigations evaluating articular cartilage healing with BMPs have involved osteochondral defect repair models. These studies have demonstrated that certain BMPs can improve the healing response in both the bone and cartilage tissue in comparison to the untreated controls. The models have involved deep defects of the knee and have utilized a variety of animal species including rabbit, dog, sheep or goat models. For the most part type I collagen has been used to locally deliver the BMP into the defect sites.

The earliest studies were done using various models in the rabbit knee. The most comprehensive study was conducted using recombinant BMP-2 delivered in a type I collagen sponge to defects created in the trochlear groove [70–72]. Repair of defects (3 mm in diameter, 3 mm deep) was evaluated at 4, 8, 24 and 52 weeks post implantation. In comparison with control defects, treatment with BMP-2 greatly accelerated the formation of new subchondral bone and improved the articular cartilage with the formulation of a tidemark between the tissues. At 6 months the thickness of the cartilage was 70% that of the normal adjacent cartilage and this was maintained to 12 months. In this regard it is interesting that this repair did not require the presence of BMP-2 after the initial induction process since the residence time of the BMP-2 was demonstrated to be 14 days in the defect site. The BMP-2-treated defects showed an improvement in the integration of the repair cartilage with the adjacent cartilage, an improvement in cellular morphology and significantly less type I collagen when compared with untreated defects. In conclusion, the authors clearly stated that the repair cartilage was not identical to normal articular cartilage, but BMP-2 stimulated a faster repair with more abundant and more hyalin-like cartilage over that seen in the control defects.

Similar results to those seen with BMP-2 were described for BMP-7. Multiple studies were also done in rabbit models using recombinant protein delivered on bone-derived type I collagen particles [73, 74]. In one study defects were created in femoral condyles (6 mm long, 3 mm wide and 3 mm deep) and histological examination of the repair at 2 months showed a significant difference in healing of the defects treated with BMP-7 compared to those left empty or treated with the collagen only [74]. Defects that were not treated with BMP-7 were filled with primarily fibrous tissue and fibrocartilage. Defects treated with BMP-7 showed extensive regeneration of both the subchondral bone and a hyaline-like cartilage layer on top. These defects were completely bridged with abundant tissue that consisted of small rounded cells organized in columns and embedded in compact extracellular matrix. In a study reported using a chitosan matrix to deliver BMP-7, no significant repair of either bone or cartilage was demonstrated [75], suggesting that the matrix is an important component of the repair process.

More recently, a rabbit study has been reported evaluating GDF-5 in combination with a collagen carrier [76] for repair of osteochondral defects. Defects (3.2

mm wide and 5 mm deep) implanted with GDF-5 showed an accelerated regeneration over controls at 4 and 8 weeks. However, at 6 months there was no difference between the groups with fibrocartilage being the predominate tissue, suggesting that GDF-5 is not as potent for cartilage repair as BMP-2 and -7.

In addition to applications of recombinant BMPs, multiple rabbit studies have been reported evaluating the local application of a BMP gene to repair osteochondral defects [77]. In the initial study, a *BMP* gene was evaluated for the ability to augment repair with stem cells. In this study, the *BMP-7* gene was introduced into periosteal-derived allogenic mesenchymal stem cells using a retroviral vector. The cells were then seeded onto a polyglycolic acid scaffold and the device implanted into defects (3 mm in diameter, 2 mm deep) in the femoropatellar groove. Regeneration of the bone and cartilage was evaluated at 4, 8 and 12 weeks. The data showed that defects treated with the *BMP-7* gene modified cells showed complete or near complete bone and cartilage regeneration at 8 and 12 weeks, while the controls with empty defects or allogenic cells only had poor repair of both bone and cartilage. It was concluded that the cell-based approach was not successful in this model without the introduction of the *BMP-7* gene.

In another rabbit study, bone marrow-derived mesenchymal cells were used after transient transfection by lipofection with the GDF-5 gene to enhance repair of osteochondral defects [78]. The cells were embedded in a type I collagen gel and press-fitted into the defects. The 8-week evaluation showed that repair was improved in the GDF-5-treated defects in comparison to cell-only implants.

Finally, the combination of a naked DNA construct for *BMP-2* and a type I collagen sponge was also investigated and directly compared with recombinant BMP-2 protein delivered on the same collagen sponge [79]. At 12 weeks after implantation in rabbit osteochondral defects, the repair tissue in sites treated with naked DNA showed substantial hyaline-like cartilage repair that was nearly equivalent in quality to the BMP-2 protein-induced repair. The results apparently demonstrated that mesenchymal cells at or near the site could successfully incorporate the naked DNA and, once transfected, produce the BMP-2 protein.

The encouraging results from multiple rabbit studies provided the impetus to extend the studies on osteochondral defect repair to larger animal models. Only a few of these studies have been done, but have included dog, sheep and goat models [80–84]. The most extensive evaluation has been done in dogs using BMP-7 and the data have demonstrated that significant repair could also be observed in a large animal model [80]. Variables such as protein dose, delivery materials and containment methods have been investigated in defects (5 mm in diameter, 6 mm deep) created in the femoral condyles. BMP-7 was delivered on bone-derived type I collagen particles, with carboxymethylcellulose as an additive to make a putty-like formulation, and compared to defects left empty or treated with the delivery material alone. Similar to the rabbit data, at 16 weeks both the subchondral bone and cartilage showed statistically significant improvement in BMP-7-treated sites, although there was a higher degree of animal-to-animal variability than seen in small animals. Histologically in the best BMP-7-treated sites, maturing cartilage was present that appeared similar to the intact articular cartilage, had thickness similar to the surrounding intact cartilage, was minimally degraded at the defect interface and was generally continuous with the repair cartilage. The control defects were filled primarily with fibrous tissue and/or what appeared to be fibrocartilage and a moderate degeneration of the cartilage at the defect interfaces were noted with large clusters of chondrocytes observed at the interface and fissures separating the intact cartilage from the repair tissue.

In a pilot study done in sheep, BMP-7 delivered on type I collagen particles was evaluated for use in augmentation of the mosaicplasty (osteochondral autograft) procedure [84]. In this study, two osteochondral defects were created in each femoral condyle. One of the defects on each condyle served as the graft donor site, (5.5 mm in diameter, 10 mm deep) and the other as the graft recipient site (5.4 mm in diameter, 10 mm deep). In the treatment knees, the BMP-7 was press-fitted into the donor site as well as the recipient site prior to the plug implantation. The contralateral knees served as controls without BMP-7 and all the animals were evaluated at 3, 6 and 12 weeks. This investigation determined that the combination of bone-derived type I collagen particles and carboxymethylcellulose could improve the histological outcome of: (1) the interface between transplanted and host cartilage, (2) the interface between the transplanted and host bone, and (3) donor site healing. In regard to the graft sites, differences between BMP-7-treated and control healing were significant. The graft sites treated with BMP-7 appeared to have smoother articular surfaces, less severe chondrocyte clustering, significant integration with the normal adjacent tissue and, most importantly, seamless subchondral bone integration without marked sclerotic transformation. The control graft sites demonstrated much poorer cartilage integration and subchondral bone that was sclerotic and did not integrate with the existing bone. It was concluded from the data that mosaicplasty augmentation was a promising clinical application for BMP-7.

One large animal study has also been reported using BMP-2 [82]. This interesting study tested the ability of an anti-angiogenesis factor to inhibit the upgrowth of blood vessels from the marrow into the cartilage compartment in shallow osteochondral defects. The model utilized narrow defects (1.2 mm in width, 7–8 mm in length and 0.8 mm in depth) in the femoral patellar groove of minipigs. The lower (bony) portion of the defects was filled with a fibrin-gelatin matrix containing either free TGF-β or IGF-1 and liposome-encapsulated BMP-2. The upper (cartilage) portion was filled with the same matrix and factors, but in addition contained the anti-angiogenic factor, suramin, in free or liposome-encapsulated form. At 8-week sacrifice the time-released suramin was highly efficacious in suppressing bone tissue upgrowth into the cartilage compartment and allowed tissue transformation into cartilage within the cartilage compartment at 50–80%. Without the anti-angiogenic factor, a significant amount of bone upgrowth occurred into the cartilage compartment, suppressing the amount of cartilage tissue present. It was suggested that such a factor is required in osteochondral defects to block the deleterious effects of blood/marrow on cartilage tissue formation and that preferentially stem cells for the cartilage repair should be derived from the synovium.

Finally, a study has also reported the use of GDF-5 for healing osteochondral defects in a large animal; this study used a minipig model [83]. Recombinant GDF-5 was combined with a type I collagen/hyaluronic acid matrix and press-fitted into defects (6.3 mm wide, 10 mm deep) created in medial femoral condyles. Sacrifice at 3 or 12 months demonstrated that the cartilage repair was not improved by the GDF-5/matrix implants or by the matrix alone. It was speculated that the matrix may have inhibited the repair process and other matrixes should be investigated.

Focal chondral defects models

Chondral defect repair studies have more clearly demonstrated that certain BMPs have the ability to be a cartilage anabolic factor *in vivo*. These studies have shown that hyaline-like tissue repair can be induced in a non-repairing model where the defects are not exposed to stem cells and factors from the blood/marrow. However, the models are technically more difficult than the osteochondral defect models. Creation of defects that do not penetrate the subchondral bone is challenging and methods for containment of the BMP in the shallow defects have proven difficult. As a result these studies have utilized large animals where the cartilage is much thicker than in the rabbit and have involved novel techniques to deliver the BMP.

The most important study evaluated BMP-7 for repairing focal chondral defects in sheep [85]. In this study, large 10-mm defects were created such that the subchondral bone was not damaged and liquid BMP-7 in acetate buffer was delivered to the SF *via* an extra-articularly positioned mini-osmotic pump connected to the joint by a polyethylene tubing. Two defects were created in each knee; one on the medial condyle and the other on the trochlea of the femur, and infused with BMP-7 over a 2-week period. At 3 months following surgery, defects treated with BMP-7 were partially filled with newly formed cartilage, precartilagineous tissue and connective tissue at the top of the defect. The cartilage formation initially took place at the bottom progressing towards the surface of the defect. In control knees, there was no sign of cell ingrowth into the defect area. In BMP-7-treated knees, condylar defects showed a 40–62% fill and the trochlear defects showed a 56–81% fill. At 6 months, defects treated with BMP-7 showed increased new cartilage over that seen at 3 months and this cartilage was well fused to the old cartilage and stained positive for type II collagen. All of these defect sites showed significant filling in both condyle (57–71%) and trochlea (74–92%) locations, but none of the control defects showed healing. In this regard it is interesting that the repair process continued to progress between 3 and 6 months, even though BMP-7 was delivered for only the first 2 weeks and without a scaffolding material. It was hypothesized that the continuous presence of BMP-7 throughout the initial weeks following surgery may have attracted sufficient mesenchymal-like cells originating from the synovium into the defect area. Subsequently, BMP-7 could stimulate these cells to differentiate into chondrocytes, which would then produce the appropriate extracellular matrix for filling the defect site. The actual filling of the defect site continues at a slow process long after the BMP-7 has been cleared from the joint.

A chondral repair study has also been reported comparing recombinant BMP-2 and BMP-13 (also called GDF-6 or CDMP-2) in a much smaller defect in the minipig knee [86]. The defects were long and narrow (0.5 mm wide, 5–6 mm long and 0.5 mm deep) and the BMPs, encapsulated in liposomes, were contained at the defect site by a fibrin clot. Both BMPs were efficacious in inducing cartilage-like tissue fill in approximately 90% of the defect site 6 weeks after surgery in this model. It should be noted that this study also included TGF-β, IGF-1, EGF, TGF-A and Tenascin-C for comparison, and of this group only TGF-β showed the ability to induce cartilage tissue in this model. However, the amount of tissue was significantly less with TGF-β than that seen with the BMPs.

Recently, a study was reported evaluating chondral defect repair with a combination of the microfracture procedure and a BMP [87]. In this rabbit study, defects (2 mm × 7 mm) were created in the patellar groove and treated by (1) no treatment, (2) liquid BMP-7 painted onto the defects, (3) microfracture alone, (4) microfracture with a collagen sponge (press-fitted into the microfracture holes), and (5) microfracture with the collagen sponge containing BMP-7. Repair was evaluated qualitatively and quantitatively at 24 weeks by histological and gross appearance. Liquid BMP-7 increased the quantity but not the quality of repair tissue and microfracture improved both the quantity and surface smoothness of repair tissue. However, the combination of microfracture and BMP-7 increased both the quantity and quality (more hyaline-like) of the repair tissue. It was concluded that microfracture and BMP-7 act synergistically to stimulate cartilage repair and, although the procedure converts a chondral defect into an osteochondral defect, it is technically simple and could easily be adopted into clinical practice.

In addition to the studies using recombinant BMPs, studies have also been reported using a *BMP* gene to repair a focal chondral defect [88]. A large animal study evaluated the repair of large 15-mm-diameter defects extending down to, but not though, the calcified cartilage layer in the horse patellofemoral joints. The *bmp-7* gene on an adenoviral vector was delivered to the defect site *via* transfected allogenic chondrocytes embedded in a fibrin clot. In comparison with the control cells without BMP-7, the BMP-treated defects showed accelerated healing at 4 weeks and markedly more hyalin-like morphology. However, by 8 months, both the control and BMP-7-treated defects had similarly healed with cartilage repair tissue and it was concluded that the advantage of the BMP appears to be limited to an acceleration of cartilage healing in this model using modified allogenic chondrocytes.

In regard to the type of stem cell that could be used for delivering a BMP, one study [89] has reported a comparison of stromal cells from perichondrium/periosteal, bone marrow or fat. Using a rat model, cells stimulated with adenoviral nectors carrying the *bmp-2* gene were transplanted into patellar groove defects. Histochemical evaluation of the repair tissue of 8 weeks demonstrated a hyaline-like cartilaginous tissue in defects treated with the perichondrial/periosteal or bone marrow-derived cells. BMP-2-activated fat stromal cells formed mainly fibrous tissue and were similar to cells of each population that were not activated. It was concluded that BMP activation of stem cells was required, but fat-derived cells were significantly inferior.

In summary, the animal studies have demonstrated that certain BMPs can improve both cartilage and bone repair in osteochondral defects. However, the repaired tissues are not perfect and some studies show the repair cartilage may not be stable over extended periods. Few studies have been done with chondral defects, but the data show that the repair cartilage is more hyaline than has been obtained using osteochondral defects. It has been suggested that the presence of blood/marrow at the defect site is not only unnecessary, but also undesirable for achieving optimal regeneration of the cartilage.

OA models

A role for BMPs in the management of articular degeneration is based on the aforementioned *in vitro* evidence that BMPs regulate chondrocyte metabolism in normal, osteoarthritic and aged cartilage as well as playing a synergistic role with IGF-1 in reversing the effect of catabolic cytokines such as IL-1 and TNF. The anti-catabolic, anabolic and reparative functions of these molecules make them an ideal candidate in OA prevention and treatment, but the establishment and progression of this disease is multifactorial. OA has many manifestations in the knee, hip, and hands and other joints. The progression of degenerative changes is often driven by joint injury, supra-physiological loading (obesity, misalignment), instability or incongruency, so the potential for any new therapy needs to be considered in this context. While the chondroprotective potential of BMPs in arthritis can be estimated by *in vitro* studies, additional proof in animal models is needed where the entire synovial environment and immune system are intact. The ideal therapy would enhance cell viability, block catabolic chondrocyte and synoviocyte metabolism as well as facilitate repair of early degeneration or damaged cartilage. The following outlines evidence for BMP activity in this context.

Necrosis, apoptosis and areas of acellular matrix in human cartilage have been observed adjacent to femoral condyle bone bruises [90], and these lesions may progress to acellular areas that delaminate from the subchondral bone. The enlarging chondral defect creates abnormal surface strains and drives OA progression.

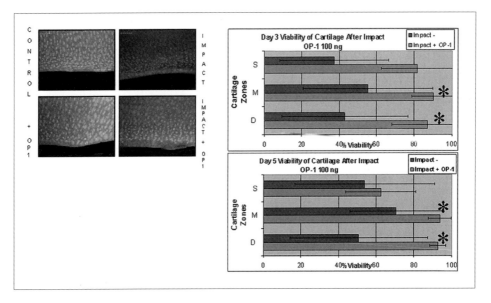

Figure 3
In vitro *viability studies of bovine explant cultures subjected to 30 MPa demonstrated enhanced preservation of all cartilage layers 3 days post injury and the middle and deep cartilage zones on day 5 (*p<0.05). There were no significant differences at time=0 or 1 day after injury. Photomicrographs on left show calcein and ethidium bromide labeled vibratome sections and demonstrate full thickness chondrocyte injury in control sections, whereas BMP-7 treated explants had injury limited to the superficial zone.*

Enhanced cell viability in the face of high mechanical loads is particularly important where mechanical or neuromuscular instability allows repetitive injury to already compromised cartilage. BMP-7 with IGF-1 enhanced the viability and PG production of aged and osteoarthritic chondrocytes in alginate cultures [11, 16–19], and BMP-2 improved the viability of fibroblasts in collagen scaffolds [91] and suppressed apoptosis through the NF-κB activation system in cultured chondrocytes [92]. Gene array data [34] suggests suppression of some members of the caspase family and apoptosis-related proteins. When bovine cartilage explants were exposed to a 30-MPa impact injury, 100 ng/mL BMP-7 improved chondrocyte viability (Fig. 3) and suppressed PG release 3 and 5 days after injury [93]. All these data suggest that BMPs may promote chondrocyte viability after sub-lethal injury arising from joint injury.

Nishida et al. [42] used hyaluronan hexasaccharide to deplete the CD44 receptor and cause PG depletion in cartilage explants. This leads to the loss of cell-associated matrix and activates gelatinase and aggrecanase and paradoxical up-regulation of PG synthesis that is not retained. BMP-7 (100 ng) dramatically improved cell associated PG deposition and prevented further matrix degradation. Fahlgren et al. [43] showed that the precursor pro- and mature forms of BMP-7 are up-regulated after an arthrotomy incision in rabbits. Rabbit knees that underwent meniscectomy developed OA and demonstrated high levels of mature BMP-7 in the superficial zone of articular cartilage as compared to non-operated controls and to rabbits that underwent a joint capsule incision only. The inactive pro-BMP-7 form of the protein was decreased implying that injury resulted in increased conversion to the active form during the 12-week study.

In a chymopapain-induced model of OA in rabbits [66], mature BMP-7 was low in the superficial zone but higher in the middle and deep zones, whereas there was strong immunostaining for pro-BMP-7 in the superficial zone at 28 days. The apparent differences in pro-BMP-7 localization between these two experiments may be related to time course. Increased levels of pro-BMP-7 were also found as long as 12 weeks after injection of human recombinant BMP-7 in model of post-traumatic OA in sheep [93]. Undoubtedly, other members of the TGF-β and BMP families such as BMP-2 have also been demonstrated after injury and are part of the reparative process [56]. These experiments provide confirmation that BMPs may be part of normal response to injury, and the reduced production of BMP-7 in aged and osteoarthritic chondrocytes [16, 22] allows speculation that supplementation with exogenous BMPs might be beneficial. Furthermore, Kaps et al. [94] showed that BMP-7-expressing chondrocytes suppress ingrowth of fibroblasts and elaboration of fibrous connective tissue (pannus) in nude mice. This is an important aspect of the repair response because re-direction of extrinsic repair after injury may be needed in some forms of arthritis, particularly if there is a strong inflammatory component.

Bone edema and microdamage to the subchondral bone plate may create pain, but can be catastrophic if the bone is sufficiently compromised to allow the cartilage surface to collapse [95]. The evidence for BMP-7 promoting bone remodeling and repair in long bone fractures is strong [96], but improvement of osteochondral defect healing has also been observed [80–85]. Repair and remodeling at bone-tendon or bone-ligament interfaces may also pay a role in arthritis, and both BMP-2 and BMP-7 have been shown to play a role in an anterior cruciate ligament (ACL) model in rabbits where these proteins were associated with osseous integration of the transplanted tendons or ligaments.

Since treatments are seldom administered at the time of injury, most patients with impending or early OA will have areas of hypocellular cartilage matrix, superficial zone delamination and fibrillation. Thus, all the information relating to repair of focal chondral and osteochondral defects described in the previous sections of this chapter supports the use of BMPs in early OA.

Reliable delivery of therapeutic proteins to the synovial environment is difficult, but the stability of the BMP-7 protein confers an advantage over more labile agents. Recent studies in the authors' laboratories have shown that intra-articular injections of 1 mg BMP-7 protein into the sheep knee yielded between 10 and 100 ng/mL in the SF for 10 days after injection. Since *in vitro* studies show protective effects of BMP-7 against the catabolic influence of IL-1 and TNF at these concentrations, a series of appropriately timed injections, with or without slow release carriers, may be able to maintain therapeutic levels. This may be quite useful in the context of sports injuries where the index injury time, such as a tear of the ACL, is known and surgical reconstruction is anticipated. Bracing, reduced activity and a chondroprotective agent would be logical until stability and a population of competent chondrocytes can be restored.

With this in mind, recent data from an ACL transection model in rabbits demonstrated that BMP-7 had a protective effect on cartilage degeneration [97]. In the ACL model, there is repeated daily injury leading to damage to the meniscus and articular cartilage that culminates in rapid progression and severe degeneration, so chondroprotection can be difficult to demonstrate. Badlani et al. [97] administered BMP-7 by osmotic pumps implanted into the knee joint at the time of ACL transection. Significant improvements in histological and morphometric scores and expression of type II collagen were found in addition to suppression of aggrecanase activity.

In a model of direct contusive injury to the knee joint [98–100], focal cartilage lesions developed in the medial femorotibial joint compartment of horses, dogs and sheep by 3 months, and progressed to severe medial femorotibial compartment arthritis by 6 months. The ability of BMP-7 to pre-empt lesion development and progression was tested in six sheep by intra-articular injection of 300 µg recombinant BMP-7 in a collagen carrier (BMP-7 Putty®, Stryker Biotech, Hopkinton, MA) at the time of injury and 1 week later [38]. The contralateral limb received carrier alone. BMP-7 injections resulted in significant improvements in histological scores, cell viability, and PG content of the injured cartilage 12 weeks after injury. Lesions at the impact site were absent or very subtle in three of six animals and consisted of minor superficial zone delamination in the remaining three; however, all six contralateral joints had severe cartilage degeneration in the medial condyle (Fig. 4). A similar experiment using a single injection 1 week post injury also suppressed PG loss and progression of histological degeneration in the medial femoral condyle.

In subsequent experiments, the ability of BMP-7 to reverse an established injury was studied by delayed administration of BMP-7 (300 µg, intra-articularly) 3 and 4 weeks after injury to the femoral condyle. Assessments were made 12 weeks after the last injection [93]. Macroscopic and histological damage to the femoral condyle was reduced, as was the C3/4 short collagen epitope immunostaining, the later indicating that there was protection against metalloproteinase-mediated collagen degradation. In this study, repletion of PG in the injured cartilage was incomplete, leading to speculation additional treatments or a longer study was required to fully characterize the window of opportunity to treat such joint injuries. Another experi-

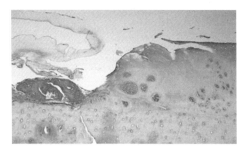 

Figure 4
Histological sections of the impact area of BMP-7 treated (right) and control sheep (left) femoral condyles 12 weeks after injury. In control sheep there is loss of most of the thickness of the cartilage due to delamination of necrotic cartilage. In BMP-7-treated knees there was partial loss of the superficial zone and small fissures. H&E (×20 mag) and Safranin-O stained (×100 mag), 5-μm-thick sections.

mental group received the same dose of BMP-7 at 12 and 13 weeks after injury, but when these animals were sacrificed 12 weeks after the last intra-articular injection, there was only minor improvement in histological scores and no other indications of efficacy. This was not surprising given that well-established lesions were present in previous experiments 12 weeks post injury and any improvement would have required extensive regeneration and repair.

It is important to note that, in all the aforementioned sheep experiments, there was no intra-articular bone formation or osteophytes in joints that received intra-articular BMP-7. An extensive analysis of 27 joints available for microCT imaging revealed four joints with linear mineralized tracts in the joint capsule and subcutaneous tissue associated with needle injections. This underscores the need for accurate delivery of BMPs and the effect of microenvironment on activity.

This body of preclinical evidence supports a role for administration of BMP-7 in prevention and treatment of early post-traumatic injuries and OA. Since there was little or no evidence of intrinsic or extrinsic cartilage repair reported in these studies, it must be assumed that the beneficial effects observed are due to enhanced survival of a more functional chondrocyte population that is resistant to the environment associated with injury and inflammation that interferes with chondrocyte metabolism. The duration of exposure and concentration of BMP-7 needed to create this effect in patients is unknown, but, based on the above evidence, 50–100 ng/ml over 2 weeks may be sufficient to reverse or significantly reduce the size of early traumatic lesions and prevent progression of degeneration to OA. The window of opportunity to address developing lesions may be inversely proportional to the energy absorbed and size of the impact zone at the time of injury, but in many cases, early treatment within the first 4–6 weeks should be beneficial.

Table 3 -. Non-articular cartilage animal studies

Citation	Year	Species	BMP	Carrier	Model
Katic et al. [114]	2000	Dog	BMP-7	Allograft	Thyroid cartilage (Larynx) repair BMP coated thyroid allografts; 4-month evaluation
Okamoto et al. [115]	2004	Dog	BMP-2	Gelatin sponge	Tracheal cartilage repair contained with periosteum; 1-, 6-month evaluation
Walsh et al. [111]	2004	Mice	GDF-5	Liquid	Intervertebral disc repair - compression degeneration model; 1-, 4-week evaluation
Tcacencu et al. [116]	2006	Rabbit	BMP-2	Collagen sponge	Cricoid cartilage (Larynx) repair; 2-, 4-week evaluation
Masuda et al. [108]	2006	Rabbit	BMP-7	Liquid	Intervertebral disc repair -puncture degeneration model; 4-, 24-week evaluation
Chujo et al. [104]	2006	Rabbit	GDF-5	Liquid	Intervertebral disc repair -puncture degeneration model; 4-, 16-week evaluation
Wallach et al. [113]	2006	Rabbit	BMP-2	Liquid	Intradiscal gene transfer as safety study- Intradural injection of DNA; 3-, 7-week evaluation
Imai et al. [110]	2007	Rabbit	BMP-7	Liquid	Intervertebral disc repair – chondroitinase ABC degeneration model; 6-, 16-week evaluation

If biological therapies are to succeed in OA prevention and treatment, the use of TNF-blocking antibodies and soluble receptors in inflammatory arthritis [101] may be a good model. A similar paradigm might be possible in OA provided the staging, temporal application, dose, and complementary combinations of therapies can be elucidated. BMPs have many of the required anti-catabolic, anabolic and reparative properties to be chondroprotective treatments, but as the pressure to discover new therapies mounts, the additional evidence for BMP efficacy in preclinical studies will accumulate.

Non-articular cartilage repair

A role for BMPs in repairing non-articular cartilage has also been emerging in the literature (Tab. 3). Most exciting is the injection of liquid formulations of BMPs into the intervertebral disc to induce repair and these studies are reviewed in this section. In addition, animal studies evaluating the use of BMPs for repairing larynx and trachea cartilage have also been reported and are reviewed.

Intervertebral disc models

The intervertebral disc (IVD) provides articulation between adjoining vertebral bodies and acts as a weight-bearing cushion that dissipates axially applied spinal loads. These biomechanical functions are made possible by the unique structure of the IVD, which is composed of an outer collagen-rich annulus fibrosus (AF) surrounding a central hydrated PG-rich gelatinous nucleus pulposus (NP). The AF is composed of concentric layers that contain variable proportions of type I and II collagen. The outer annular layers are anchored firmly to the rim of the vertebral bodies by Sharpey's fibers. The NP, in contrast, is a gelatinous tissue that provides weight-bearing properties to the disc. Endogenous cells in both the NP and AF are responsible for the synthesis and degradation of these matrix components to maintain tissue homeostasis. Apart from a limited blood supply to the periphery of the AF, the remainder of the normal IVD is avascular. The IVD undergoes profound cellular and matrix changes with aging and degenerates much earlier than other weight-bearing cartilaginous tissues. With the onset of age, cells in the NP progressively die as the tissue becomes depleted of PGs, less hydrated and more fibrous. These changes cause a decrease in biomechanical functions and trauma can lead to further structural degeneration.

Certain BMPs, as discussed above, have been known for some time to have potent anabolic effects *in vitro* on articular chondrocytes, particularly for the stimulation of the synthesis of PGs and type II collagen. In similar studies using IVD cells, BMPs, such as BMP-7 and GDF-5, have also been shown to stimulate the production of extracellular matrix components and the effect is observed using cultures of cells derived from either the NP or the AF [102–104]. Furthermore, after depletion of the extracellular matrix by exposing IVD cells to IL-1 or chondroitinase ABC, BMP-7 was also found to be effective in the replenishment of the PGs and collagen [105, 106] and the stimulatory effect of this BMP was also demonstrated *in vivo* after the intradiscal administration in normal rabbits [107]. At 2 weeks after a single injection of 2 µg BMP-7, the mean disc height index was 15% greater than the control group and the increase was sustained for up to 8 weeks. Biochemically, a significant increase in the PG content of the NP was observed in the BMP-7-treated discs.

Based on the above *in vitro* and *in vivo* data, models using degenerated discs were developed to evaluate the therapeutic effects of BMPs. Initially, a rabbit model was used where a needle puncture of the AF induced degeneration of the disc [108, 109]. At 4 weeks after puncture and development of disc degeneration, 100 µg BMP-7 in a solution of 5% lactose was injected into the NP. At 6 weeks after injection, a restoration of the disc height was observed and maintained for up to 24 weeks. Restoration did not occur in the control animals injected with lactose solution. Histological scoring after Safranin-O staining demonstrated less degeneration for the BMP-7-treated discs and biomechanical measurements demonstrated that BMP-7 also restored the viscoelastic properties of the disc to the level of non-punctured control discs. Comparable results were seen in another rabbit model using chondroitinase-ABC to produce degeneration of the disc; a single injection of 100 μg into the disc 4 weeks after the initial insult was able to restore the disc height after 8 weeks [110].

A similar rabbit study has been reported examining the effects of GDF-5 after injection into a degenerated disc [104]. Using the same model described above, 10 ng, 1 μg or 100 μg GDF-5 was injected 4 weeks after annular puncture. During the 16-week follow-up, the data demonstrated that GDF-5 restored disc height, improved MRI and histological grading scores. Restoration was achieved with the 1 and 100 μg doses; the low GDF-5 dose was not significantly different from the control discs, which maintained the degenerated disc to sacrifice at week 12. The results comparing the 100-μg dose for GDF-5 with that for BMP-7 showed very similar disc regeneration. A mouse model [111] where a compressed disc was evaluated after GDF-5 injection also showed a significant increase in disc height compared to controls.

A rat model has also been used to study the therapeutic efficacy of intradiscal injection of BMP-7 [37, 112, 113]. This model used compression of tail vertebrae with an Illizarov-type apparatus to induce degeneration. The apparatus was used to apply chronic compression to the tail for 4 weeks after surgery. Subsequently, either BMP-7 in 5% lactose or a control solution was injected into the instrumented discs and the discs were harvested at 4 weeks. The ability of BMP-7 to reverse the damage as well as to reduce discogenic pain was evaluated. To evaluate pain the harvested tissue was applied to the left lumbar nerve roots of the rats and pain assessed by behavioral measurements for 3 weeks. The data showed that pain was observed in the sham and control group, but not in the BMP-7-treated group. Histological evaluation was used to determine the degree of disc degeneration and the appearance of the extracellular matrix in the intervertebral discs. In the control group, NP cells became spindle-shaped, while in the BMP-7 group, the NP cells became swollen with vacuolated cytoplasm, and the content of the extracellular matrix was markedly increased. Immunohistochemical analysis demonstrated an anti-catabolic effect through reduced levels of aggrecanase, MMP-13, TNF-α, IL-1β and substance P. Because substance P is a neuropeptide linked with inflammation and pain, the observed reduction supports the lower level of pain-related behavior of the BMP-7-treated rats. Thus, BMP-7 injection into the degenerated IVD appeared to not only enhance extracellular matrix production, as observed in the rabbit studies described above, but also to inhibit pain. These results were particularly important because this was the first demonstration of pain inhibition by BMP-7.

Larynx and trachea

In addition to the intervertebral disc, a few animal studies have been reported for other non-articular cartilage tissues and these have involved larynx and trachea cartilage [114–116]. The larynx has nine different cartilage tissues and two, thyroid and cricoid, have been evaluated. *In vitro* studies have reported stimulatory effects on nose and meniscus cartilage, but no animal studies have yet been reported [117, 118]; it has been shown that cells derived from either of these tissues could be stimulated by BMPs to produce extracellular matrix.

Thyroid cartilage is the largest cartilage structure making up the larynx and is hyaline. One study has been reported evaluating the ability of BMP-7 to regenerate the thyroid cartilage using a dog model [114]. In this study defects were treated with thyroid allografts that included allograft alone or allograft coated with BMP-7. Dogs were sacrificed 4 months after surgery. The results showed that BMP-7 regenerated the cartilage defects by inducing new bone, cartilage and ligament-like structures, acting as a multiple tissue morphogen in this specific environment. The regenerated cartilage was well integrated with the pre-existing cartilage, suggesting that BMP-7 can promote stable remodeling of the cartilage matrix. The data demonstrated the potential for the reconstruction of the larynx through regeneration of the cartilage.

A second larynx study described repair of a cricoid cartilage defect in a rabbit model [116]. Cricoid cartilage is also a prominent cartilage of the larynx and is responsible for adjoining the thyroid cartilage with the trachea. BMP-2 in combination with a type I collagen sponge was used to repair a 2-mm-wide defect created in the cartilage. Evaluation was at 1, 2 or 4 weeks after surgery and demonstrated the formation of bone, not cartilage at the defect site. Although cartilage formed initially, it was not stable and was replaced by bone in the endochondral bone formation process as seen in subcutaneous or bony sites. It was concluded that BMP-2 alone could not result in a terminal cartilage repair tissue.

Finally, one study has also described the use of a BMP for repair of tracheal cartilage [115]. A gap, 1×5 cm, was made in the anterior cervical trachea in a dog model. BMP-2 in combination with a collagen sponge was implanted in the defect and sacrifice was at 1, 3 or 6 months after surgery. The results demonstrated the induction of cartilage in the defect sites treated with BMP-2, but not in control defects. This regenerated cartilage was stable for at least 6 months and was strong enough to maintain the integrity of the repaired trachea.

In summary, the animal studies reported thus far have demonstrated that BMPs also have therapeutic potential in non-articular cartilage tissues. The disc regeneration studies are the most promising and represent the simplest form of application of a BMP, that is, direct injection of the liquid formulation, but the results have not yet been reproduced in large animals. Large animal studies have demonstrated the ability of BMPs to repair larynx and trachea with some early success. However, these types of cartilage will require extensive research into procedures and matrices.

Discussion

The purpose of this chapter is to review the current knowledge on BMPs in cartilage biology from the standpoint of both *in vitro* and animal repair studies. The data clearly show that BMPs have an important role in cartilage, both in normal homeostasis and in repair, and predict a bright future for the use of certain BMPs in the engineering of cartilage.

In vitro studies have demonstrated that many BMPs are endogenously expressed in cartilage and some act as anabolic factors for chondrocytes in culture. BMP-2, -3, -4, -5, -6, -7 and GDF-5 and -6 have been localized to cartilage and BMP-7 has also been localized to SF, synovium, ligament, tendon and meniscus. In regard to the anabolic activity, the role BMPs play in stimulating chondrocyte differentiation, extracellular matrix production and maintenance of the adult chondrocytic phenotype is well documented. However, few direct comparisons of BMPs have been reported and the only extensively studied BMP has been BMP-7. BMP-7 has been shown to stimulate the synthesis of all the major cartilage extracellular matrix proteins and to counteract the degenerative effect of numerous catabolic mediators. Furthermore, recent antisense studies have demonstrated that down-regulation of *BMP-7* mRNA induced a significant decrease in PG synthesis in articular chondrocytes in culture. Thus, the data from *in vitro* studies have clearly demonstrated that at least one BMP, BMP-7, is very important in articular cartilage homeostasis. Certainly a detailed analysis of the importance of other BMPs needs to be done.

Data from numerous studies in animals show that at least three BMPs, BMP-7, BMP-2 and GDF-5, have therapeutic potential for cartilage repair. Although most studies have used the recombinant proteins, a few have described the potential use of BMP gene therapy. In several large osteochondral defect studies, both BMP-7 and BMP-2 were observed to induce a significant improvement in repair of both the cartilage and bone compartment over that observed in control defects; the BMP-treated sites exhibited less fibrocartilage and more hyaline-like cartilage. In a large chondral defect study in sheep, BMP-7 was shown to induce significant repair in a model where no repair takes place in the controls; the repair was hyaline-like and well bonded with the surrounding cartilage. In studies evaluating models of OA, BMP-7 was shown to prevent development of damage and in some models reverse the damage. Finally, in animal models of degenerated IVD, the injection of liquid formulations of BMP-7 or GDF-5 has been shown to stimulate restoration of the tissue.

Although BMP-induced repair can be significant in some animal models, the goal of perfectly repaired cartilage remains to be achieved. In the future, long-term stability of the repair tissue needs to be better evaluated, and studies using models of OA need to be significantly expanded. At the same time, there will have to be extensive evaluations of a variety of formulations, scaffolds, methods of administration and possibly combinations with other factors. Most likely each cartilage site will have unique requirements for the optimal use of BMPs. In addition, a wider range of BMPs needs to be evaluated since in the limited number of BMPs tested thus far there appear to be differences between BMPs and particular BMPs may be better suited for different cartilage applications. However, the relevance of animal repair models to the repair of human cartilage is a frequently asked question and thus thoughtful design of future studies is mandatory to adequately prepare for the initial clinical studies.

In regard to using BMPs in clinical studies, the animal studies have clearly demonstrated that one BMP, BMP-7 is efficacious and can be safely administered to both the articular joint and the disc. In addition, recombinant BMP-7, and to a lesser extent BMP-2, have been evaluated in various formulations, concentrations, frequencies of dosing and delivery routes in a variety of animal models involving rabbits, dogs, goats, and sheep. In these studies, there have been no reports of side effects, such as bone formation on the synovial or disc surface or free floating objects in the SF or inflammation such as synovitis, pannus formation or joint effusion. The only conflicting data have come from a study evaluating BMP-2 injected into a normal mouse joint [119] where it was reported that, after multiple injections of BMP-2, chondrogenesis (termed chondrophytes) was induced in the region where the growth plates meet the joint space and at later times these chondrophytes developed into osteophytes. However, given the numerous reported repair studies done with BMP-2 in multiple animal species where no such side effects were observed, this result may to be specific to the mouse.

In conclusion, a BMP-based therapy for damaged cartilage would appear to have significant clinical potential. The clinical demand is so great for new cartilage repair procedures that simply an improvement over the repair currently achieved with microfracture or mosaicplasty can be an acceptable first goal [120]. The BMP protein could be delivered locally to a focal defect site on an appropriate scaffold material, preferably by arthroscopic administration, which would increase the clinical attractiveness of such a therapeutic. However, the use of an injectable BMP, perhaps in a slow release formulation, would seem to be the ideal route of administration to extend the therapeutic potential dramatically, particularly in the area of treatment and, ultimately, prevention of OA and degenerative disc disease.

References

1   Hogan BLM (1996) Bone morphogenetic proteins: multifunctional regulators of vertebrate development. *Genes Dev* 10: 1580–1594

2   Wozney JM, Rosen V, Celeste AJ, Mitsock LM, Whitters MJ, Kriz RW, Hewick RM, Wang EA (1988) Novel regulators of bone formation: Molecular clones and activities. *Science* 242: 1528–1534
3   Itoh S, Itoh F, Gourmans M-J, ten Dijke P (2000) Signaling of transforming growth factor-β family members through Smad proteins. *Eur J Biochem* 267: 6954–6967
4   Massagué J, Chen YG (2000) Controlling TGF-β signaling. *Genes Dev* 14: 627–644
5   Roberts AB, Sporn MB (1990) The transforming growth factor-βs. In: Sporn MB, Roberts AB (eds): *Peptide growth factors and their receptors, Part I*. Springer, Berlin 95, 419–472
6   Goumans MJ, Mummery C (2000) Functional analysis of the TGFβ receptor/Smad pathway through gene ablation in mice. *Int J Dev Biol* 44: 253–265
7   Schier AF, Shen MM (2000) Nodal signaling in vertebrate development. *Nature* 403: 385–389
8   Sampath TK, Reddi AH (1981) Dissociative extraction and reconstitution of extracellular matrix components involved in local bone differentiation. *Proc Natl Acad Sci USA* 78: 7599–7603
9   Urist MR, Mikulski A, Lietze A (1979) Solubilized and insolubilized bone morphogenetic protein. *Proc Natl Acad Sci USA* 76: 1828–1832
10 Urist MR (1997) Bone morphogenetic protein: The molecularization of skeletal system development. *J Bone Miner Res* 12: 343–346
11 Chubinskaya S, Kumar B, Merrihew C, Heretis K, Rueger D, and Kuettner KE (2002) Age-related changes in cartilage endogenous BMP-7. *Biochim Biophys Acta* 1588: 126–134
12 Merrihew C, Kumar B, Heretis K, Rueger DC, Kuettner KE, Chubinskaya S (2003) Alterations in endogenous osteogenic protein-1 (BMP-7) with degeneration of human articular cartilage. *J Orthop Res* 21: 899–907
13 Soeder S, Hakimiyan A, Rueger D, Kuettner KE, Aigner T, Chubinskaya S (2005) Antisense inhibition of osteogenic protein-1 disturbs human articular cartilage integrity. *Arthritis Rheum* 52: 468–478
14 Flechtenmacher J, Huch K, Thonar EJ-MA, Mollenhauer JA, Davies SR, Schmid TM, Puhl W, Sampath TK, Aydelotte MB, Kuettner KE (1996) Recombinant human osteogenic protein 1 is a potent stimulator of the synthesis of cartilage proteoglycans and collagens by human articular chondrocytes. *Arthritis Rheum* 39: 1896–1904
15 Nishida Y, Knudson CB, Eger W, Kuettner KE, Knudson W (2000) Osteogenic protein-1 stimulates cell-associated matrix assembly by normal human articular chondrocytes: Upregulation of hyaluronan synthase, CD 44 and aggrecan. *Arthritis Rheum* 43: 206–214
16 Loeser RF, Pacione CA, Chubinskaya S (2003) The combination of insulin-like growth factor-1 and osteogenic protein-1 promotes increased survival of and matrix synthesis by normal and osteoarthritic human articular chondrocytes. *Arthritis Rheum* 48: 2188–2196
17 Fan Z, Chubinskaya S, Rueger DC, Bau B, Haag J, Aigner T (2004) Regulation of ana-

bolic and catabolic gene expression in normal and osteoarthritic adult human articular chondrocytes by BMP-7 (BMP-7). *J Clin Exp Rheum* 22: 103–106

18 Loeser R, Chubinskaya S, Pacione C, Im H-J (2005) Basic fibroblast growth factor inhibits the anabolic activity of insulin-like growth factor-1 and osteogenic protein-1 in adult human articular chondrocytes. *Arthritis Rheum* 52: 3910–3917

19 Chubinskaya S, Hakimiyan A, Pacione C, Yanke A, Rappoport L, Aigner T, Rueger D, Loeser RF (2007) Synergistic Effect of IGF-1 and BMP-7 on matrix formation by normal and OA chondrocytes cultured in alginate beads. *Osteoarthritis Cart* 15: 421–430

20 Lietman S, Yanagishita M, Sampath TK, Reddi AH (1997) Stimulation of proteoglycan synthesis in explants of porcine articular cartilage by recombinant osteogenic protein-1 (bone morphogenetic protein-7). *J Bone Joint Surg* 79–A: 1132–1137

21 Nishida Y, Knudson CB, Kuettner KE, Knudson W (2000) Osteogenic protein-1 promotes the synthesis and retention of the extracellular matrix within bovine articular cartilage and chondrocyte cultures. *Osteoarthritis Cartilage* 8: 127–136

22 Im, HJ, Pacione C, Chubinskaya S, Van Wijnen AJ, Sun Y, and Loeser RF (2003) Inhibitory effects of insulin-like growth factor-1 and osteogenic protein-1 on fibronectin fragment- and interleukin-1beta-stimulated matrix metalloproteinase-13 expression in human chondrocytes. *J Biol Chem* 278: 25386–25394

23 Chubinskaya S, Hakimiyan A, Otten L, Rappoport L, Rueger DC, Sobhy M, Cole B (2008) Response of human chondrocytes prepared for autologous chondrocyte implantation to growth factors. *J Knee Surg* (in press)

24 Vukicevic S, Luyten FP, Reddi AH (1989) Stimulation of the expression of osteogenic and chondrogenic phenotypes *in vitro* by osteogenin. *Proc Natl Acad Sci USA* 86: 8793–8797

25 Luyten FP, Yu YM, Yanagishita M, Vukicevic S, Hammonds RG, Reddi AH (1992) Natural bovine osteogenin and recombinant human bone morphogenetic protein-2B are equipotent in the maintenance of proteoglycans in bovine articular cartilage explant culture. *J Biol Chem* 267: 3691–3695

26 Luyten FP, Chen P, Paralkar V, Reddi AH (1994) Recombinant bone morphogenetic protein-4, transforming growth factor-b, and activin A enhance the cartilage phenotype of articular chondrocytes *in vitro*. *Exp Cell Res* 210: 224–229

27 Morris E (1996) Differential effects of TGF-beta superfamily members on articular cartilage metabolism: Stimulation by rhBMP-9 and rhBMP-2 and inhibition by TGF-beta. *Trans Orthop Res Soc* 42: 175

28 van Susante JLC, Buma P, van Beuningen HM, van den Berg WB, Veth RPH (2000) Responsiveness of bovine chondrocytes to growth factors in medium with different serum concentrations. *J Orthop Res* 18: 68–77

29 Sailor LZ, Hewick RM, Morris EA (1996) Recombinant human bone morphogenetic protein-2 maintains the articular chondrocyte phenotype in long-term culture. *J Orthop Res* 14: 937–945

30 Stewart MC, Saunders KM, Burton-Wurster N, Macleod JN (2000) Phenotypic stabil-

ity of articular chondrocytes *in vitro*: The effect of culture models, bone morphogenetic protein 2, and serum supplementation. *J Bone Miner Res* 15: 166–174

31 Bobacz K, Gruber R, Soleiman A, Erlacher L, Smolen JS, Graninger WB (2003) Expression of bone morphogenetic protein 6 in healthy and osteoarthritic human articular chondrocytes and stimulation of matrix synthesis *in vitro*. *Arthritis Rheum* 48: 2501–1508

32 Chubinskaya S, Rueger DC, Berger RA, Kuettner KE (2002) Osteogenic protein-1 and its receptors in human articular cartilage. In: Hascall V, Kuettner KE (eds): *The many faces of osteoarthritis*. Birkhäuser, Basel, 81–89

33 Chubinskaya S, Segalite D, Enockson C, Pikovsky D, Gattuso V, Rueger DC (2006) Anabolic response induced by BMPs in human articular chondrocytes: Comparative studies. *Proceedings of the 6th ICRS Symposium*, San Diego, CA, Jan 8–11

34 Chubinskaya S, Otten L, Soeder S, Aigner T, Loeser RF, Rueger DC (2007) Regulation of anabolic and catabolic pathways by osteogenic protein-1: Gene array data. *Trans Orthop Res Soc* 53: 546

35 Vinall RL, Lo SH, Reddi AH (2002) Regulation of articular chondrocyte phenotype by bone morphogenetic protein 7, interleukin 1, and cellular context is dependent on the cytoskeleton. *Exp Cell Res* 272: 32–44

36 Yao J, Cole AA, Huch K, Kuettner KE (1996) The effect of BMP-7 on IL-1beta induced gene expressions of matrix metalloproteinases and TIMP in human articular cartilage. *Trans Orthop Res Soc* 42: 305

37 Chubinskaya S, Kawakami M, Rapoport L, Matsumoto T, Migita N, Rueger DC (2007) Anti-catabolic effect of BMP-7 in chronically compressed intervertebral discs. *J Ortho Res* 25: 517–530

38 Hurtig MB, Chubinskaya S (2004) The protective effect of BMP-7 in early traumatic osteoarthritis-animal studies. *Trans 5th Combined Orthop Res Soc*, Banff, Canada, October 10–13, 70

39 Huser CAM, Peacock M, Davies ME (2006) Inhibition of caspase-9 reduces chondrocyte apoptosis and proteoglycan loss following mechanical trauma. *Osteoarthritis Cartilage* 14: 1002–1010

40 Huch K, Wilbrink B, Flechtenmacher J, Koepp HE, Aydelotte MB, Sampath TK, Kuettner KE, Mollenhauer JA, Thonar EJ-MA (1997) Effects of recombinant human osteogenic protein 1 on the production of proteoglycan, prostaglandin E2, and interleukin-1 receptor antagonist by human articular chondrocytes cultured in the presence of interleukin-1β. *Arthritis Rheum* 40: 2157–2161

41 Koepp HE, Sampath KT, Kuettner KE, Homandberg GA (1999) Osteogenic protein-1 (BMP-7) blocks cartilage damage caused by fibronectin fragments and promotes repair by enhancing proteoglycan synthesis. *Inflamm Res* 47: 1–6

42 Nishida Y, Knudson CB, Knudson W (2004) Osteogenic protein-1 inhibits matrix depletion in a hyaluronan hexasaccharide-induced model of osteoarthritis. *Osteoarthritis Cartilage* 12: 374–382

43 Fahlgren A, Chubinskaya S, Messner K, Aspenberg P (2006) A capsular incision leads to

a fast osteoarthrotic response, but also elevated levels of activated osteogenic protein-1 in rabbit knee joint cartilage. *J Scand J Med Sci Sports* 16: 456–462

44 Blanco FJ, Geng Y, Lotz M (1995) Differentiation-dependent effects of IL-1 and TGF-beta on human articular chondrocyte proliferation are related to inducible nitric oxide synthase expression. *J Immunol* 154: 4018–4026

45 Berg WB (2000) Role of nitric oxide in the inhibition of BMP-2-mediated stimulation of proteoglycan synthesis in articular cartilage. *Osteoarthritis Cartilage* 8: 82–86

46 Elshaier AM, Hakimiyan A, Margulis A, Rueger DC, Chubinskaya S (2005) Effect of IL-1β on BMP-7 signaling in human adult articular chondrocytes. *Trans Orthop Res Soc* 51: 4

47 Ostrowski K, Rohde T, Asp S, Schjerling P, Pedersen BK (1999) Pro- and anti-inflammatory cytokine balance in strenuous exercise in humans. *J Physiol* 515: 287–291

48 Beg AA, Finco TS, Nantermet PV, Baldwin AS Jr (1993) Tumor necrosis factor and interleukin-1 lead to phosphorylation and loss of I kappa B alpha: A mechanism for NF-kappa B activation. *Mol Cell Biol* 13: 3301–3310

49 Lee MJ, Yang CW, Jin DC, Chang YS, Bang BK, Kim YS (2003) Bone morphogenetic protein-7 inhibits constitutive and interleukin-1 beta-induced monocyte chemoattractant protein-1 expression in human mesangial cells: Role for JNK/AP-1 pathway. *J Immunol* 170: 2557–2563

50 Grimm OH, Gurdon JB (2002) Nuclear exclusion of Smad2 is a mechanism leading to loss of competence. *Nat Cell Biol* 4: 519–522

51 Kretzschmar M, Doody J, Timokhina I, Massague J (1999) A mechanism of repression of TGFbeta/ Smad signaling by oncogenic Ras. *Genes Dev* 13: 804–816

52 Bobacz K, Sunk IG, Hofstaetter JG, Amoyo L, Toma CD, Akira S, Weichhart T, Saemann M, Smolen JS (2007) Toll-like receptor and chondrocytes: The lipopolysaccharide-induced decrease in cartilage matrix synthesis is dependent on the presence of toll-like receptor 4 and antagonized by bone morphogenetic protein 7. *Arthritis Rheum* 56: 1880–1893

53 Heldin C-H, Miyazono K, ten Dijke P (1997) TGF-beta signaling from cell membrane to nucleus through SMAD proteins. *Nature* 390: 465–471

54 Hirota Y, Tsukazaki T, Yonekura A, Miyazaki Y, Osaki M., Shindo H., Yamashita S (2000) Activation of specific MEK-ERK cascade is necessary for TGF-beta signaling and crosstalk with PKA and PKC pathways in cultured rat articular chondrocytes. *Osteoarthritis Cartilage* 8: 241–247

55 Hu MC, Wasserman D, Hartwig S, Rosenblum ND (2004) p38MAPK acts in the BMP7-dependent stimulatory pathway during epithelial cell morphogenesis and is regulated by Smad1. *J Biol Chem* 279: 12051–12059

56 Fukui N, Zhu Y, Maloney WJ, Clohisy J, Sandell L (2003) Stimulation of BMP-2 expression by pro-inflammatory cytokines IL-1 and TNF-alpha in normal and osteoarthritic chondrocytes. *J Bone Joint Surg Am* 85-A (Suppl 3): 59–66

57 Chen AL, Fang C, Liu C, Leslie MP, Chang E, Di Cesare PE (2004) Expression of bone

morphogenetic proteins, receptors, and tissue inhibitors in human fetal, adult, and osteoarthritic articular cartilage. *J Orthop Res* 22: 1188–1192

58 Erlacher L, Ng C-K, Ullrich R, Krieger S, Luyten FP (1998) Presence of cartilage-derived morphogenetic proteins in articular cartilage and enhancement of matrix replacement *in vitro*. *Arthritis Rheum* 41: 263–273

59 Cui Y, Jean F, Thomas G, Christian JL (1998) BMP-4 is proteolytically activated by furin and/or PC6 during vertebral embryonic development. *EMBO J* 17: 4735–4743

60 Chubinskaya S, Oakes B, Shimmin A, Rueger DC, Kildey R (2004) Anabolic response in the articular joint induced by BMP-7 in the goat model of osteochondral defects. *Proceedings of the 5th ICRS Symposium*, Gent, Belgium, May 26–29, 365

61 Anderson HC, Hodges PT, Aguilera XM, Missana L, Moylan PE (2000) Bone morphogenetic protein (BMP) localization in developing human and rat growth plate, metaphysis, epiphysis, and articular cartilage. *J Histochem Cytochem* 48: 1493–1502

62 Rueger DC, Chubinskaya S (2004) BMPs in articular cartilage repair. In: Vukicevic S, Sampath KT (eds): *Bone morphogenetic proteins: Regeneration of bone and beyond*. Birkhäuser, Basel, 109–132

63 Chubinskaya S, Frank BS, Michalska M, Kumar B, Merrihew CA, Thonar EJ-MA, Lenz ME, Otten L, Rueger DC, Block JA (2006) Osteoarthritic protein-1 in synovial fluid from patients with rheumatoid arthritis or osteoarthritis: Relationship to disease and levels of hyaluronan and antigenic keratan sulfate. *Arthritis Res Ther* 8: R73

64 Miossec R, Naviliat M, Dupuy DA, Sany J, Banchereau J (1990) Low levels of interleukin-4 and high levels of transforming growth factor beta in rheumatoid synovitis. *Arthritis Rheum* 33: 1180–1187

65 Taketazu F, Kato M, Gobl A, Iichijo H, ten Dijke P, Itoh J, Kyogoku M, Ronnelid J, Miyazono K, Heldin CH (1994) Enhanced expression of transforming growth factor-beta s and transforming growth factor-beta type II receptor in the synovial tissues of patients with rheumatoid arthritis. *Lab Invest* 70: 620–630

66 Muehleman C, Kuettner KE, Rueger DC, ten Dijke P, Chubinskaya S (2002) Immunohistochemical localization of osteogenic protein-1 and its receptors in rabbit articular cartilage. *J Histochem Cytochem* 50: 1341–1350

67 Patwari P, Chubinskaya S, Hakimiyan A, Kumar B, Cole AA, Kuettner KE, Rueger DC, Grodzinsky AJ (2003) Injurious compression of adult human donor cartilage explants: Investigation of anabolic and catabolic processes. *Trans Orthop Res Soc* 49: 695

68 Chu CQ, Field M, Allard S, Abney E, Feldman M, Maini RN (1992) Detection of cytokines at the cartilage/pannus junction in patients with rheumatoid arthritis: Implication for the role of cytokines in cartilage destruction and repair. *Br J Rheumatol* 31: 653–661

69 Fava R, Olsen N, Keski Oja J, Moses H, Pincus T (1989) Active and latent forms of transforming growth factor beta activity in synovial effusions. *J Exp Med* 169: 291–296

70 Sellers RS, Peluso D, Morris EA (1997) The effect of recombinant human bone mor-

phogenetic protein-2 (rhBMP-2) on the healing of full-thickness defects of articular cartilage. *J Bone Joint Surg Am* 79-A: 1542–1463

71 Sellers RC, Zhang R, Glasson SS, Kim HD, Peluso D, D'Augusta DA, Beckwith K, Morris EA (2000) Repair of articular cartilage defects one year after treatment with recombinant human bone morphogenetic protein-2 (rhBMP-2). *J Bone Joint Surg Am* 82-A: 151–160

72 Frenkel SR, Saadeh PB, Mehrar BJ, Chin GS, Steinbrech DS, Brent B, Gittes JK, Longaker MT (2000) Transforming growth factor beta superfamily members: Role in cartilage modeling. *Plast Recon Surg* 105: 980–990

73 Cook SD, Rueger DC (1996) Osteogenic protein-1: Biology and applications. *Clin Orthop Relat Res* 324: 29–38

74 Grgic M, Jelic M, Basic V, Basic N, Pecina M, Vukicevic S (1997) Regeneration of articular cartilage defects in rabbits by osteogenic protein-1 (bone morphogenetic protein-7). *Acta Med Croatia* 51: 23–27

75 Mattioli-Belmonte M, Gigante A, Muzzarelli RAA, Politano R, De Benedittis A, Specchia N, Buffa A, Biagini G, Greco F (1999) N,N-Dicarboxymethyl chitosan as delivery agent for bone morphogenetic protein in the repair of articular cartilage. *Med Biol Eng Comp* 37: 130–134

76 Simank HG, Sergi C, Jung M, Adolf S, Eckhardt C, Ehemann V, Ries R, Lill C, Richter W (2004) Effects of local application of growth and differentiation factor-5 (GDF-5) in a full-thickness cartilage defect model. *Growth Factors* 22: 35–43

77 Mason JM, Breibart AS, Barcia M, Porti D, Pergolizzi RG, Grande DA (2000) Cartilage and bone regeneration using gene-enhanced tissue engineering. *Clin Orthop Relat Res* (379S): S171–S178

78 Katayama R, Wakitani S, Tsumaki N, Morita Y, Matsushita I, Gejo R, Kimura T (2004) Repair of articular cartilage defects in rabbits using CDMP1 gene-transfected autologous mesenchymal cells derived from bone marrow. *Rheumatology (Oxford)* 43: 980–985

79 Di Cesare PE, Frenkel SR, Carlson CS, Fang C, Liu C (2006) Regional gene therapy for full-thickness articular cartilage lesions using naked DNA with a collagen matrix. *J Orthop Res* 24: 1118–1127

80 Cook SD, Patron LP, Salkeld SL, Rueger DC (2003) Repair of articular cartilage defects with osteogenic protein-1 (BMP-7) in dogs. *J Bone Joint Surg* 85-A (Supp 3): 116–123

81 Louwerse RT, Iheyligers IC, Klein-Nulend J, Sugiihara S, van Kampen GPJ, Semeins CM, Goei SW, de Koning MHMT, Wuisman PIJM, Burger EH (2000) Use of recombinant human osteogenic protein-1 for the repair of subchondral defects in articular cartilage in goats. *J Biomed Mater Res* 49: 506–516

82 Hunziker EB, Driesang MK (2003) Functional barrier principle for growth-factor-based articular cartilage repair. *Osteoarthritis Cartilage* 11: 320–327

83 Jung M, Tuischer JS, Sergi C, Gotterbarm T, Pohl J, Richter W, Simank HG (2006) Local application of a collagen type I/hyaluronate matrix and growth and differentiation fac-

tor 5 influences the closure of osteochondral defects in a minipig model by enchondral ossification. *Growth Factors* 24: 225–232

84 Shimmin A, Young D, O'Leary S, Shih MS, Rueger DC Walsh WR (2002) Growth factor augmentation of an ovine mosaicplasty model. *Proceedings of the 4th ICRS Symposium*, Toronto, Canada, June 15–18, 16

85 Jelic M, Pecina M, Haspl M, Kos J, Taylor K, Maticic D, McCartney J, Yin S, Rueger D, Vukicevic S (2001) Regeneration of articular cartilage chondral defects by osteogenic protein-1 (bone morphogenetic protein-7) in sheep. *Growth Factors* 19: 101–113

86 Hunziker EB, Dreisang IMK, Morris EA (2001) Chondrogenesis in cartilage repair is induced by members of the transforming growth factor-beta superfamily. *Clin Orthop Relat Res* (391S): S171–S181

87 Kuo AC, Rodrigo JJ, Reddi AH, Curtiss S, Grotkopp E, Chiu M (2006) Microfracture and bone morphogenetic protein 7 (BMP-7) synergistically stimulate articular cartilage repair. *Osteoarthritis Cartilage* 14: 1126–1135

88 Hidaka C, Goodrich LR, Chen CT, Warren RF, Crystal RG, Nixon AJ (2003) Acceleration of cartilage repair by genetically modified chondrocytes over expressing bone morphogenetic protein-7. *J Orthop Res* 21: 573–583

89 Park J, Gelse K, Frank S, von der Mark K, Aigner T, Schneider H (2006) Transgene-activated mesenchymal cells for articular cartilage repair: A comparison of primary bone marrow-, perichondrium/periosteum- and fat-derived cells. *J Gene Med* 8: 112–125

90 Colwell CW Jr, D'Lima DD, Hoenecke HR, Fronek J, Pulido P, Morris BA, Chung C, Resnick D, Lotz M (2001) *In vivo* changes after mechanical injury. *Clin Orthop Relat Res* 391 Suppl: S116–123

91 Kim HD, Valentini RF (2002) Retention and activity of BMP-2 in hyaluronic acid-based scaffolds *in vitro*. *J Biomed Mater Res* 59: 573–578

92 Sugimori K, Matsui K, Motomura H, Tokoro T, Wang J, Higa S, Kimura T, Kitajima I (2005) BMP-2 prevents apoptosis of the N1511 chondrocytic cell line through PI3K/Akt-mediated NF-kappa B activation. *J Bone Miner Metab* 23: 411–419

93 Hurtig MB (2006) Delayed administration of BMP-7 reduces articular degeneration after contusive impact injury. *Trans Orthop Res Soc* 52: 1338

94 Kaps C, Bramlage C, Smolian H, Haisch A, Ungethum U, Burmester GR, Sittinger M, Gross G, Haupl T (2002) Bone morphogenetic proteins promote cartilage differentiation and protect engineered artificial cartilage from fibroblast invasion and destruction. *Arthritis Rheum* 46: 149–162

95 Costa-Paz M, Muscolo DL, Ayerza M, Makino A, Aponte-Tinao L (2001) Magnetic resonance imaging follow-up study of bone bruises associated with anterior cruciate ligament ruptures. *Arthroscopy* 17: 445–449

96 Cook SD, Barrack RL, Patron LP, Sakeld SL (2004) Osteogenic protein-1 in knee arthritis and arthroplasty. *Clin Orthop Relat Res* 428: 140–145

97 Badlani N, Inoue A, Healey R, Coutts R, Amiel D (2007) The protective effect of BMP-7 on articular cartilage in the development of osteoarthritis. *Proceedings of the 7th ICRS Symposium*, Warsaw, Sept 29–Oct 2, 150

98 Bolam C, Hurtig M, Cruz A, McEwen B (2006) Characterization of a model of post-traumatic osteoarthritis in the equine femorotibial joint. *Am J Vet Res* 67: 433–447
99 Hurtig M, Akens M (2004) A comparison of the contusive impact and ACL transection models of osteoarthritis. *Trans Orthop Res Soc* 50: 927
100 Hurtig MB, Runciman J, Dickey J, Newbound G (2002) A standardized model of knee injury. *Trans Orthop Res Soc* 48: 104
101 Gartlehner G, Hansen RA, Jonas BL, Thieda P, Lohr KN (2006) The comparative efficacy and safety of biologics for the treatment of rheumatoid arthritis: A systematic review and metaanalysis. *J Rheumatol* 33: 2398–2408
102 Masuda K, Takegami K, An H, Kumano F, Chiba K, Andersson GBJ, Schmid T, Thonar, E (2003) Recombinant osteogenic protein-1 upregulates extracellular matrix metabolism by rabbit annulus fibrosus and nucleus pulposus cells cultured in alginate beads. *J Orthop Res* 21: 922–930
103 Zhang Y, An HS, Song S, Toofanfard M, Masuda K, Andersson GB, Thonar EJ (2004) Growth factor osteogenic protein-1: Differing effects on cells from three distinct zones in the bovine intervertebral disc. *Am J Phys Med Rehabil* 83: 515–521
104 Chujo T, An HS, Akeda K, Miyamoto K, Muehleman C, Attawia M, Andersson G, Masuda K (2006) Effects of growth differentiation factor-5 on the intervertebral disc – *In vitro* bovine study and *in vivo* rabbit disc degeneration model study. *Spine* 31: 2909–2917
105 Takegami K, An HS, Kumano F, Chiba K, Thonar EJ, Singh K, Masuda K (2005) Osteogenic protein-1 is most effective in stimulating nucleus pulposus and annulus fibrosus cells to repair their matrix after chondroitinase ABC-induced *in vitro* chemonucleolysis. *Spine J* 5: 231–238
106 Takegami K, Thonar EJMA, An HS, Kamada H, Masuda K (2002) Osteogenic protein-1 enhances matrix replenishment by intervertebral disc cells previously exposed to interleukin-1. *Spine* 27: 1318–1325
107 An HS, Takegami K, Kamada H, Nguyen CM, Thonar E, Singh K, Andersson GB, Masuda K (2005) Intradiscal administration of Osteogenic Protein-1 increases intervertebral disc height and proteoglycan content in the nucleus pulposus in normal adolescent rabbits. *Spine* 30: 25–32
108 Masuda K, Imai Y, Okuma M, Muehleman C, Nakagawa K, Akeda K, Thonar E, Andersson G, An HS (2006) Osteogenic protein-1 injection into a degenerated disc induces the restoration of disc height and structural changes in the rabbit anular puncture model. *Spine* 31: 742–754
109 Miyamoto K, Masuda K, Kim JG, Inoue N, Akeda K, Andersson GB, An HS (2007) Intradiscal injections of osteogenic protein-1 restore the viscoelastic properties of degenerated intervertebral discs. *Spine J* 6: 692–703
110 Imai Y, Okuma M, An HS, Nakagawa K, Yamada M, Muehleman C, Thonar E, Masuda K (2007) Restoration of disc height loss by recombinant human osteogenic protein-1 injection into intervertebral discs undergoing degeneration induced by an intradiscal injection of chondroitinase ABC. *Spine* 32: 1197–1205

111 Walsh AJ, Bradford DS, Lotz JC (2004) *In vivo* growth factor treatment of degenerated intervertebral discs. *Spine* 29: 156–163
112 Kawakami M, Matsumoto T, Hashizume H, Kuribayashi K, Chubinskaya S, Yoshida M (2005) Osteogenic protein-1 (osteogenic protein-1/bone morphogenetic protein-7) inhibits degeneration and pain-related behavior induced by chronically compressed nucleus pulposus in the rat. *Spine* 30: 1933–1939
113 Wallach CJ, Kim JS, Sobajima S, Lattermann C, Oxner WM, McFadden K, Robbins PD, Gilbertson LG, Kang JD (2006) Safety assessment of intradiscal gene transfer: A pilot study. *Spine J* 6: 107–112
114 Katic V, Majstorovic L, Maticic D, Pirkic B, Yin S, Kos J, Martinovic S, McCartney JE, Vukicevic S (2000) Biological repair of thyroid cartilage defects by osteogenic protein-1 (bone morphogenetic protein-7) in dog. *Growth Factors* 17: 221–232
115 Okamoto T, Yamamoto Y, Gotoh M, Huang CL, Nakamura T, Shimizu Y, Tabata Y, Yokomise H (2004) Slow release of bone morphogenetic protein 2 from a gelatin sponge to promote regeneration of tracheal cartilage in a canine model. *J Thorac Cardiovasc Surg* 127: 329–334
116 Tcacencu I, Carlsöö B, Stierna P, Hultenby K (2006) Local treatment of cricoid cartilage defects with rhBMP-2 induces growth plate-like morphology of chondrogenesis. *Otolaryngol Head Neck Surg* 135: 427–433
117 Hicks DL, Sage AB, Shelton E, Schumacher BL, Sah RL, Watson D (2007) Effect of bone morphogenetic proteins 2 and 7 on septal chondrocytes in alginate. *Otolaryngol Head Neck Surg* 136: 373–379
118 Lietman SA, Hobbs W, Inoue N, Reddi AH (2003) Effects of selected growth factors on porcine meniscus in chemically defined medium. *Orthopedics* 26: 799–803
119 van Beuningen HM, Glansbeek HL, van der Kraan PM, van den Berg W (1998) Differential effects of local application of BMP-2 or TGF-β1 on both articular cartilage composition and osteophyte formation. *Osteoarthritis Cartilage* 6: 306–317
120 Jordan KM, Arden NK, Doherty M, Bannwarth B, Bijlsma JW, Dieppe P, Gunther K, Hauselmann H, Herrero-Beaumont G, Kaklamanis P et al (2003) EULAR Recommendations 2003: An evidence based approach to the management of knee osteoarthritis: Report of a Task Force of the Standing Committee for International Clinical Studies Including Therapeutic Trials (ESCISIT). *Ann Rheum Dis* 62: 1145–1155

Systemic administration of bone morphogenetic proteins

Slobodan Vukicevic[1], Petra Simic[1], Lovorka Grgurevic[1], Fran Borovecki[2] and Kuber Sampath[3]

[1]Laboratory for Mineralized Tissues, School of Medicine, University of Zagreb, 10000 Zagreb, Croatia; [2]Center for Functional Genomics, School of Medicine, University of Zagreb, 10000 Zagreb, Croatia; [3]Genzyme Corporation, Framingham, MA 01701-9322, USA

Systemic administration of bone morphogenetic proteins

Apart from the local application, bone morphogenetic proteins (BMPs) have been systemically applied in rats for the following indications: bone formation in a model of osteoporosis [1], kidney regeneration in models of acute and chronic renal failure [2–7], liver regeneration [8], ischemic coronary infarction [6, 9] and stroke [10], and in a nude mouse model of human prostate, breast, brain and melanocyte cancer [11–14]. BMPs induced organ regeneration recapitulating embryonic development without recorded systemic side effects, except transient bone formation at the site of injection around the tail vein. This was not unexpected, since around the blood vessels the existing progenitor cells can differentiate into cartilage and bone in the presence of a recombinant BMP. No ossification of other tissues has been recorded even after a long period of daily BMP injections. The idea of using BMPs systemically for promoting tissue regeneration and repair has existed for years. However, do we have enough information on their pharmacodynamic and pharmacokinetic properties and a reproducible production of pure preparations for clinical use, and on reproducible results of their efficacy? Do BMPs circulate under physiological conditions? These questions are discussed in this chapter in an attempt to support evidence for their systemic use.

For example, using Western blot analysis it has been claimed that BMP-7 is present in the rat plasma [15] or CNS liquor [16]. However, characterization of BMPs in the plasma is impossible without a mass spectrometry analysis of individual proteins due to cross-reactivity of antibodies against various members of the transforming growth factors (TGF)-β superfamily. In search for BMPs in biological fluids, we detected BMP-6 and GDF-15 in the plasma of normal individuals and patients with chronic renal failure, respectively (Grgurevic et al., unpublished observations).

BMP-6 increases bone volume in osteoporotic rats

Numerous studies have unequivocally demonstrated that recombinant human BMPs induce new bone formation when administered locally in animals [17] and in men [18]. Recently, we showed that systemically administered BMP-6 restores the bone volume and mechanical characteristics of both the trabecular and cortical bone in aged ovariectomized (OVX) rats [1]. The effect is more pronounced on the peripheral skeleton as compared to the lumbar spine (Fig. 1). BMP-6 applied to rats 12 months following OVX significantly increases the osteoblast surface, serum osteocalcin and osteoprotegerin levels, and decreases the osteoclast surface, serum C-telopeptide and IL-6 after 3 months. BMP-6 exerted its effects on bone *via* promoting bone formation and reducing bone resorption thereby providing a distinct means to accumulate the bone tissue mass. No known therapeutic agent achieves both effects by *in vivo* systemic administration. BMP-6 increases *in vivo* the bone expression of ALK-2, ALK-6, Smad5, alkaline phosphatase, collagen type I and decreases the expression of *BMP-3* and BMP antagonists, chordin and cerberus [1]. BMP-6 thus restores bone volume and bone quality in osteoporotic rats making it a feasible candidate for treating osteoporosis in patients with pronounced bone loss. In the same studies, 17β-estradiol (E_2) is significantly less effective *in vivo* and is not synergistic with BMP-6. We and others [1, 19, 20] have previously suggested that the effect of E_2 on bone is associated with increased levels of *BMP-6* mRNA *in vitro* and in mouse bone marrow stromal cells adjacent to active bone formation surface *in vivo*. It has also been demonstrated that E_2 specifically up-regulates *BMP-6* mRNA in different osteoblastic cells [21]. The exact relationship between E_2 and BMP-6 signaling pathways still remains to be elucidated.

BMP-7 in cancer biology

During bone metastases, tumor cells interact with bone cells to disrupt the normal bone remodeling, causing abnormal new bone formation or bone destruction, characteristic of osteoblastic and osteocytic metastases. Tumor cells in the bone microenvironment, stimulated by TGF-β, secrete parathyroid hormone-related protein (PTHrP), interleukin(IL)-6, and IL-11 that can further stimulate osteoclastic resorption and increase more TGF-β release from bone. TGF-β plays a central role in this forward stimulation of osteoclastic bone resorption [22]. Therefore, inhibiting the TGF-β signaling and the subsequent osteolytic metastases attributable to prostate cancer is a novel therapeutic option. The role of BMPs in tumor biology is controversial. It has been suggested that BMP-2 and BMP-4 promote tumor invasion and metastasis of lung and other carcinomas [11]. BMP-2 has also been demonstrated to induce, in a self-autonomous manner, a signaling pathway that promotes malignant transformation, while high levels of BMP-2 at the invasion front enhance the migra-

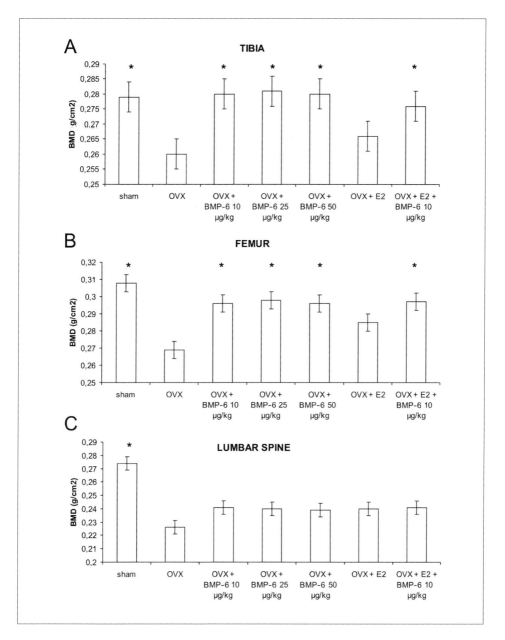

Figure 1
Bone mineral density (BMD) of ex vivo tibia (A), femur (B) and lumbar spine (C) of rats treated with BMP-6 at doses of 10, 25 and 50 μg/kg 3× a week i.v, estradiol (E2) at dose of 50 μg/kg 3× a week i.p. and a combination of E2 and BMP-6 (10 μg/kg 3× week) i.v. for 3 months, 12 months following OVX. *$p<0.05$ vs OVX, ANOVA Dunnett test.

tory and invasive properties of breast cancer in a xenograft model, as well as promote vascularization and tumor angiogenesis [23]. It has been reported that BMPs also promote melanoma cell invasion, angiogenesis and vasculogenic mimicry, and therefore might have important roles in the progression of malignant melanoma [24]. Recently, the role of BMP-7 in tumor expansion has been elaborated in several animal models utilizing various malignant cell lines of prostate, breast brain and melanocyte [12–14, 25]. By counteracting the epithelial-to-mesenchymal transition (EMT), inhibiting the growth of tumor cells and *via* inhibition of TGF-β, BMP-7 prevents the acquisition of an invasive metastatic phenotype (Fig. 2).

Specifically, BMP-7 is a putative regulator of epithelial homeostasis in the human prostate and inhibits prostate cancer bone metastasis *in vivo* [12]. It has been found that *BMP-7* expression in laser-microdissected primary human prostate cancer tissue is strongly down-regulated compared to normal prostate luminal epithelium. Exogenous addition of BMP-7 to human prostate cancer cells dose-dependently inhibits TGF-β-induced activation of nuclear Smad3/4 complexes *via* ALK5 and induced E-cadherin expression. Moreover, BMP-7-induced activation of nuclear Smad1/4/5 signaling transduced *via* BMP type I receptors is synergistically stimulated in the presence of TGF-β. Daily BMP-7 administration to nude mice inhibits the growth of cancer cells in bone. In contrast, no significant growth inhibitory effects of BMP-7 are observed in the human prostate underscoring a decisive role of the tumor microenvironment in mediating the therapeutic response of BMP-7. BMP-7 counteracts the EMT process in the metastatic tumor [12]. The fact that BMPs can reverse the action of TGF-β is demonstrated for the first time *in vitro* when osteoblastic cells are grown in the presence of TGF-β losing their potential to produce alkaline phosphatase and converse to the fibroblastic phenotype [26]. However, when a highly purified BMP preparation is added to the same cells they regain their original osteoblastic phenotype. Buijs et al. [13] found decreased *BMP-7* expression in the primary breast cancer and its association with the formation of clinically overt bone metastases in patients with ≥ 10 years of follow-up. In line with clinical observations, *BMP-7* expression is inversely related to tumorigenicity and invasive behavior of human breast cancer cell lines. Moreover, BMP-7 decreased the expression of vimentin, a mesenchymal marker associated with invasiveness and poor prognosis, in human MDA-MB-231-B/Luc (+) breast cancer cells under basal and TGF-β-stimulated conditions. In addition, exogenous addition of BMP-7 to TGF-β-stimulated MDA-231 cells inhibited Smad-mediated TGF-β signaling. In a well-established bone metastasis model using whole-body bioluminescent reporter imaging, stable overexpression of *BMP-7* in MDA-231 cells inhibited *de novo* formation and progression of osteolytic bone metastases and, hence, their metastatic capability. In line with these observations, daily intravenous administration of BMP-7 (100 μg/kg per day) inhibits orthotopic and intrabone growth of cells in nude mice. Decreased BMP-7 expression during carcinogenesis in the human breast contributes to the acquisition of a bone metastatic phenotype. Because exogenous BMP-7 can

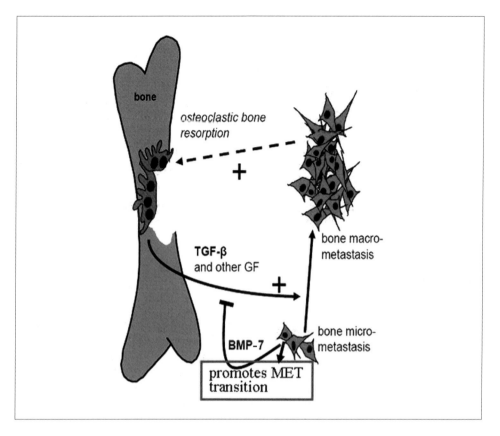

Figure 2
Schematic drawing of the roles of TGF-β and BMP-7 in bone metastases. BMP-7 promotes mesenchymal-to-epithelial transition (MET) and inhibits the growth of bone metastases that induce osteoclastic bone resorption. TGF-β and other growth factors (GF) inhibit MET and promote the growth of bone metastases.

still counteract the breast cancer growth at the primary site and in bone, BMP-7 may serve as a novel therapeutic molecule for repression of local and bone metastatic growth of breast cancer [13] (Fig. 2). Since BMP-7 is essential for the early eye morphogenesis, and lack of BMP-7 causes epithelial developmental abnormalities of the eye and subsequently birth of blind mice [27], Notting et al. [14] tested the role of BMP-7 in the progression of the human uveal melanoma. Primary tumor tissue of patients with uveal melanoma had low or no detectable expression of *BMP-7*. *BMP-7* mRNA was also low or undetectable in cultured human uveal melanoma cell lines when compared to normal cultured melanocytes [14]. The association of tumorigenicity and malignant uveal melanoma was further tested in OCM-1 cell line overexpressing BMP-7 using targeted homologous recombination. When OCM-1 cells overexpressing BMP-7 were inoculated into the anterior chamber of the eye in nude mice, tumor progression was inhibited, confirming that decreased expression of BMP-7 contributes to the progression of uveal melanoma. Surprisingly, BMP-7 was effective in inhibiting growth of both primary and metastatic tumors of various tissues using exact *in vivo* models with precise efficacy determination.

BMPs and diabetes

Diabetes is a major health problem, with more than 5% of the population having impaired insulin function [28]. Despite the variety of anti-diabetic drugs, new treatments that have prolonged effect and act more specifically to reduce the occurrence of side effects, such as weight gain or hypoglycemia, are still required. An integrated functional genomics screening program revealed a role for BMP-9 in glucose homeostasis [29]. BMP-9 gave a positive response in two independent assays: reduction of phosphoenolpyruvate carboxykinase (PEPCK) expression in hepatocytes and activation of Akt kinase in differentiated myotubes [29]. Purified recombinant BMP-9 potently inhibited hepatic glucose production and activated the expression of key enzymes of lipid metabolism. In freely fed diabetic mice, a single subcutaneous injection of BMP-9 reduced glycemia to near-normal levels, with a maximal reduction observed 30 h after treatment [29]. Celeste [11] revealed that livers of mice treated systemically with BMP-9 showed microvesicular changes and necrosis, thereby questioning the potential of BMP-9 for the treatment of diabetes. It has also been shown that transgenic expression of Bmp-4 in beta cells enhances glucose-stimulated insulin secretion and glucose clearance, and that systemic administration of BMP-4 protein to adult mice significantly stimulates glucose-stimulated insulin secretion and ameliorates glucose tolerance in a mouse model of glucose intolerance [30]. On the other hand, despite the identification of many BMPs during development of the pancreas, only the mis-expression of BMP-6 results in a complete agenesis of the pancreas and leads to reduction of the size of the stomach and spleen, causing fusion of the liver and duodenum [31]. BMP signaling also has a key role in mediating insulin-positive differentiation of the acinar-like AR42J cells into insulin-secreting cells through the intracellular Smad signaling pathway [32]. We explored the function of endogenous BMP-6 in the development and homeostasis of the pancreas by using BMP-6 knockout ($^{-/-}$) mice. First, we have noticed that Bmp-6$^{-/-}$ mice have significantly higher daily water consumption as compared to wild type (WT) (Fig. 3A). Bmp-6$^{-/-}$ mice have agenesis of the pancreas. Immunohistochemistry of the pancreas revealed a reduced number of Langerhans islands as compared to WT mice, resulting in decreased levels of blood insulin and increased levels of blood glucose (Fig. 3B–D). Surprisingly, i.v. injection of 10 µg/kg BMP-6 reduced blood glucose to normal within 2 h. Single i.v. injection of 60 µg/kg BMP-6 gradually reduced glycemia after the glucose tolerance

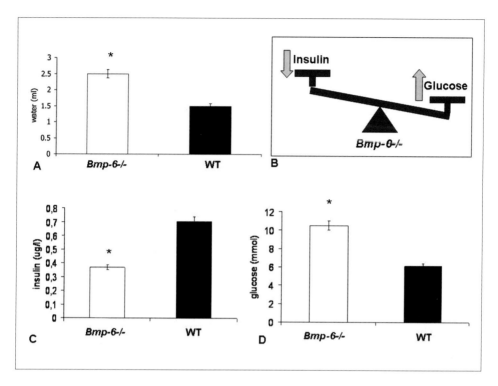

Figure 3
*Bmp-6$^{-/-}$ mice have significantly higher daily water consumption as compared to wild type (WT) (A). Bmp-6$^{-/-}$ mice have disturbed balance of insulin and glucose levels (B) with decreased levels of insulin (C) and increased levels of blood glucose (D). *p<0.05 vs WT, t-test*

test to near-normal values in both WT rats and mice, as well as in bmp-6$^{-/-}$ mice, with a maximal reduction observed 2 h following treatment. Alloxan-induced diabetic rats, with no detectable serum insulin showed a 57–63% reduction of blood glucose within 30 min to 48 h following BMP-6 i.v. injection. Hyperglycemic, insulin-free non-obese diabetic (NOD) mice also showed a significant reduction of blood glucose levels for a prolonged period of time after a single i.v. 60 μg/kg BMP-6 administration. More importantly, BMP-6 therapy improved survival of NOD mice as compared to insulin therapy, with 90% survival in BMP-6-treated diabetic mice after 7 days, while all insulin-treated mice died within 48 h. Next, by using ^{18}F-labeled deoxyglucose (^{18}FDG), we tested its blood and urine levels for 180 min in BMP-6-treated normal and diabetic rats. BMP-6 reduced blood and urine ^{18}FDG levels up to 44% and 47% in both normal and diabetic animals, respectively. Molecular analyses of tissue samples revealed that BMP-6 inhibited hepatic glucose production and activated the expression of the key enzymes of lipid metabolism. BMP-6 reduced the expression of PEPCK, 3-hydroxy-

3methylglutamyl-CoA lyase and fatty acid synthase in the liver [33]. Improper regulation of fatty acid metabolism is believed to have a central function in mediating insulin resistance in the peripheral tissues [34]. BMP-6 increased the expression of insulin-like growth factor-I (IGF-I) in the pancreas, liver and bone, and increased the level of IGF-I protein in serum. IGF-I is necessary for the normal insulin sensitivity, and impairment of IGF-I synthesis results in worsening of the insulin resistance [35]. IGF-I also enhances glucose metabolism by controlling both the endogenous glucose output and the peripheral glucose uptake [36]. We showed that coadministration of IGF-I antibody and BMP-6 significantly reduces its glucose-lowering effect. BMP-6 reduces the blood glucose level, at least in part, *via* an IGF-I-related pathway [33]. We suggest that BMP-6 reduces glycemia in rodent models *via* an insulin-independent pathway, and could therefore provide a novel therapy for diabetes.

BMPs in regeneration of the liver

Bmp-2 is expressed in the normal adult rat liver and negatively regulates hepatocyte proliferation [37]. The observed down-regulation of *Bmp-2* following partial hepatectomy suggests that such down-regulation may be necessary for the hepatocyte proliferation [37]. On the other hand, *Bmp-7* expression is absent in the liver, but the receptors for BMP-7 are present on adult hepatocytes, which could enable BMP-7 to function as an endogenous regulator of adult hepatocyte proliferation and liver homeostasis. Neutralization of circulating endogenous BMP-7 results in significantly impaired regeneration of the liver after partial hepatectomy, while systemic administration of BMP-7 significantly enhances liver regeneration associated with accelerated improvement of liver function [8]. The role of BMPs has also been investigated in hemochromatosis, a disease when too much iron builds up in the liver. Systemic iron balance is regulated by hepcidin, a peptide hormone secreted by the liver. Hepcidin deficiency plays a key role in the pathogenesis of hemochromatosis. Hemojuvelin is a coreceptor for BMP signaling and BMP signaling positively regulates hepcidin expression in the liver cells *in vitro* [38]. Hemojuvelin mutants are associated with hemochromatosis and have an impaired BMP signaling ability [38].

BMP-2 administration increases hepcidin expression and decreases serum iron levels *in vivo* [39]. These data support a role for modulators of the BMP signaling pathway in treating diseases of iron overload and anemia of chronic disease.

BMPs in regeneration of the heart

BMP-2 has been shown to induce ectopic expression of cardiac transcription factors, beating cardiomyocytes in non-precardiac mesodermal cells, suggesting that BMP-2 is an inductive signaling molecule that participates in cardiac development.

BMP-2 promotes survival of isolated neonatal rat cardiac myocytes without any hypertrophic effect [40]. Furthermore, embryonic stem cells used for transplantation in a diseased heart are directed to differentiate into cardiomyocytes by signaling mediated through TGF-β/BMP-2, a cardiac paracrine pathway [41].

BMP-7 was studied for its anti-ischemic properties in rats subjected to myocardial ischemia and reperfusion. At 10 min after ligation of the left coronary artery, 2 or 20 μg/rat BMP-7 or its vehicle, was given i.v. The 20-μg BMP-7 dose significantly reduced reperfusion injury 24 h later compared to rats receiving vehicle alone. BMP-7 also preserved rat coronary endothelial function in perfused hearts exerting significant anti-ischemic effects [9]. Cardiac fibrosis is associated with the emergence of fibroblasts originating from endothelial cells, suggesting an endothelial-mesenchymal transition. TGF-β1 induces endothelial cells to undergo EMT, whereas BMP-7 preserves the endothelial phenotype. Systemic administration of BMP-7 significantly inhibits EMT and the progression of cardiac fibrosis in mouse models of pressure overload and chronic allograft rejection [42].

BMPs in stroke

BMP-7 selectively induces dendritic outgrowth from cultured neurons *in vitro* [43]. BMP-7 (1 or 10 μg) was injected into the cisterna magna of Sprague-Dawley rats 1 and 4 days following focal cerebral infarction induced by middle cerebral artery occlusion (MCAO) [10]. BMP-7 treatment was associated with a marked enhancement of recovery of sensorimotor function of the impaired fore limb and hind limb (contralateral to infarcts). There was no difference in the infarct volume between BMP-7 and vehicle-treated rats [43]. In a similar study, BMP-7 enhanced glucose utilization in the basal ganglia ipsilateral to stroke and improved local cerebral blood flow in ipsilateral subthalamus, but decreased the local cerebral blood flow and the glucose utilization in contralateral cortical regions [44]. Recent studies have indicated that receptors for BMP are up-regulated after brain ischemia [45]. It is possible that this up-regulation may facilitate endogenous neurorepair in the ischemic brain. After ischemia/reperfusion injury, animals receiving BMP-7 i.v. show a decrease in the body asymmetry from days 7 to 14, and an increase in the locomotor activity on day 14 following MCAO. BMP-7 given parenterally after stroke passes through the blood-brain barrier on the ischemic side and induces a behavioral recovery in stroke animals at longer testing times [45].

BMPs and kidney regeneration

The kidney has been identified as a major site of BMP-7 synthesis during embryonic and post-natal development [46–49]. Gene knockout [27, 50] and *in vitro* experiments [49, 51] demonstrated the importance of BMP-7 in the kidney development. Many developmental features are recapitulated during renal injury, and BMPs may be important in both preservation of function and resistance to injury [3, 4]. BMP-7 has a cytoprotective and anti-inflammatory effect in both acute and chronic renal failure [3, 4].

Acute renal failure

The finding that *BMP-7* kidney expression remains high both in the fetal and in postnatal life suggests that BMP-7 may have a systemic function and a role in the continued repair and regeneration of the adult kidney [3, 48].

Acute renal failure represents a clinical condition with persistently high mortality (40–80%), despite technical advances in both critical care medicine and dialysis [52]. The damaged kidney is capable of undergoing complete repair and regeneration after an acute injury and the process recapitulates features that occur during the development. It is assumed that regenerating cells take a step back, towards an earlier ontogenic stage, which makes the cells sensitive to embryonic stimuli [53, 54]. BMP-7 may be important in both preservation of function and resistance to injury [3].

The mechanisms controlling the cascade of cellular migration, growth and proliferation following acute renal failure undoubtedly comprise a number of autocrine and paracrine growth factors [55, 56], such as IGFs, epidermal growth factor (EGF), fibroblast growth factor (FGF), TGF-α, TGF-β, and hepatocyte growth factor (HGF) [57–60]. Animal studies dealing with acute renal failure due to an ischemic-reperfusion insult proved that administration of BMP-7 is beneficial to the extent of injury and the regeneration of kidney function [3]. Bioavailability studies have shown that human BMP-7 has a serum half-life of about 30 min, and that significant amounts of ^{125}I-labeled BMP-7 can be found in both the kidney cortex and medulla shortly after i.v. administration [3].

Apart from being protective in ischemic acute renal failure, BMP-7 also influences the course of toxic kidney injury *in vitro*, as well as in acute nephrotoxic animal models utilizing administration of mercuric chloride and cisplatinum [61]. Both prophylactic and therapeutic systemic administration of BMP-7 to rats given mercuric chloride protects the kidney function and significantly extends the survival rate. Similarly, BMP-7 protects the kidney function in rats treated with a high dose of cisplatinum. Since we have shown that prostaglandin E2 (PGE2) regulates the expression of *BMP-7* [62], we tested the efficacy of three selective agonists of the PGE2 receptor, EP2, EP2/4 and EP4, in two models of acute renal failure [63]. In the nephrotoxic $HgCl_2$ rat model of acute kidney failure systemically administered EP4 agonist reduced the serum creatinine values and increased the survival rate. Although the EP2 or the EP2/4 agonist did not change the serum creatinine values, the EP2 receptor agonist increased the survival rate. Histological evaluation of kidneys from EP4-treated rats indicated less proximal tubular necrosis and fewer apoptotic cells [63].

Chronic renal failure

Progressive and permanent reduction in the glomerular filtration rate (GFR), which is associated with the loss of functional nephron units, leads to chronic renal failure (CRF). The subject progresses to the end-stage renal disease (ESRD) when the GFR continues to decline to less than 10% of normal values (5–10 ml/min). At this point, renal failure will rapidly progress to cause death unless the subject receives renal replacement therapy, i.e., chronic hemodialysis, continuous peritoneal dialysis or kidney transplantation, or therapy that delays the progression of the chronic renal disease.

Diabetic nephropathy

Morphological traits of diabetic nephropathy include progressive thickening of the glomerular basement membrane and expansion of the mesangial compartment due to accumulation of extracellular matrix (ECM). Transdifferentiation of epithelial cells to fibroblastic phenotype has been implicated as the major pathological event that promotes tubulointerstitial fibrosis, a hallmark of progressive renal failure [64]. In the model of streptozotocin (STZ)-induced diabetes in rats, BMP-7 systemic therapy partially reversed the renal hypertrophy, restored GFR to normal and decreased proteinuria 32 weeks following a single dose of diabetes-inducing agent [65]. Recent studies have shown that renal BMP-7 may act as a protective agent in mouse models of diabetes, in part by preserving the podocytes [66]. Diabetic mice expressing the phosphoenolpyruvate caboxykinase promoter-driven BMP-7 transgene exhibited a marked reduction in podocyte dropout, as well as reduction in glomerular fibrosis and interstitial collagen accumulation. In another study, three different mouse strains, CD1, namely inbred C57BL/6 and 129Sv strains, were subjected to intraperitoneal application of STZ, leading to ESRD associated with prominent tubulointerstitial nephritis and fibrosis [67]. Application of STZ in CD1 mice led to the development of ESRD within 3 months, with ensuing death by month 6–7 due to diabetic complication. BMP-7 and inhibitors of advanced glycation end products, i.e., aminoguanidine and pyridoxamine, were next applied to test their therapeutic ability to inhibit and ameliorate the progression of renal disease in diabetic CD1 mice. The results indicate marked efficacy of all three compounds, with BMP-7 being the most effective in the inhibition of tubular inflammation and tubulointerstitial fibrosis.

We tested the efficacy of three selective agonists of the PGE2 receptor EP2, EP2/4 and EP4 in a rat model of chronic renal failure and the three receptor agonists decreased the serum creatinine and increased the glomerular filtration rate at 9 weeks following therapy. Kidneys treated with the EP4 agonist have less glomerular sclerosis, better preservation of proximal and distal tubules and blood vessels, increased convoluted epithelium proliferation and fewer apoptotic cells [63].

Renal osteodystrophy

Chronic renal disease influences the skeletal system through an extensive range of skeletal abnormalities, which are the result of impaired kidney function and secondary hyperparathyroidism. Adynamic bone disorder, a low-turnover variant of the renal osteodystrophy, is an important complication of chronic renal disease. In the animal model of chronic renal disease accompanied by adynamic bone disorder, which is characterized by absence of secondary hyperparathyroidism, treatment with BMP-7 reversed the abnormal bone histomorphometry to normal values [68]. Untreated animals developed hyperphosphatemia, secondary hyperparathyroidism, and a mild osteodystrophy. Significant adynamic bone disorder developed in the untreated animals. The animals treated with BMP-7 exhibited a significant decrease in the phosphorus plasma levels when compared to the untreated animals [68]. More recent studies have shown that LDL receptor knockout mice fed high-fat diets exhibit vascular calcification [69]. Careful analysis of the skeletal system in these animals revealed significant reductions in bone formation rates, associated with increased vascular calcification and hyperphosphatemia. LDL receptor knockout mice with superimposed chronic kidney disease developed a low turnover osteodystrophy, with more pronounced vascular calcification. BMP-7 treatment in these animals led to the decreased plasma phosphate, as well as the inhibition of vascular calcification and correction of osteodystrophy.

Renal fibrosis

Scarring of the kidney due to progressive fibrosis occurs during a variety of diseases, such as hypertension, primary glomerulopathies, autoimmune disease, diabetes mellitus, toxic injury or congenital malformations [70, 71].

TGF-β1 has been identified as the main inducer of EMT in the kidney. TGF-β1 contributes to progressive glomerular and interstitial fibrosis by increasing gene expression of several proteins of the ECM and acts to reduce degradation of these proteins [72, 73]. *In vitro* studies on cultured mesangial cells have shown that TGF-β1 increased cell-associated collagen type IV and fibronectin, soluble collagen type IV, thrombospondin and connective tissue growth factor (CTGF) [74]. Treatment with recombinant human BMP-7 impaired the increase in the aforementioned ECM proteins and CTGF. It has been shown that BMP-7 reverses the TGF-β1-induced EMT by reinduction of E-cadherin, an important cell adhesion molecule. Due to the fact that TGF-β1 and BMP-7 mediate their effect through the Smad signaling pathway, interactions on the level of these second messenger molecules has also been implicated in the reversal of TGF-β1-induced EMT. While TGF-β1 directly inhibits E-cadherin expression and induces EMT in a Smad3-dependent manner, BMP-7 enhances E-cadherin expression *via* Smad5 and restores the epithelial phenotype [75]. *In vivo* studies on the TGF-β1-induced *de novo* EMT model have shown that systemic administration of human recombinant BMP-7 results in repair of severely damaged renal tubular epithelial cells and ultimately leads to reversal of the chronic renal injury [6].

Inhibition of TGF-β1-mediated EMT has been the focus of several recent studies. It has been shown that a novel protein, kielin/chordin-like protein (KCP), enhances BMP signaling in a paracrine manner. This is done through binding to BMP-7 and subsequent increased binding to the type I receptor [76]. This, in turn, leads to increased levels of Smad1-dependent transcription with phosphorylated Smad1 levels being elevated. Additionally, $Kcp^{-/-}$ mice exhibit heightened susceptibility to renal interstitial fibrosis, are more sensitive to tubular injury, and show substantial pathology after recovery.

Among other promising compounds, histone deacetylase (HDAC) inhibitors have been shown to possess an anti-fibrogenic effect, and may thus be effective in preventing the EMT observed in chronic renal disease. Histone acetylation plays an important role in regulation of gene expression through a direct effect on the chromatin structure. *In vitro* studies using trichostatin A (TSA), an HDAC inhibitor, have shown that treatment of human renal proximal tubular epithelial cells with TSA completely prevents TGF-β1-induced EMT [77]. Application of TSA leads to up-regulation of E-cadherin and down-regulation of collagen type I, but fails to affect phosphorylation of Smad2 and Smad3. BMP-7 exhibits marked up-regulation, indicating that beneficial effects of TSA might in part be mediated through the activation of BMP-7 transcription. A subsequent study, using a nephrotoxic serum nephritis mouse model, exhibited efficacy of TSA in prevention of the progression of proteinuria, glomerulosclerosis and interstitial fibrosis by inducing BMP-7 expression [78]. TSA was effective both in prophylactic and therapeutic models of the disease.

BMPs in biological fluids

There is a controversy regarding the presence of BMPs in biological fluids. The potential use of their remarkable biology in organ regeneration will greatly depend on their presence in the circulation. For example, it will be of great support prov-

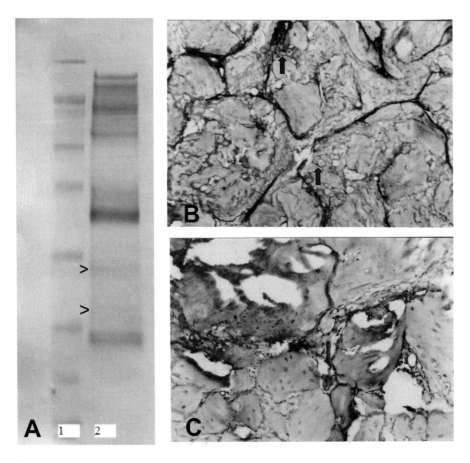

Figure 4
(A) Characterization of BMP-7/6 hybrid molecule in rat urine. The fraction following purification of a 24-h urine collection with heparin-Sepharose column was precipitated and subjected to gel electrophoresis and immunoblotting. Lane 1: Molecular mass markers; lane 2: fraction 1 M NaCl. Bands were visualized with specific BMP-6 and BMP-7 antibodies. Arrowheads indicate the prodomain and mature domain of BMP-7/6 hybrid molecule. (B, C) Induction of new bone formation by BMP-7/6 hybrid molecule protein isolated from the urine. Eluted protein solution was mixed with rat type I collagen, demineralized bone matrix and chondroitin sulfate. (B) Eluted protein solution from the band of 35 kDa related to BMP-7/6 hybrid molecule shows new cartilage and bone formation (arrows); (C) control implants with fibrous tissue between the implanted particles. Toluidine blue staining (×25).

ing that renal osteodystrophy develops due to the lack of BMP-7 in the blood as a consequence of CRF [79]. Whether BMP-7 is released to serum from the kidney is not known. Systemic administration of a hybrid BMP-7/6 molecule composed of a BMP-7 prodomain and a BMP-6 mature domain made it possible to test whether a

BMP passes the kidney glomerular barrier and appears in the urine following an i.v. injection [80]. The blood was collected from the rats' orbital plexus and purified using a heparin-Sepharose column. Protein identity was confirmed by Western blot and by liquid chromatography–mass spectrometry (LC-MS) of the resulting peptides. BMP-7/6 remained intact in the rat plasma and could still be visualized 30 min after its systemic administration. Two protein bands at 35 and 39 kDa were detected with anti-BMP antibodies in the urine of rats, corresponding to the mature active BMP-6 dimer and the prodomain of BMP 7, respectively (Fig. 4A). LC-MS analysis detected only peptides derived from the BMP-7/6 molecule. Histological analysis of implanted pellets, composed of the protein derived from the 35-kDa band and the inactive demineralized bone matrix, revealed the formation of a new endochondral bone 14 days following implantation at a subcutaneous site (Fig. 4B, C). These results show that a systemically administered BMP-7/6 hybrid molecule is secreted into the urine and that its biological activity is preserved, suggesting that analysis of BMP in urine might reflect its presence in serum where the detection is limited due to low amounts. These findings suggest that the level of BMPs in serum, mainly of BMP-7 and BMP-6, could reflect the ability of the kidney to regenerate in various acute and chronic kidney diseases [81]. Full understanding of the availability and biodistribution of naturally released and systemically administered BMPs will help in the design of new approaches in treating BMP-deficient symptoms and in developing new biomarkers for kidney- and bone-related diseases, as well as assays for other organs such as liver, kidney and gut, where BMPs play an important role in preventing and regenerating injuries [7, 82].

Conclusion

BMPs upon systemic delivery circulate and promote repair and regeneration of organs in rodent models of osteoporosis, renal failure, liver diseases, ischemic myocardial infarction, stroke and in human cancers. In all the models used, recombinant BMP-7 induces organ repair and regeneration without systemic side effects, except transient bone formation at the injection site. These studies have opened the possibilities of novel therapeutic options based on BMP systemic biology.

References

1. Simic P, Buljan Culej J, Orlic I, Grgurevic L, Draca N, Spaventi R, Vukicevic S (2006) Systemically administered bone morphogenetic protein-6 (BMP-6) restores bone in aged ovariectomized rats by increasing bone formation and suppressing bone resorption. *J Biol Chem* 281: 25509–25521
2. Vukicevic S, Grgic M, Stavljenic A, Sampath TK (1996) Recombinant human OP-1

(BMP-7) prevents rapid loss of glomerular function and improves mortality associated with chronic renal failure. *J Am Soc Nephrol* 7: A3102

3   Vukicevic S, Basic V, Rogic D, Basic N, Shih MS, Shepard A, Jin D, Dattatreyamurty B, Jones W, Dorai H et al (1998) Osteogenic protein-1 (bone morphogenetic protein-7) reduces severity of injury after ischemic acute renal failure in rat. *J Clin Invest* 102: 202–214

4   Hruska KA, Guo G, Wozniak M, Martin D, Miller S, Liapis H, Loveday K, Klahr S, Sampath TK, Morrissey J (2000) Osteogenic protein-1 prevents renal fibrogenesis associated with ureteral obstruction. *Am J Physiol Renal Physiol* 279: F130–F143

5   González EA, Lund RJ, Martin KJ, McCartney JE, Tondravi MM, Sampath TK, Hruska KA (2002) Treatment of a murine model of high-turnover renal osteodystrophy by exogenous BMP-7. *Kidney Int* 61: 1322–1331

6   Zeisberg M, Hanai J, Sugimoto H, Mammoto T, Charytan D, Strutz F, Kalluri R (2003) BMP-7 counteracts TGF-beta1-induced epithelial-to-mesenchymal transition and reverses chronic renal injury. *Nat Med* 9: 964–968

7   Borovecki F, Simic P, Grgurevic L, Vukicevic S (2004) BMPs in regeneration of kidney. In: Vukicevic S, Sampath TK (eds): *Bone morphogenetic proteins: Regeneration of bone and beyond*. Birkhäuser, Basel, 213–244

8   Sugimoto H, Yang C, LeBleu VS, Soubasakos MA, Giraldo M, Zeisberg M, Kalluri R (2007) BMP-7 functions as a novel hormone to facilitate liver regeneration. *FASEB J* 21: 256–264

9   Lefer AM, Tsao PS, Ma XL, Sampath TK (1992) Anti-ischaemic and endothelial protective actions of recombinant human osteogenic protein (hOP-1). *J Mol Cell Cardiol* 24: 585–593

10 Ren J, Kaplan PL, Charette MF, Speller H, Finklestein SP (2000) Time window of intracisternal osteogenic protein-1 in enhancing functional recovery after stroke. *Neuropharmacology* 39: 860–865

11 Simic P, Vukicevic S (2007) Bone morphogenetic proteins: From developmental signals to tissue regeneration. Conference on bone morphogenetic proteins. *EMBO Rep* 8: 327–331

12 Buijs JT, Rentsch CA, van der Horst G, van Overveld PG, Wetterwald A, Schwaninger R, Henriquez NV, Ten Dijke P, Borovecki F, Markwalder R et al (2007) BMP7, a putative regulator of epithelial homeostasis in the human prostate, is a potent inhibitor of prostate cancer bone metastasis *in vivo*. *Am J Pathol* 171: 1047–1057

13 Buijs JT, Henriquez NV, van Overveld PG, van der Horst G, Que I, Schwaninger R, Rentsch C, Ten Dijke P, Cleton-Jansen AM, Driouch K et al (2007) Bone morphogenetic protein 7 in the development and treatment of bone metastases from breast cancer. *Cancer Res* 67: 8742–8751

14 Notting I, Buijs J, Mintardjo R, van der Horst G, Vukicevic S, Lowik C, Schalij-Delfos N, Keunen J, van der Pluijm G (2007) Bone morphogenetic protein 7 inhibits tumor growth of human uveal melanoma *in vivo*. *Invest Ophthalmol Vis Sci* 48: 4882–4889

15 Irie A, Habuchi H, Kimata K, Sanai Y (2003) Heparan sulfate is required for bone morphogenetic protein-7 signaling. *Biochem Biophys Res Commun* 308: 858–865
16 Dattatreyamurty B, Roux E, Horbinski C, Kaplan PL, Robak LA, Beck HN, Lein P, Higgins D, Chandrasekaran V (2001) Cerebrospinal fluid contains biologically active bone morphogenetic protein-7. *Exp Neurol* 172: 273–281
17 Cook SD, Rueger DC (2002) Preclinical models of recombinant BMP induced healing of orthopedic defects. In: S Vukicevic, TK Sampath (eds): *Bone morphogenetic proteins: From laboratory to clinical practice*. Birkhäuser, Basel, 121–144
18 Friedlaender GE (2004) Clinical experience of osteogenic protein 1 (OP-1) in the repair of bone defects and fractures of long bones. In: S Vukicevic, TK Sampath (eds): *Bone morphogenetic proteins: Regeneration of bone and beyond*. Birkhäuser, Basel, 157–162
19 Martinovic S, Basic N, Vukicevic S (1999) BMP-4 expression decreases during differentiation of osteoblastic MC3T3-E1 cells. *Bone* 24: 4a
20 Plant A, Tobias JH (2002) Increased bone morphogenetic protein-6 expression in mouse long bones after estrogen administration. *J Bone Miner Res* 17: 782–790
21 Rickard DJ, Hofbauer LC, Bonde SK, Gori F, Spelsberg TC, Riggs BL (1998) Bone morphogenetic protein-6 production in human osteoblastic cell lines Selective regulation by estrogen. *J Clin Invest* 101: 413–442
22 Fournier PG, Guise TA (2007) BMP7: A new bone metastases prevention? *Am J Pathol* 171: 739–743
23 Raida M, Clement JH, Leek RD, Ameri K, Bicknell R, Niederwieser D, Harris AL (2005) Bone morphogenetic protein 2 (BMP-2) and induction of tumor angiogenesis. *J Cancer Res Clin Oncol* 131: 741–750
24 Rothhammer T, Poser I, Soncin F, Bataille F, Moser M, Bosserhoff AK (2005) Bone morphogenic proteins are overexpressed in malignant melanoma and promote cell invasion and migration. *Cancer Res* 65: 448–456
25 Klose A, Waerzeggers Y, Klein M, Monfared P, Vukicevic S, Kaijzel EL, Winkeler A, Löwik CWGM, Jacobs AH (2007) Imaging BMP-7–induced cell cycle arrest in experimental gliomas. *Cancer Research* 67: 8742–8751
26 Vukicevic S, Luyten FP, Reddi AH (1990) Osteogenin inhibits proliferation and stimulates differentiation in mouse osteoblast-like cells (MC3T3–E1). *Biochem Biophys Res Commun* 166: 750–756
27 Luo O, Hofmann A, Bronckers JJ, Sohocki M, Bradley A, Karsenty G (1995) BMP-7 is an inducer of nephrogenesis and is also required for eye development and skeletal patterning. *Genes Dev* 9: 2808–2820
28 Adeghate E, Schattner P, Dunn E (2006) An update on the etiology and epidemiology of diabetes mellitus. *Ann NY Acad Sci* 1084: 1–29
29 Chen C, Grzegorzewski KJ, Barash S, Zhao Q, Schneider H, Wang Q, Singh M, Pukac L, Bell AC, Duan R et al (2003) An integrated functional genomics screening program reveals a role for BMP-9 in glucose homeostasis. *Nat Biotechnol* 21: 294–301
30 Goulley J, Dahl U, Baeza N, Mishina Y, Edlund H (2007) BMP4–BMPR1A signaling in

beta cells is required for and augments glucose-stimulated insulin secretion. *Cell Metab* 5: 207–219

31 Dichmann DS, Miller CP, Jensen J, Scott Heller R, Serup P (2003) Expression and misexpression of members of the FGF and TGFbeta families of growth factors in the developing mouse pancreas. *Dev Dyn* 226: 663–674

32 Yew KH, Hembree M, Prasadan K, Preuett B, McFall C, Benjes C, Crowley A, Sharp S, Tulachan S, Mehta S et al (2005) Cross-talk between bone morphogenetic protein and transforming growth factor-beta signaling is essential for exendin-4-induced insulin-positive differentiation of AR42J cells. *J Biol Chem* 280: 32209–32217

33 Simic P, Zuvic M, Rogic D, Dodig D, Stavljenic-Rukavina A, Vukicevic S (2004) BMP-6 regulates blood glucose level *via* an insulin independent pathway. In: *Abstracts of the 6th International Conference on Bone Morphogenetic Proteins*, 12–16 September, Nagoya, Japan, 70

34 Watkins BA, Lippman HE, Le Bouteiller L, Li Y, Seifert MF (2001) Bioactive fatty acids: Role in bone biology and bone cell function. *Prog Lipid Res* 40: 125–148

35 Clemmons DR (2004) Role of insulin-like growth factor in maintaining normal glucose homeostasis. *Horm Res* 62 (Suppl 1): 77–82

36 Saukkonen T, Shojaee-Moradie F, Williams RM, Amin R, Yuen KC, Watts A, Acerini L, Umpleby AM, Dunger DB (2006) Effects of recombinant human IGF-I/IGF-binding protein-3 complex on glucose and glycerol metabolism in type 1 diabetes. *Diabetes* 55: 2365–2370

37 Xu CP, Ji WM, van den Brink GR, Peppelenbosch MP (2006) Bone morphogenetic protein-2 is a negative regulator of hepatocyte proliferation downregulated in the regenerating liver. *World J Gastroenterol* 12: 7621–7625

38 Babitt JL, Huang FW, Wrighting DM, Xia Y, Sidis Y, Samad TA, Campagna JA, Chung RT, Schneyer AL, Woolf CJ et al (2006) Bone morphogenetic protein signaling by hemojuvelin regulates hepcidin expression. *Nat Genet* 38: 531–539

39 Babitt JL, Huang FW, Xia Y, Sidis Y, Andrews NC, Lin HY (2007) Modulation of bone morphogenetic protein signaling *in vivo* regulates systemic iron balance. *J Clin Invest* 117: 1933–1939

40 Izumi M, Fujio Y, Kunisada K, Negoro S, Tone E, Funamoto M, Osugi T, Oshima Y, Nakaoka Y, Kishimoto T et al (2001) Bone morphogenetic protein-2 inhibits serum deprivation-induced apoptosis of neonatal cardiac myocytes through activation of the Smad 1 pathway. *J Biol Chem* 276: 31133–31141

41 Behfar A, Zingman LV, Hodgson DM, Rauzier JM, Kane GC, Terzic A, Pucéat M (2002) Stem cell differentiation requires a paracrine pathway in the heart. *FASEB J* 16: 1558–1566

42 Zeisberg EM, Tarnavski O, Zeisberg M, Dorfman AL, McMullen JR, Gustafsson E, Chandraker A, Yuan X, Pu WT, Roberts AB et al (2007) Endothelial-to-mesenchymal transition contributes to cardiac fibrosis. *Nat Med* 13: 952–961

43 Kawamata T, Ren J, Chan TC, Charette M, Finklestein SP (1998) Intracisternal osteo-

genic protein-1 enhances functional recovery following focal stroke. *Neuroreport* 9: 1441–1445

44 Liu Y, Belayev L, Zhao W, Busto R, Saul I, Alonso O, Ginsberg MD (2001) The effect of bone morphogenetic protein-7 (BMP-7) on functional recovery, local cerebral glucose utilization and blood flow after transient focal cerebral ischemia in rats. *Brain Res* 905: 81–90

45 Chang CF, Lin SZ, Chiang YH, Morales M, Chou J, Lein P, Chen HL, Hoffer BJ, Wang Y (2003) Intravenous administration of bone morphogenetic protein-7 after ischemia improves motor function in stroke rats. *Stroke* 34: 558–564

46 Helder MN, Ozkaynak E, Sampath TK, Luyten FP, Latin V, Oppermann H, Vukicevic S (1995) Expression pattern of osteogenic protein-1 (bone morphogenetic protein-7) in human and mouse development. *J Histochem Cytochem* 43: 1035–1044

47 Vukicevic S, Stavljenic A, Pecina M (1995) Discovery and clinical applications of bone morphogenetic proteins. *Eur J Clin Chem Clin Biochem* 33: 661–671

48 Ozkaynak E, Schnegelsberg PN, Opperman H. (1991) Murine osteogenic protein -1 (OP-1): High levels of mRNA in kidney. *Biochem Biophys Res Commun* 179: 116–123

49 Vukicevic S, Kopp JB, Luyten FB, Sampath TK (1996) Induction of nephrogenic mesenchyme by osteogenic protein-1 (bone morphogenetic protein 7). *Proc Natl Acad Sci USA* 93: 9021–9026

50 Dudley AT, Lyons KM, Robertson EJ (1995) A requirement for bone morphogenetic protein-7 during development of the mammalian kidney and eye. *Genes Dev* 9: 2795–2807

51 Simon M, Maresh JG, Harris SE, Hernandez JD, Arar M, Olson MS, Abboud HE (1999) Expression of bone morphogenetic protein-7 mRNA in normal and ischemic adult rat kidney. *Am J Physiol* 276: 382–389

52 Michos O, Gonçalves A, Lopez-Rios J, Tiecke E, Naillat F, Beier K, Galli A, Vainio S, Zeller R (2007) Reduction of BMP4 activity by gremlin 1 enables ureteric bud outgrowth and GDNF/WNT11 feedback signalling during kidney branching morphogenesis. *Development* 134: 2397–2405

53 Thadhani R, Pascual M, Bonventre JV (1996) Acute renal failure. *N Engl J Med* 334: 1448–1460

54 Humes HD, MacKay SM, Funke AJ, Buffington DA (1997) Acute renal failure: Growth factors, cell therapy and gene therapy. *Proc Assoc Am Physicians* 109: 547–557

55 Hirschberg R, Ding H (1998) Growth factors and acute renal failure. *Semin Nephrol* 18: 191–207

56 Humes DH, Liu S (1994) Cellular and molecular basis of renal repair in acute renal failure. *J Am Soc Nephrol* 5: 1–11

57 Witzgall R, Brown D, Schwarz C, Bonventre JV (1994) Localization of proliferating cell number antigen, vimentin, c-Fos and clusterin in the post-ischemic kidney: Evidence for a heterogenous genetic response among nephron segments and a large pool of mitotically active and differentiated cells. *J Clin Invest* 93: 2175–2188

58 Hirschberg R, Kopple JD (1989) Evidence that insulin-like growth factor I increases renal plasma flow and glomerular filtration rate in fasted rats. *J Clin Invest* 83: 326–330

59 Andersson G, Jennische E (1988) IGF-I immunoreactivity is expressed by regenerating renal tubule cells after ischaemic injury in the rat. *Acta Physiol Scand* 132: 453–457

60 Sugimura K, Goto T, Kasai S, Tsuchida K, Takemoto Y, Yamagami S (1998) The activation of serum hepatocyte growth factor in acute renal failure. *Nephron* 76: 364–365

61 Coimbra TM, Cieslinski DA, Humes HD (1990) Epidermal growth factor accelerates renal repair in mercuric chloride nephrotoxicity. *Am J Physiol* 259: 438–443

62 Paralkar VM, Grasser WA, Mansolf AL, Baumann AP, Owen TA, Smock SL, Martinovic S, Borovecki F, Vukicevic S, Ke HZ et al (2002) Regulation of BMP-7 expression by retinoic acid and prostaglandin E(2). *J Cell Physiol* 190: 207–217

63 Vukicevic S, Simic P, Borovecki F, Grgurevic L, Orlic I, Grasser W, Thompson DD, Paralkar V (2006) Role of EP2 and EP4 receptor-selective agonists of prostaglandin E-2 in acute and chronic kidney failure. *Kidney Int* 70: 1099–1106

64 Dolan V, Hensey C, Brady HR (2003) Diabetic nephropathy: Renal development gone awry? *Pediatr Nephrol* 18: 75–84

65 Wang S, Chen Q, Simon TC, Strebeck F, Chaudhary L, Morrissey J, Liapis H, Klahr S, Hruska KA (2003) Bone morphogenic protein-7 (BMP-7), a novel therapy for diabetic nephropathy. *Kidney Int* 63: 2037–2049

66 Wang S, de Caestecker M, Kopp J, Mitu G, Lapage J, Hirschberg R (2006) Renal bone morphogenetic protein-7 protects against diabetic nephropathy. *J Am Soc Nephrol* 17: 2504–2512

67 Sugimoto H, Grahovac G, Zeisberg M, Kalluri R (2007) Renal fibrosis and glomerulosclerosis in a new mouse model of diabetic nephropathy and its regression by bone morphogenic protein-7 and advanced glycation end product inhibitors. *Diabetes* 56: 1825–1833

68 Lund RJ, Davies MR, Brown AJ, Hruska KA (2004) Successful treatment of an adynamic bone disorder with bone morphogenetic protein-7 in a renal ablation model. *J Am Soc Nephrol* 15: 359–369

69 Davies MR, Lund RJ, Mathew S, Hruska KA (2005) Low turnover osteodystrophy and vascular calcification are amenable to skeletal anabolism in an animal model of chronic kidney disease and the metabolic syndrome. *J Am Soc Nephrol* 16: 917–928

70 Zeisberg M, Muller GA, Kalluri R (2004) Are there endogenous molecules that protect kidneys from injury? The case for bone morphogenic protein-7 (BMP-7). *Nephrol Dial Transplant* 19: 759–761

71 Remuzzi G, Bertani T (1998) Pathophysiology of progressive nephropathies. *N Engl J Med* 339: 1448–1456

72 Brenner BM (2002) Remission of renal disease: Recounting the challenge, acquiring the goal. *J Clin Invest* 110: 1753–1758

73 Tomooka S, Border WA, Marshall BC, Noble NA (1992) Glomerular matrix accumulation is linked to inhibition of the plasmin protease system. *Kidney Int* 42: 1462–1469

74　Wilson HM, Reid FJ, Brown PA, Power DA, Haites NE, Booth NA (1993) Effect of transforming growth factor-beta 1 on plasminogen activators and plasminogen activator inhibitor-1 in renal glomerular cells. *Exp Nephrol* 1: 343–350

75　Wang S, Hirschberg R (2003) BMP7 antagonizes TGF-beta-dependent fibrogenesis in mesangial cells. *Am J Physiol Renal Physiol* 284: F1006–1013

76　Lin J, Patel SR, Cheng X, Cho EA, Levitan I, Ullenbruch M, Phan SH, Park JM, Dressler GR (2005) Kielin/chordin-like protein, a novel enhancer of BMP signaling, attenuates renal fibrotic disease. *Nat Med* 11: 387–393

77　Yoshikawa M, Hishikawa K, Marumo T, Fujita T (2007) Inhibition of histone deacetylase activity suppresses epithelial-to-mesenchymal transition induced by TGF-beta1 in human renal epithelial cells. *J Am Soc Nephrol* 18: 58–65

78　Imai N, Hishikawa K, Marumo T, Hirahashi J, Inowa T, Matsuzaki Y, Okano H, Kitamura T, Salant D, Fujita T (2007) Inhibition of histone deacetylase activates side population cells in kidney and partially reverses chronic renal injury. *Stem Cells* 25: 2469–2475

79　Li T, Surendran K, Zawaideh MA, Mathew S, Hruska KA (2004) Bone morphogenetic protein 7: A novel treatment for chronic renal and bone disease. *Curr Opin Nephrol Hypertens* 13: 417–422

80　Grgurevic L, Macek B, Erjavec I, Mann M, Vukicevic S (2007) Urine release of systemically administered bone morphogenetic protein hybrid molecule. *J Nephrol* 20: 311–319

81　Hruska KA, Saab G, Chaudhary LR, Quinn CO, Lund RJ, Surendran K (2004) Kidney-bone, bone-kidney, and cell-cell communications in renal osteodystrophy. *Semin Nephrol* 24: 25–38

82　Maric I, Poljak L, Zoricic S, Bobinac D, Bosukonda D, Sampath KT, Vukicevic S (2003) Bone morphogenetic protein-7 reduces the severity of colon tissue damage and accelerates the healing of inflammatory bowel disease in rats. *J Cell Physiol* 196: 258–264

Index

absorbable collagen sponge 72
acromesomelic chondrodysplasia
    (AMCD) 143, 147, 148
ACVR1 (ALK2, BMP type1 receptor) 149
adult chondrocyte phenotype,
    maintenance 287
AKT 85
allograft bone dowel 73
alveolar bone regeneration 241, 249
alveolar cleft 61
alveolar ridge 46
ameloblast 201, 205–207
ameloblastin 206
amelogenesis 206, 207
amelogenin 206
anabolic activity 278–283
annulus fibrosus (AF) 302
anterior cervical interbody fusion 74
anterior lumbar interbody fusion 72–74
anti-catabolic activity 281
apoptosis 89, 200
articular cartilage animal study 287–290
articular cartilage repair 288
autoinduction, bone formation by 233

bacterial artificial chromosome (BAC)
    transgenesis 119, 125
bacterial homologous recombination
    (BHR) 125
Barx1, homeobox gene 200
BDA2 148

beta-catenin 164, 200, 264, 265
biological fluids, BMPs in 329–331
biological matrix 71
biphasic calcium phosphate 8
BMP and activin membrane-bound inhibitor
    (BAMBI) 95, 216
BMP antagonist 216–219, 264, 267, 269
BMP discovery 1
BMP dissection 115–134
BMP receptor 83, 91, 161–163
BMP receptor type I (BMPR I) 161–163
BMPR1B 148
BMP responsive transcription factor 90
BMP signaling 88, 117–120, 141–152, 163,
    164, 200–203, 263, 269
BMP signaling network 117–120
BMP signaling route 88
BMP, endogenous expression 283–287
BMP-2 25–38, 163–168, 179–195, 201, 277
BMP-2, conditional allele 128, 129
Bmp-2 targeting vector 126
BMP-2cKO 189, 190
BMP-4 179–195, 199–208, 214, 279
BMP-4, expression 199, 201, 207
BMP-4, protein 205
BMP-4 knockout 123, 203, 204
BMP-4 signaling 200, 201, 203
BMP-4cKO 186
BMP-4cKO mouse 205, 206
BMP-4-mRNA 201
BMP-6 318

BMP-7   129–132, 201, 213, 277, 284–288, 303, 318-322
BMP-7, conditional allele   129–132
BMP-7, intradiscal injection   303
BMP-7, pro- and mature domain   284
BMP-7/6 hybrid molecule   331
BMP-9   279
BMP-induced signaling complex (BISC)   86
bone allograft, (de-)mineralized   64
bone formation, phases   71
bone formation by autoinduction   233
bone graft   8
bone graft replacement   71
bone graft substitute   2, 8
bone loss   28, 32
bone mineral density   183
bone regeneration   51, 53, 73, 241, 249
bone remodeling   166
bone volume   318, 319
bone volume to total volume (BV/TV)   185
bone-derived type I collagen particle   291
brachydactyly   82, 143, 150, 151
brachydactyly type A2   143
brachydactyly type B   150, 151
brachydactyly type C   143
brachypodism (bp) mouse   152
Brorin   95

cancer biology   317, 320–322
cancer cell   167
cartilage, anabolic response   278
cartilage homeostasis   286
cartilage repair   277, 278, 288, 301
cartilage repair, *in vitro* study   278
cartilage repair, non-articular   301
cartilage-derived morphogenetic protein (CDMP)   279
β-catenin   164, 200, 264, 265
CCN family   95
CD2-Cre   122
cementoblast   204
cementogenesis   240, 241

cementum   201, 204
ceramic granule   75
CFU-F   193
Cgraft®   64
chondrocyte differentiation   287
Chordin   92, 94, 95
c-Jun-N-terminal kinase (JNK)   89
clathrin-coated pit   86
collagen sponge, absorbable   72
composite graft   64
compression resistant matrix (CRM)   8, 76, 77
conditional allele   128–132
Cre-loxP system   203
crim1   223
crossveinless 2 (Cv2), BMP agonist   224
cystine knot   82, 92, 217
cytodifferentiation   199, 201–203, 206

Dan family   92
de novo bone regeneration   53
degenerative disc disease   277
dental implant   43, 56
dental sac   199
dentin   201, 203, 204, 206, 207, 241, 243
dentin matrix protein DMP1   204
dentin sialophosphoprotein (DSPP)   204
dentinal tubule   206
dentinogenesis   201
diabetes   322–324, 327
Dickkopf (DKK) protein   264–267
Dlx5   201
DuPan syndrome   147

embryonic development, recapitulation of   234, 237, 241
enamel   205–207
enamel knot   199–201
endochondral ossification   94
endocytosis   87
endofin   87
epithelial signaling   201
epithelial-mesenchymal interaction   199, 200

epithelium, dental   200, 201
epithelium, oral   199
ERK kinase   87
extracellular matrix production   287

facial bone   43–67
facial cleft   61
fibroblast growth factor (FGF)   199, 280
fibrodysplasia ossificans progressiva (FOP)   84, 149
fluorescein di-β-D-galactopyranoside (FDG)   122
focal chondral defect model   294–296
focal osteochondral defect   288
focal osteochondral defect model   289–294
fracture, open   28

β-galactosidase   119
GDF-5   143, 279
GDF-6   279
GDF-7   279
gene array   286
gene-deletion, phenotype   117, 118
gene-targeting construct   122
genome engineering   115–134
Grebe syndrome   147
Gremlin   221, 222
Growth differentiation factor (GDF)   81, 84, 141–143, 147, 148
GS-box   82

hemojuvelin   96
heparin sulfate proteoglycans   217
hepcidin   96
heterotopic bone, induction of   234
high bone mass (HBM)   261, 266, 267
homeobox gene   199, 200
Hunter-Thompson syndrome   147

in situ hybridization   181
induction of cementogenesis   240
induction of heterotopic bone   234, 235

induction of tissue and morphogenesis   234, 236, 240
INFUSE® Bone graft   44
insulin-like growth factor-1 (IGF-1)   280
intertransverse posterolateral spinal fusion   21
intervertebral disc (IVD)   302
ischemic coronary infarction   317

c-Jun-N-terminal kinase (JNK)   89
juvenile polyposis syndrome   84

kidney regeneration   317, 325, 326
kielin/chordin-like protein (KCP)   223, 224
knockout, conditional   202
knockout mouse   123, 200, 203–206

lamina, dental   200
larynx   303, 304
ligament, periodontal   201
liver regeneration   317, 324
low-density lipoprotein (LDL) receptor-related protein-5 or -6 (LRP5/6)   264–267
LRP5 gene   261, 266, 267
LRP5 HBM   261, 266, 267
lymphoid-enhanced-factor 1 (Lef1)   199, 200

Mad homology   87
mandible, development   200
mandible, osteopenia   203
mandibular defect   61
marker gene   201
Mastergraft Matrix®   64
MASTERGRAFT® Granules   75
maxillary defect   46, 61
maxillofacial skeleton   60
maxillofacial surgery   46
mesenchymal signaling   201
mesenchymal stem cell   71
mesenchyme, dental   199
mesenchyme, oral   200
metalloprotease   94

metalloproteinase (TIMP), tissue inhibitor of   280
microfracture procedure, chondral defect repair with   295
mineral quality   186
mitogen-activated protein kinase (MAPK)   89, 162, 163, 165
morpho-differentiation within tooth germ   200
mosaicplasty procedure, augmentation of   293
mouse model, 3.6 Col1a1-Cre   203
multiple synostosis syndrome (SYNS)   82, 147, 151

nephropathy, diabetic   327, 328
NF-κB   168, 282
Noggin   82, 92, 94, 151, 222
non-obese diabetic (NOD) mouse   323
nonunions, treatment   26–28, 35–38
nucleus pulposus (NP)   302

OAZ   162
odontoblast   201–207
odontogenesis   201, 204, 206
osteoarthritis (OA)   147, 277
OA model   290, 296–300
osteoblast   71, 180, 202, 203, 258, 259, 263
osteocalcin expression   192
osteoclast   187
osteocyte   258, 260, 268
osteodystrophy, renal   328
osteogenic protein-1 (OP-1)   213, 277
osteogenic soluble molecular signals, redundancy of   237
osteopenia   203
osteopenic phenotype   183
osteoporosis   257, 269, 270, 317
osterix   93, 168, 201, 203, 204, 259

p83 pathway   206
paired box gene 9 (Pax9)   199, 200
papilla, dental   199–201
pericyte   62

periodontal ligament   241
periodontum   203, 206
phenotypes, associated with gene deletion   117, 118
phosphatase   87
phosphatidylinositiol 3 kinase (PI3K)   165–168
PI3K signaling   166
PI3K/Akt signaling   166–168
Phosphor-Smad1/5/8 (P-Smad1/5/8)   188, 203, 206
platelet-derived growth factor (PDGF)   280
posterolateral fusion   74, 75
posterior maxillary site, bone regeneration   51
pre-ameloblast   201, 205
predentin   201, 203
preformed complex (PFC)   87
preodontoblast   201
prosthesis   60
protein kinase B (PKB/Akt)   165, 166
proteoglycans synthesis   279
proximal symphalangism (SYM1)147, 151
pulmonary arterial hypertension (PAH)   85
pulp, dental   201, 207

RACK1   85
recombinant human bone morphogenetic protein-2 (rhBMP-2)   7, 8, 25–38, 45–67, 71–78
RedE/T recombineering   125
regenerative medicine   1
renal failure   317, 326, 327
renal fibrosis   328, 329
renal osteodystrophy   328
reporter gene   119
repulsive guidance molecule (RGM)   95
rhBMP-2 carrier   8
rhBMP-2/ACS   45–67
Robinow syndrome   96, 150
Ror2   96, 150
Runx2   89, 168, 200, 259

sclerosteosis   93, 151, 260, 261, 269

sclerostin 257–270
short ear mouse 152
signaling 88, 117–120, 141–152, 163, 164, 200–203, 263, 269
sinus lift 64
sinus lift bone graft 46
site/tissue specificity 240
Smad 84, 87, 161–169, 185, 191, 203, 206, 262–264, 267
Smad/CBP interaction 168
Smad1 87, 162
Smad4 84, 161, 167
Smad4-Smad5 complex 162
Smad6 162
Smad-binding element (SBE) 87
Smurf 89
SOST 151
SOST gene 257, 260–262
SOST mRNA 258, 259, 266
SOST/sclerostin expression 269, 270
Speman organizer 95
stroke 317, 325
symphalangism 82, 147, 151
synostosis syndrome (SYNS), multiple 82, 147, 151

TAK1 89
targeted deletion, in osteoblast 180
tarsal carpal coalition syndrome 151
threaded titanium cage 73
thymus 119–122
tissue and morphogenesis, induction of 234, 236, 240

tissue engineering 1
tissue inhibitor of metalloproteinase (TIMP) 280
tissue regeneration 97
tooth development 199, 200, 203
tooth germ 200, 201, 207
trabecular bone 71
trachea 303, 304
transforming growth factor β (TGF-β) 161, 278
trap staining 187
tumor necrosis factor-α (TNF-α) 280
twisted gastrulation (Tsg) 93
tyrosine kinase 96

uterine sensitization-associated gene-1 (USAG-1) 218, 257, 267

Van Buchem disease 93, 151, 261, 262, 269
Vav-Cre 122
VelociGene® 126

WISP 95
Wnt pathway 200
Wnt response 268
Wnt signaling 93, 163, 200, 264–268
Wnt/β-catenin signaling 164

x-chromosomal linked inhibitor of apoptosis (XIAP) 89
x-ray 204

The PIR-Series
Progress in Inflammation Research

Homepage: http://www.birkhauser.ch

Up to-date information on the latest developments in the pathology, mechanisms and therapy of inflammatory disease are provided in this monograph series. Areas covered include vascular responses, skin inflammation, pain, neuroinflammation, arthritis cartilage and bone, airways inflammation and asthma, allergy, cytokines and inflammatory mediators, cell signalling, and recent advances in drug therapy. Each volume is edited by acknowledged experts providing succinct overviews on specific topics intended to inform and explain. The series is of interest to academic and industrial biomedical researchers, drug development personnel and rheumatologists, allergists, pathologists, dermatologists and other clinicians requiring regular scientific updates.

Available volumes:
T Cells in Arthritis, P. Miossec, W. van den Berg, G. Firestein (Editors), 1998
Medicinal Fatty Acids, J. Kremer (Editor), 1998
Cytokines in Severe Sepsis and Septic Shock, H. Redl, G. Schlag (Editors), 1999
Cytokines and Pain, L. Watkins, S. Maier (Editors), 1999
Pain and Neurogenic Inflammation, S.D. Brain, P. Moore (Editors), 1999
Apoptosis and Inflammation, J.D. Winkler (Editor), 1999
Novel Inhibitors of Leukotrienes, G. Folco, B. Samuelsson, R.C. Murphy (Editors), 1999
Metalloproteinases as Targets for Anti-Inflammatory Drugs,
    K.M.K. Bottomley, D. Bradshaw, J.S. Nixon (Editors), 1999
Gene Therapy in Inflammatory Diseases, C.H. Evans, P. Robbins (Editors), 2000
Cellular Mechanisms in Airways Inflammation, C. Page, K. Banner, D. Spina (Editors), 2000
Inflammatory and Infectious Basis of Atherosclerosis, J.L. Mehta (Editor), 2001
Neuroinflammatory Mechanisms in Alzheimer's Disease. Basic and Clinical Research,
    J. Rogers (Editor), 2001
Inflammation and Stroke, G.Z. Feuerstein (Editor), 2001
NMDA Antagonists as Potential Analgesic Drugs,
    D.J.S. Sirinathsinghji, R.G. Hill (Editors), 2002
Mechanisms and Mediators of Neuropathic pain, A.B. Malmberg, S.R. Chaplan (Editors), 2002
Bone Morphogenetic Proteins. From Laboratory to Clinical Practice,
    S. Vukicevic, K.T. Sampath (Editors), 2002
The Hereditary Basis of Allergic Diseases, J. Holloway, S. Holgate (Editors), 2002
Inflammation and Cardiac Diseases, G.Z. Feuerstein, P. Libby, D.L. Mann (Editors), 2003
Mind over Matter – Regulation of Peripheral Inflammation by the CNS,
    M. Schäfer, C. Stein (Editors), 2003
Heat Shock Proteins and Inflammation, W. van Eden (Editor), 2003
Pharmacotherapy of Gastrointestinal Inflammation, A. Guglietta (Editor), 2004
Arachidonate Remodeling and Inflammation, A.N. Fonteh, R.L. Wykle (Editors), 2004
Recent Advances in Pathophysiology of COPD, P.J. Barnes, T.T. Hansel (Editors), 2004
Cytokines and Joint Injury, W.B. van den Berg, P. Miossec (Editors), 2004

Cancer and Inflammation, D.W. Morgan, U. Forssmann, M.T. Nakada (Editors), 2004
Bone Morphogenetic Proteins: Bone Regeneration and Beyond, S. Vukicevic, K.T. Sampath (Editors), 2004
Antibiotics as Anti-Inflammatory and Immunomodulatory Agents, B.K. Rubin, J. Tamaoki (Editors), 2005
Antirheumatic Therapy: Actions and Outcomes, R.O. Day, D.E. Furst, P.L.C.M. van Riel, B. Bresnihan (Editors), 2005
Regulatory T-Cells in Inflammation, L. Taams, A.N. Akbar, M.H.M Wauben (Editors), 2005
Sodium Channels, Pain, and Analgesia, K. Coward, M. Baker (Editors), 2005
Turning up the Heat on Pain: TRPV1 Receptors in Pain and Inflammation, A.B Malmberg, K.R. Bley (Editors), 2005
The NPY Family of Peptides in Immune Disorders, Inflammation, Angiogenesis and Cancer,
Z. Zukowska, G. Z. Feuerstein (Editors), 2005
Toll-like Receptors in Inflammation, L.A.J. O'Neill, E. Brint (Editors), 2005
Complement and Kidney Disease, P. F. Zipfel (Editor), 2006
Chemokine Biology – Basic Research and Clinical Application, Volume 1: Immunobiology of Chemokines, B. Moser, G. L. Letts, K. Neote (Editors), 2006
The Hereditary Basis of Rheumatic Diseases, R. Holmdahl (Editor), 2006
Lymphocyte Trafficking in Health and Disease, R. Badolato, S. Sozzani (Editors), 2006
In Vivo *Models of Inflammation, 2nd Edition, Volume I*, C.S. Stevenson, L.A. Marshall, D.W. Morgan (Editors), 2006
In Vivo *Models of Inflammation, 2nd Edition, Volume II*, C.S. Stevenson, L.A. Marshall, D.W. Morgan (Editors), 2006
Chemokine Biology – Basic Research and Clinical Application. Volume II: Pathophysiology of Chemokines, K. Neote, G.L. Letts, B. Moser (Editors), 2007
Adhesion Molecules: Function and Inhibition, K. Ley (Editor), 2007
The Immune Synapse as a Novel Target for Therapy, L. Graca (Editor), 2008
The Resolution of Inflammation, A. G. Rossi, D. A. Sawatzky (Editors), 2008

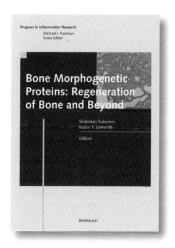

Bone Morphogenetic Proteins: Regeneration of Bone and Beyond

Vukicevic, S., Laboratory for Mineralized Tissues, Zagreb, Croatia / **Sampath, K.T.**, Genzyme Corporation, Framingham, USA (eds)

Content:
Bone morphogenetic proteins and their role in regenerative medicine.- Bone morphogenetic protein receptors and their nuclear effectors in bone formation.- Biology of bone morphogenetic proteins.- BMPs in development.- BMPs in articular cartilage repair.- Craniofacial reconstruction with BMP.- Clinical experience of osteogenic protein-1 (OP-1) in the repair of bone defects and fractures of long bones.- Development of the first commercially available recombinant human bone morphogenetic protein (rhBMP-2) as an autograft replacement for spinal fusion and ongoing R & D direction.- Bone morphogenetic proteins and the synovial joints.- The role of bone morphogenetic proteins in developing and adult kidney.- Bone morphogenetic proteins in the nervous system.- BMP and cancer.- Bone healing: BMPs and beyond.

2004. XI, 310 p. 96 illus.,Hardcover
ISBN 978-3-7643-7139-5
PIR — Progress in Inflammation Research

See also
**Bone Morphogenetic Proteins:
From Laboratory to Clinical Practice**
Vukicevic, S., Zagreb, Croatia / **Sampath, K.T.**, Framingham, MA, USA (eds)
2002. XII, 326 p. Hardcover
ISBN 978-3-7643-6509-7
PIR — Progress in Inflammation Research

BIRKHÄUSER

www.birkhauser.ch

The Resolution of Inflammation

Rossi, A.G. / Sawatzky, D.A. (eds)
University of Edinburgh, UK

Much progress has been made in elucidating the mechanisms regulating the induction and progression of inflammatory diseases. More recently, however, it has become apparent that the manipulation of mechanisms governing the resolution of inflammation is an important way to develop novel strategies for the therapy of such diseases. This book provides an up-to-date and comprehensive view on the resolution of inflammation and on new developments in this area (e.g. pro-resolution mediators, apoptosis, macrophage clearance of apoptotic cells, possible novel drug developments). The chapters are written by leading scientists in the field. This topical volume is of interest to medical and scientific researchers in academia and industry working on inflammatory and cardiovascular diseases such as arthritis, asthma, bronchitis, atherosclerosis, and more.

2008. XII, 238 pp. 25 illus., 2 in color. Hardcover
ISBN 978-3-7643-7505-8
PIR - Progress in Inflammation Research

www.birkhauser.ch